Interfacial Engineering in Functional Materials for Dye-Sensitized Solar Cells

Interfacial Engineering in Functional Materials for Dye-Sensitized Solar Cells

Edited by

Alagarsamy Pandikumar
Academy of Scientific and Innovative Research (AcSIR), Ghaziabad, India
Functional Materials Division, CSIR-Central Electrochemical Research Institute
Karaikudi, India

Kandasamy Jothivenkatachalam
Department of Chemistry, Bharathidasan Institute of Technology (UCE- BIT Campus)
Anna University, Tiruchirappalli, India

Karuppanapillai B Bhojanaa
CSIR-Central Electrochemical Research Institute
Karaikudi, India

Registered Office
John Wiley & Sons, Inc., 111 River Street, Hoboken, NJ 07030, USA

Editorial Office
111 River Street, Hoboken, NJ 07030, USA

For details of our global editorial offices, customer services, and more information about Wiley products visit us at www.wiley.com.

Wiley also publishes its books in a variety of electronic formats and by print-on-demand. Some content that appears in standard print versions of this book may not be available in other formats.

Library of Congress Cataloging-in-Publication Data is applied for

Hardback ISBN: 9781119557333

Cover Design: Wiley
Cover Image: © MILANTE/Getty Images

Set in 10/12pt WarnockPro by SPi Global, Chennai, India

Printed in United States of America

V10014993_102119

Contents

List of Contributors

S. Akshaya

Electrochemical Materials and Devices Lab
Department of Chemistry
Bharathiar University
Coimbatore
Tamil Nadu
India

S. Anandhi

Jeppiaar Maamallan Engineering College
Department of Physics
Sriperumbudur
India

A. Arulraj

Department of Physics
University College of Engineering –
Bharathidasan Institute of Technology
(BIT) Campus
Anna University
Tiruchirappalli
India

Suresh Kannan Balasingam

Department of Materials Science and
Engineering
Faculty of Natural Sciences
Norwegian University of Science and
Technology (NTNU)
Trondheim
Norway

Giovana R. Cagnani

São Carlos Institute of Physics
University of São Paulo
Department of Physics
São Paulo
Brazil

F. Manik Clinton

Electrochemical Materials and Devices Lab,
Department of Chemistry
Bharathiar University
Coimbatore
Tamil Nadu
India

K.V. Hemalatha

Department of Chemistry
Coimbatore Institute of Technology (CIT)
Coimbatore
Tamil Nadu
India

Nirav Joshi

São Carlos Institute of Physics
University of São Paulo
Department of Physics
São Paulo
Brazil

C.R. Kalaiselvi

Department of Physics
Erode Sengunthar Engineering College
Perundurai
India

S.S. Kanmani

Hindusthan College of Engineering and
Technology
Coimbatore
India

S.N. Karthick

Electrochemical Materials and Devices Lab
Department of Chemistry
Bharathiar University
Coimbatore
Tamil Nadu
India

C. Karthik Kumar

Sathyabama Institute of Science and
Technology
Centre of Excellence for Energy Research
Chennai
India

A. Muthu Kumar

Nanostructure Lab, Department of Physics
The Gandhigram Rural Institute-Deemed to
be University
Gandhigram
Tamil Nadu
India

Hee-Je Kim

School of Electrical and Computer Science
Engineering
Pusan National University (PNU)
Gumjeong-Ku
Jangjeong-Dong
Busan
Republic of Korea

Su Pei Lim

Xiamen University Malaysia
School of Energy and Chemical Engineering
Jalan Sunsuria
Bandar Sunsuria
Sepang Selangor Darul Ehsan
Malaysia

and

Xiamen University
College of Chemistry and Chemical
Engineering
Xiamen
China

R.V. Mangalaraja

Department of Materials Engineering
Faculty of Engineering
Advanced Ceramics and Nanotechnology
Laboratory
University of Concepcion
Concepcion
Chile

and

Technological Development Unit (UDT)
University of Concepcion
Coronel Industrial Park
Coronel
Chile

Hari Murthy

Department of Electronics and Communica-
tion Engineering
CHRIST University
Kanminike
Bengaluru
India

G. Murugadoss

CSIR- Central Electrochemical Research
Institute
Karaikudi
India

P. Nithiananthi

Nanostructure Lab, Department of Physics
The Gandhigram Rural Institute-Deemed to
be University
Gandhigram
Tamil Nadu
India

Subhendu K. Panda

CSIR- Central Electrochemical Research
Institute
Karaikudi
India

M. Paulraj

University of Concepcion
Department of Physics
Faculty of Physical and Mathematical Sciences
Concepcion
Chile

I. John Peter

Nanostructure Lab, Department of Physics
The Gandhigram Rural Institute-Deemed to
be University
Gandhigram
Tamil Nadu
India

T. Raguram

Department of Sciences
Amrita School of Engineering, Amrita Vishwa
Vidyapeetham
Coimbatore
India

C. Raja Mohan

Nanostructure Lab, Department of Physics
The Gandhigram Rural Institute-Deemed to
be University
Gandhigram
Tamil Nadu
India

Rajkumar C

Department of Electronics and
Communication
University of Allahabad
Allahabad
India

K.S. Rajni

Department of Sciences
Amrita School of Engineering
Amrita Vishwa Vidyapeetham
Coimbatore
India

K. Ramachandran

Nanostructure Lab Department of Physics
The Gandhigram Rural Institute-Deemed to
be University
Gandhigram
India

A. Dennyson Savariraj

Department of Materials Engineering
Faculty of Engineering
Advanced Ceramics and Nanotechnology
Laboratory
University of Concepcion
Concepcion
Chile

and

Department of Chemical Engineering
Khalifa University of Science and Technology
Abu Dhabi
United Arab Emirates

T.S. Senthil

Department of Physics
Erode Sengunthar Engineering College
Perundurai
India

Flavio M. Shimizu

Laboratório Nacional de Nanotecnologia
Centro Nacional de Pesquisa em Energia e
Materiais
São Paulo
Brazil

T.S. Shyju

Sathyabama Institute of Science and
Technology
Centre for Nanoscience and Nanotechnology
Chennai
India

and

University of Concepcion
Department of Physics
Faculty of Physical and Mathematical Sciences
Concepcion
Chile

and

Sathyabama Institute of Science and
Technology
Centre of Excellence for Energy Research
Chennai
India

R. Thangamuthu

CSIR- Central Electrochemical Research
Institute
Karaikudi
India

P. Vengatesh

Sathyabama Institute of Science and
Technology
Centre of Excellence for Energy Research
Chennai
India

Preface

Solar energy has paved the alternative way to fossil fuels for present and future energy demands. Dye-sensitized solar cells (DSSCs) are one of the most promising technique to harvest solar energy and convert in to electrical energy because of their ease of production, low cost, flexibility, relatively high conversion efficiency, and low toxicity to the environment. DSSC consists of components such as electrolyte, dye, counter electrode, and photoanode. Among these, photoanode plays a vital role and serves as a support for dye molecules and transport photo-excited electrons. Performance of the devices is greatly affected by charge generation, collection, and charge recombination occurs at interfaces (TiO_2 photoanode/electrolyte and FTO substrate/electrolyte) and are influenced by properties of the interfacial materials.

This new book gathers and surveys a variety of novel ideas for improving the efficiency of photoanode from various experts of interdisciplinary fields of chemist, physicist, materials scientist, and engineers to widely explore the materials development in the field of DSSC to achieve higher solar energy conversion efficiency.

This book is very attractive for multidisciplinary researchers. Moreover, this is much useful for the beginners who are working in the multidisciplinary area of nanoscience and nanotechnology, physics, chemistry, energy science and technology, materials science, and engineering related to solar energy conversion through DSSCs. Further, this book will be helpful to upgrade their knowledge and establish their own research in the area of solar cells. Beyond that this book can be used for teaching and reference book for bachelor and master's degree level students including nanoscience and nanotechnology, physics, chemistry, energy science and technology, materials science, and engineering.

The first two chapters describe the operation principles, charge transfer dynamics, function of photoanode, challenges, and solutions for DSSC. Chapters 3 and 4 describe how the nanoarchitectures and light scattering materials are used as a photoanodes in DSSCs. Chapters 4 and 5 explain the role of compact layer and $TiCl_4$ posttreatment during the fabrication of DSSCs. The remaining six chapters focus on engineering the interface with functional materials like doped semiconductors, binary semiconductors metal oxide based plasmonic nanocomposites, carbon nanotubes-based nanocomposites, graphene-based nanocomposites, graphitic carbon nitride (g-C_3N_4) nanocomposites at the photoanode surface of DSSCs in order to achieve the higher efficiency.

Last but not least, we would like to express our thanks and gratitude to the authors for sharing their generous knowledge on photoanodes used in DSSCs for the benefits of our community. Without them, materializing of this book is impossible. We regret if any copyright is being infringed unknowingly. We acknowledge the sincere efforts of Wiley Book publishing authorities, for bringing the book in its final shape. The editors would like to dedicate this book to **Prof. Ramasamy Ramaraj FASc, FNASc, FNA**, CSIR-Emeritus Scientist, Madurai Kamaraj University for his pioneer contribution in the area of solar energy harvesting.

26 June 2019

Alagarsamy Pandikumar
India
Kandasamy Jothivenkatachalam
India
Karuppanapillai B Bhojanaa
India

1

Dye-Sensitized Solar Cells: History, Components, Configuration, and Working Principle

S.N. Karthick[1], K.V. Hemalatha[2], Suresh Kannan Balasingam[3], F. Manik Clinton[4], S. Akshaya[5], and Hee-Je Kim[6]

[1] *Electrochemical Materials and Devices Lab, Department of Chemistry, Bharathiar University, Coimbatore, Tamil Nadu, India*
[2] *Department of Chemistry, Coimbatore Institute of Technology (CIT), Coimbatore, Tamil Nadu, India*
[3] *Department of Materials Science and Engineering, Faculty of Natural Sciences, Norwegian University of Science and Technology (NTNU), Trondheim, Norway*
[4] *Electrochemical Materials and Devices Lab, Department of Chemistry, Bharathiar University, Coimbatore, Tamil Nadu, India*
[5] *Electrochemical Materials and Devices Lab, Department of Chemistry, Bharathiar University, Coimbatore, Tamil Nadu, India*
[6] *School of Electrical and Computer Science Engineering, Pusan National University (PNU), Gumjeong-Ku, Jangjeong-Dong, Busan, Republic of Korea*

1.1 Introduction

The ever-growing human population requires the consumption of energy in various forms, and therefore researchers in energy field focus on energy harvesting from various sources. The nonrenewable energy sources such as fossil fuels are running out, which cannot be replenished in our life time. The nonrenewable energy sources are carbon-based fossil fuels such as coal, petroleum, and natural gas that emits greenhouse gases (for example carbon dioxide) that cause global warming, a serious threat to the world and mankind. At present, worldwide around three-fourth of the electricity is obtained from the nonrenewable sources that cannot be reused or recycled [1]. Many countries such as Japan, China, France, Ukraine, and India depend on nuclear power stations for the production of electricity and also they are facing several harmful issues from these power plants that lead to environmental pollution [2]. Therefore, the focus of scientists mainly rely on the renewable energy-based energy conversion devices. Solar, wind, hydroelectric, biomass, and geothermal are some of the examples of renewable energy resources available in our earth. Of these, solar energy is an important source of renewable energy, which is available throughout a day all over the year, basically inexhaustible in nature. In case of solar energy, radiation obtained from the sunlight is capable of producing heat and light, causes photochemical reactions, and generates electricity. As the electricity becomes a first and foremost basic need for the mankind, this impressive energy source can be utilized for the conversion of solar to electrical energy using solar cell technology. The strength of solar energy is magnanimous as it provides us about 10 000 times more energy that is higher than the world's daily need of energy consumption [1]. The earth receives such a huge amount of energy every day, we are fortunate to harness it using suitable solar cell technologies. Regrettably, though solar energy is free of cost, the highly expensive technologies required for its conversion and storage which limit the technology to reach the wider community.

The concept of solar energy harvesting has been evolving since eighteenth century. Edmond Becquerel, a French scientist, has first discovered the photovoltaic (PV) effect in 1839 [3]. This effect has become a starting point for the solar energy harvesting applications. The

Interfacial Engineering in Functional Materials for Dye-Sensitized Solar Cells, First Edition.
Edited by Alagarsamy Pandikumar, Kandasamy Jothivenkatachalam and Karuppanapillai B. Bhojanaa.

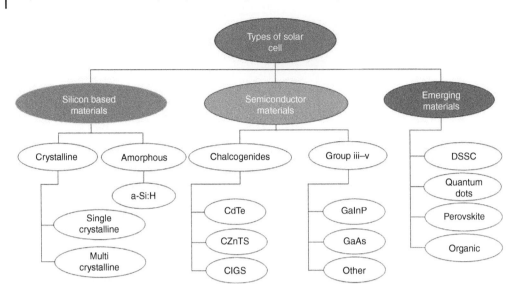

Figure 1.1 Classification of solar cells technologies. *Source:* Ibn-Mohammed et al. 2017 [7]. Reprinted with permission of Elsevier.

light energy to electrical energy conversion is being done by the special photoactive devices called photovoltaic cells. When the semiconductor material absorbs light, it produces electrical voltage and this effect is called photovoltaic effect. The first-generation solar cells use crystals of silicon to attain this effect. In 1887, German scientist Hertz has first examined the photoelectric effect and found that photons present in the light are capable of ejecting free electrons from a solid surface (usually conductor) to create power. However, based on his preliminary results, he also found that the same process produced more power when the conductor is incident with UV light rather than more intense visible light [4, 5]. Based on this phenomenon, the modern solar cells rely on the photoelectric effect to convert sunlight into electricity. Generation of solar cells are broadly classified into three different categories based on the materials properties and the period of time they evolved. For example, silicon-based materials (crystal and amorphous) belong to the first-generation solar cells, semiconductor materials (III–IV group chalcogenides/phosphide materials) belong to the second-generation solar cells, and emerging materials (Figure 1.1) belong to the third-generation solar cells. From 1953 to 1956, physicists at Bell Laboratory fabricated silicon solar cells with 6% efficiency, which is more efficient than selenium. This discovery paved a way for identifying the capability of the solar cells to power up the electrical equipment; again, the experimentation continued to improve the performance and attempted to make new devices for commercialization. Approximately, after 10 years, the second-generation thin-film solar cells were developed. The second-generation solar cells are often described as emerging thin-film solar cells that converts 30% of the solar radiation into electrical energy [6]. The semiconductor materials used in this generation is copper indium gallium selenide (CIGS), cadmium telluride (CdTe), and gallium arsenide (GaAs). The devices made out of these materials are commercially available and used in space crafts. After 1990s, the third-generation solar cell technologies have been emerged. This new generation photovoltaic technologies include dye-sensitized solar cells (DSSCs), organic/polymer solar cells, quantum dot solar cells, perovskite solar cells, etc. The power conversion efficiency of these third-generation solar cells are lower than silicon-based solar cells and thin-film solar cells, but it has its own advantages such as low processing costs and less environmental impact that induce the intensive research and development in this area.

The production cost of first- and second-generation solar cells are reported to be more than 1 US\$/W. The third-generation PV technologies aim to produce the large-scale electricity with a low cost of less than 0.5 US\$/W. This can be achieved with the module cost at the rate of 70 US\$/m^2 revealing 14% efficiency [1] for third-generation solar cells.

1.2 History of Dye-sensitized Solar Cells

During 1839, Becquerel [3] found that a voltage was produced when two platinum electrodes were immersed in the electrolyte containing a metal halide salt when it is illuminated with light. After that, the illumination of light on the organic dye is capable of producing electricity at the semiconductor electrodes in the electrochemical cell was discovered during 1960s [8]. Later, the phenomenon of photoexcitation was studied primarily at University of California during 1970s in order to simulate the mechanism of photosynthesis by extracting the chlorophyll pigment from spinach and using ZnO as semiconductor electrode material in electrochemical cells [9]. The mechanism was recognized when the photoexcitation of dye molecules injected electrons into the conduction band of the n-type semiconductor scaffold material and found that the dye molecules adsorbed on the semiconductor monolayer material was responsible for maximum yield. This study forms the basis of the bionic or biomimetic approach of sensitizing semiconductor materials for electrons excitation. Again, in 1972, the generation of electricity through dye sensitization was demonstrated by Tributsch [10]. In the study conducted by Michio Matsumura et al., in the year of 1980 [11], they inferred that the efficiency of the dye molecules could be improved by fine tuning the porosity of the semiconductor oxide material, but the stability of the dye is a mere challenge in the dye-sensitized photocell. Then the progress in this field was initiated in 1991 by Prof. Michael Grätzel who performed the sensitized electrochemical PV device made of dye sensitization on semiconductor TiO$_2$ material and named it as "dye-sensitized solar cell" [12]. They architectured the device toward the new conceptual of PV energy generation. The new conceptual approach is based on the three billion years old idea of nature's photosynthetic activity. They inferred that PV device is modeled based on the light absorption and electron transfer mechanisms in plant. When the sunlight falls on the chlorophyll molecule, the electron generated from chlorophyll is transferred from one molecule to the other till it reaches the chlorophyll reaction center. From the reaction center, the electron is transferred to the energy storage molecule. The chlorophyll lacks with one electron is being grabbed from the surrounding water molecule. This cyclical process is being imitated for harvesting energy using sun's radiation through synthetic dye material. In the first attempt, they received <2% efficiency from the DSSCs. It has its own drawback with the chlorophyll dye such as adsorption range and stability of dye molecules. To increase the performance of the device, they used the ruthenium-based dyes and polymer gel electrolyte with good thermal stability in the year of 2003 and reached around 6% maximum efficiency [13]. In 2005, researchers concentrated on improving the quantum efficiency of DSSCs, by modifying the morphology of the photoanode scaffold material and improving the electron transfer in the semiconductor, designing the particle with a small size having high surface area for more dye adsorption, etc. Then they replaced the ruthenium-based dye with low-cost dyes and achieved around 5% efficiency in 2005 [14]. In 2008 [15], 8% efficiency was achieved due to the iodine-based redox liquid electrolyte. Then the researchers attempted to solve the corrosive nature of the electrolyte, which helped to improve the photovoltage, efficiency, and the stability of devices. In this period, the efficiency of the devices increased up to 11.5% [16]. Finally, the efficiency was further improved more than 13% using porphyrin ring as dye molecules and cobalt(II/III) redox shuttle as the electrolyte [17, 18]. Later in 2009, DSSCs were first commercialized by G24 Power Limited, South Lake Drive, Imperial Park, Newport, United Kingdom.

1.3 Components of DSSCs

The basic components of DSSCs primarily consist of transparent conducting oxide (TCO) film-coated glass substrates, dye, photoanode, electrolytes, and counter electrode (CE).

1.3.1 Conductive Glass Substrate

The transparent conductive oxide-coated glass was used as substrates for both photoanode and counter electrode. The coating of this layer is required for the collection of electrons ejected from the photoanode (dye-coated TiO_2) and pass it to the counter electrode through outer circuit. Therefore, it is called current collector. Generally, TCO films are made by both fluorine-doped tin oxide (FTO) and indium-doped tin oxide (ITO). This film has a very low electrical resistance of <20 Ω/sq at room temperature. The conductive substrate should be made of highly abundant material, low cost with a maximum transparency of solar radiation [19].

1.3.2 Photoanode

A typical photoanode material consists of dye molecules-adsorbed semiconductor oxide material coated on TCO substrate, which acts as a working electrode in DSSCs. The most widely used semiconductor material is an anatase phase of titanium dioxide (TiO_2), having a band gap of 3.2 eV. It is an inexpensive, nontoxic, and abundant material present in the earth's crust. TiO_2 paste could be made using some binders such as terpineol, ethyl cellulose, polyethylene glycol, ethylene glycol, and so on, for a strong adhesion on the FTO glass plate. Screen printing or doctor blade techniques are being used for the coating of viscous TiO_2 paste onto the substrate. Removal of binder is usually performed by annealing the metal oxide paste-coated FTO glass plate at 450–500 °C for 30 minutes. Due to annealing, the metal oxide has good adhesion with TCO glass substrate, interparticle contact, and eventually forms the nanostructured porous electrode. The thickness of the formed porous electrode film is around 10–15 μm and the interconnected nanoparticle size ranges from 15 to 30 nm. Titanium dioxide has three different crystalline phases such as anatase, rutile, and brookite. The images of different TiO_2 crystal structures is shown in Figure 1.2. Apart from the titanium dioxides, the other semiconductor metal oxides such as ZnO, SiO_2, Zn_2SnO_4, CeO_2, WO_3, $SrTiO_3$, and Nb_2O_5 [20–26] were also used as the photoanode scaffold material in DSSCs. The performance of the device was poor when these materials are used in the photoanode. When ZnO is used in the photoanode doped with magnesium, it showed the efficiency of 4.11% [27]. The efficiency was 6.8% with SiO_2, 3.8% with Zn_2SnO_4, CeO_2 does not show any energy conversion efficiency, WO_3 showed the efficiency of 1.46%, $SrTiO_3$, and Nb_2O_5 showed 4.8% and 2%, respectively. TiO_2 is concluded to be the best semiconductor photoanode material because of its good efficiency and stability compared to the other materials. Among the above metal oxides, ZnO showed better efficiency because it has a band gap close to that of TiO_2 and found to contain the higher carrier mobility. Hence, it is found to be the efficient photoanode that can replace TiO_2. However, the recombination of the photoexcited charge carriers and the instability of the ZnO surface toward the acidic dyes were the major setback for the ZnO to provide high conversion efficiency. Researchers tried to improve the efficiency by modifying the structure, surface morphology, particle size, thickness optimization of photoanode, and so on.

1.3.3 Counter Electrode

The three important functions of counter electrode as a potential catalyst is (i) It is responsible for the completion of process, that is the oxidized redox couple is reduced at the surface

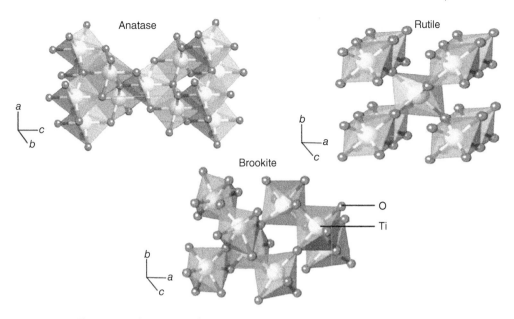

Figure 1.2 Different crystal structures of titanium dioxide (anatase, rutile, brookite). *Source:* Haggerty et al. 2017 [28]. Licensed under CC BY 4.0. https://www.nature.com/articles/s41598-017-15364-y#rightslink.

of counter electrode by accepting electrons. In solid-state DSSCs, the oxidized redox couple gets reduced by collecting the electrons from the counter electrode through ionic transport material. (ii) It collects electrons from the outer circuit and transfers them into inner cell electrolyte solution. So the ultimate role of the counter electrode is to collect electrons from the load and transfer them to the cell for circulation. (iii) The counter electrode acts as the mirror and reflects the unabsorbed light to the photoanode for the enhanced utilization of the sunlight [29]. To satisfy these basic functions, the counter electrode should possess high catalytic activity, highly conductivity and reflectivity, high surface area with porous nature, chemical corrosion resistance, high chemical and mechanical stability, energy level of counter electrode should match with the potential of the redox electrolyte, good adhesiveness toward TCO substrate. It is impossible to achieve all these parameters simultaneously with a single material. For example carbon-based counter electrode was found to achieve the high-surface area with porous nature, chemical corrosion resistance but it does not reflect the light. The counter electrode has much influence on photovoltaic parameters. The maximum photovoltage determined at zero current is due to the energy difference between the redox mediator and the semiconductor metal oxide. Under load, the output voltage seems less than the open-circuit voltage. This loss in voltage is due to the overpotential of the counter electrode that derives from the delivery of the current through electrolyte and electrolyte/CE interface. Platinum paste is used as a counter electrode material in standard cells for the completion of redox reaction. The reduction of tri-iodide was carried out by the counter electrode. The rate of reduction of tri-iodide to iodide at the counter electrode is important since after the redox reaction, formation of iodide is important to regenerate dye molecules at the anode side of the cell. For a workable cell, a slow reaction is carried out at the anode side and the fast reaction takes place at the cathode side. Apart from the platinum material, carbon, graphite, and conductive polymers are also used as counter electrode materials, platinum is still the best choice of catalysts due to its facile iodide/tri-iodide redox kinetics. Counter electrodes are being prepared by various preparation techniques such as sputtering, chemical vapor deposition, hydrothermal, electrodeposition, etc.. The platinum prepared by electrodeposition technique by Zhong et al., showed higher electrocatalytic activity than the pure Platinum [30]. Also, the light reflection value of platinum is higher than the carbon materials, which causes more reflection of light into the cell. Bimodal mesoporous carbon embedded with nickel prepared by thermal pyrolysis under nitrogen environment by Wu et al. showed the highest efficiency of 8.6% which was higher than that of Pt as counter electrode [31]. CoS films as counter electrodes prepared by Huo et al., by electrophoretic deposition and ion exchange deposition technique showed the efficiency of 7.72% [32]. Carbon nanotube (CNT) arrays were prepared by Nam et al. by the chemical vapor deposition technique. He inferred that the highly ordered structure of CNT would rapidly diffuse the electrolyte into the electrode and also helps in fast electron transport and achieved efficiency was 10.04% [33]. The conductive polymers such as polyaniline, polypyrrole, polyethylene dioxythiophene, and polythiophene derivatives were also used as counter electrodes. Anothumakkool et al. prepared polyethylene dioxythiophene-impregnated cellulose paper as a flexible counter electrode through in situ polymerization technique that achieved the efficiency of 6.1% which is on par with the efficiency achieved with Pt/FTO as counter electrode [34]. Since platinum is a very expensive material, recently researchers focus on many different metals and chalcogenide-based alternative counter electrodes for DSSCs.

1.3.4 Electrolytes

The electrolyte is yet another important compound playing an essential role in DSSCs. The presence of redox mediator in the electrolyte acts as a charge or ion transporter between the counter electrode and photoanode. The redox couple should not observe the light at the visible range. It should be chemically stable and reversible. It should have low viscosity for the rapid diffusion of charge carriers. It should not desorb the dye or disperse the semiconductor material adhered to the FTO plate. The most common, stable, and highly efficient redox mediator is iodide/tri-iodide redox couple. This redox mediator is promising due to its high penetration power into semiconductor nanoporous film, slow recombination loss, and the fast regeneration of dye molecules. This redox shuttle is used at present as an electrolyte for any standard cells. It is found to be an ideal electrolyte having a low viscosity, negligible vapour pressure, optimum boiling point, and dielectric properties. It also has important properties such as chemical inertness, environmental sustainability, and easy processing. It is believed that the redox couple encourages the charge transport between photoanode and counter electrodes. When the dye molecules inject electrons into the semiconductor metal oxide, the tri-iodide in the electrolyte helps in reducing the oxidized dye as quickly as possible to the ground state [35]. The iodide, an electron acceptor moves to the counter electrode and receives its lost electrons and regenerate as tri-iodide. The circuit is completed by the electron migration through the external load. For dye and electrolyte regeneration, the redox potential of the iodide/tri-iodide should be considered.

1.3.4.1 Types of Solvents Used in Electrolytes

The solvent used for electrolytes preparation in DSSCs are categorized into aqueous, organic and ionic liquid-based solvents.

Aqueous Electrolyte The first investigation with aqueous electrolyte (various pH ranging from 2 to 7) was done by Grätzel between pH ranging from 2 to 7 was done by Grätzel and O'Regan group by photoelectrochemical dye sensitization of porous TiO_2 electrodes. The research on water-based electrolytes was also fully concentrated for its nonflammability, nonvolatility, cost-effectiveness, and environmental compatibility. The aqueous electrolyte that is acidified for the better adsorption of dye molecules toward the TiO_2 is possible only when the pH is below the iso-electric point of TiO_2 which is 5. Next, the shift in the position of conduction band edge of TiO_2 takes place toward the negative direction of the energy scale while pH is decreasing [36].

Lindquist et al. demonstrated an experiment by adding small amount of water in nonaqueous electrolyte. The addition of water leads to elevation in photovoltage and decrease in photocurrent density [37]. It is inferred that addition of water weakens the bonding between dye molecules and the metal oxide and undergoes desorption of dye molecules. So a special dye variety should be considered when pure or mixed aqueous electrolyte is used. According to O'Regan and coworkers addition of 20% of water into nonaqueous electrolyte enhances the power energy conversion efficiency from 5.5% to 5.7% [38].

Organic Solvents-based Electrolytes Grätzel and coworkers developed the alcohol-based solvents for the preparation of lithium iodide electrolyte. It gives an efficiency of 11%, however while using aprotic solvents as electrolytes the bond between semiconducting material and dye become weaker and show instability toward aqueous solvents due to the hydrophilicity in nature [39]. Then nitrile-based solvents such as methyl cyanide, glutaronitrile, and valeronitrile were used as solvents for the electrolyte, but these are toxic in nature and expensive. So the derivative of nitrile like 3-methoxypropionitrile (MPN) and methoxyacetonitrile (MAN) were used. MPN showed comparatively good stability and exhibited 7.6% efficiency [35].

Ionic Liquids-Based Electrolytes Ionic liquids were recently found to be the most promising solvents used in electrolyte mixture because of their stability, conductivity, and low vapor pressure. The long-term stable ionic liquid electrolyte was first employed in 1996 by Papageorgiou et al. using methyl-hexyl-imidazolium iodide as solvent in DSSCs rather than other liquid electrolytes [40]. Xi and his team performed the experiments with nonimidazolium ionic liquids, i.e. *S*-ethyltetrahydrothiophenium tricyanomethide and *S*-ethyltetrahydrothiophenium iodide, showed better efficiency of 7.2% and 6.9%, respectively [41].

1.3.4.2 Alternative Redox Mediators

Though iodide/tri-iodide redox shuttle has many advantages, it also has its dark-side that they will easily corrode the substrates like silver and copper metal joints when they are used as current collectors and the photovoltage was also suppressed by the redox potential of the iodide/tri-iodide mediator. It also absorbs the visible light and some amount of photons when the dye is irradiated. So the research moves toward some other alternatives to iodide/tri-iodide mediator [42]. The bromide/tri-bromide redox couple is found to be one of the alternatives for iodide/tri-iodide electrolyte because of its high positive redox potential. This bromide/tri bromide redox couple is suitable if eosin Y dye is used as a sensitizer [43]. Pseudohalogen redox couples such as $SeCN^-/(SeCN)_3^-$ and $SCN^-/(SCN)_3^-$ were also used as the redox mediators having more positive equilibrium potential of 0.19 and 0.43 V, respectively, than I^-/I_3^- [44]. A halogen-free sulfur-based mediator like disulfide/thiolate (T^-/T_2) is also used as the redox couple with ruthenium-based dye as a sensitizer and achieved the efficiency of 6.44% [45]. Here, T^- is 5-mercapto-1-methyltetrazole anion and T_2 is its dimer. Cobalt(II/III) redox shuttle was also used as the electrolyte, which increases the overall efficiency to 13% [17]. SpiroMeOTAD, a solid hole conductor is capable of transporting charge, and dye regeneration could be used as charge mediators [42]. Polymer electrolytes are also under extensive investigation. Ionic liquids are another promising alternative electrolytes that are known for their advantages such as mutable, excellent chemical, and thermal stability, high-ionic conductivity. So finding the superior redox couple is one of the main challenges for researchers working in DSSCs field.

1.3.5 Dyes

The dye plays the centralized role in DSSCs by ejecting the electrons on irradiation and initiating the mechanism. The following are the good qualities of dyes. It should be adsorbed on photoanode and capture the sunlight at all wavelengths (λ) below 920 nm. It should have the strong anchoring groups such as $-COOH$, $-SO_3H$, $-H_2PO_3$ for the adsorption toward photoanode. The excited state of the dye should have higher energy than the conduction band of the photoanode material for the easy transport of electrons from the dye to the photoanode. For the p-type DSSCs, the HOMO level of dye should have a more positive potential than the valence band of p-type semiconductors. For the regeneration of dyes, it should have more positive oxidized state level than the redox potential of electrolyte. Optimization of the molecular structure of dye prevents the dye aggregation on the semiconductor electrode surface. The dye should have the high stability toward light and heat. Many photosensitizers such as metal complexes, porphyrins, phthalocyanines, and metal-free organic dyes were used for the past decades. Normally, the metal complexes have a central metal ion with ancillary ligand having atleast one anchoring group. Metal ligand charge transfer process is responsible for the visible light absorption in the solar spectrum. The metal ligand complex (*cis*-bis(4,4′-dicarboxy-2,2′-bipyridine)diisothiocyanato-ruthenium(II)) named N719 dye was taken as the reference dye in DSSCs due to its improved power conversion efficiency. A thiocyanato derivative *cis*-(SCN)2bis(2,2′-bipyridyl-4,4′-dicarboxylate)ruthenium(II) coded as N3 dye is an another sensitizer taken as the reference for the standard cell showed the efficiency of 10% having the longer excited state lifetime of $\approx$20 seconds. Again, to improve the efficiency of the sensitizers to near-infrared (NIR) region, ligands of the Ru complex have been modified by Grätzel and his workers. They developed N749 dye also called as black dye where Ru is bonded with three thiocyanato and one terpyridine ligands substituted with three carboxyl groups. The power conversion efficiency was 10.4% using black dye. The structure of N719, N3, black dyes are shown in Figure 1.2. A series of Ru complexes with phenanthroline ligand $Ru(dcphen)_2(NCS)_2-(TBA)_2$ as an anchor was studied as the photosensitizer showed the efficiency of 6.6%. The other metal complexes were prepared using Os, Re, Fe, Pt, and Cu as a central metal atom [46]. Some of the other sensitizers such as synthetic dyes such as N749, Z907, N712, D35, and natural dyes extracted from flowers, fruits, and vegetables were also investigated. The rockstar dyes of DSSCs are N719, N3, and black dyes structures which are shown in Figure 1.3.

Choi et al. synthesized a novel synthetic organic dye using nonplanar bis-dimethyl-fluorenylamino moiety and a terthiophene unit into the organic framework and cyanoacrylic acid moiety as an acceptor and anchoring group. With the novel synthetic dye, the achieved efficiency was 8.60%, which is on par to the efficiency of a standard cell [47]. Wang et al. used coumarin dyes based on the concept of donor-pi–conjugated bridge-acceptor, with $-CN$ groups as electron acceptor and showed the power conversion efficiency of 7.6% [48]. Hanaya and coworkers during 2015 achieved an efficiency of 11.2% using alkoxysilyl and carboxy anchor dye as cosensitizers [18]. Hemalatha et al. extracted the natural sensitizers carotenoid and anthocyanin dye from *Rosa chinensis* and *Kerria japonica* flowers, respectively, and for the dye stability, sugar molecules were added to the dye. The power conversion was very low and does not exceed 0.29% [49].

1.4 Configuration of DSSCs

1.4.1 Metal Substrates for Photoanode and Glass/TCO for Counter Electrode

The basic configuration of DSSCs consists of TCO as a glass as substrate. Dye coated with TiO_2 as photoanode. In this configuration, TCO film transmits photons and conducts electrons

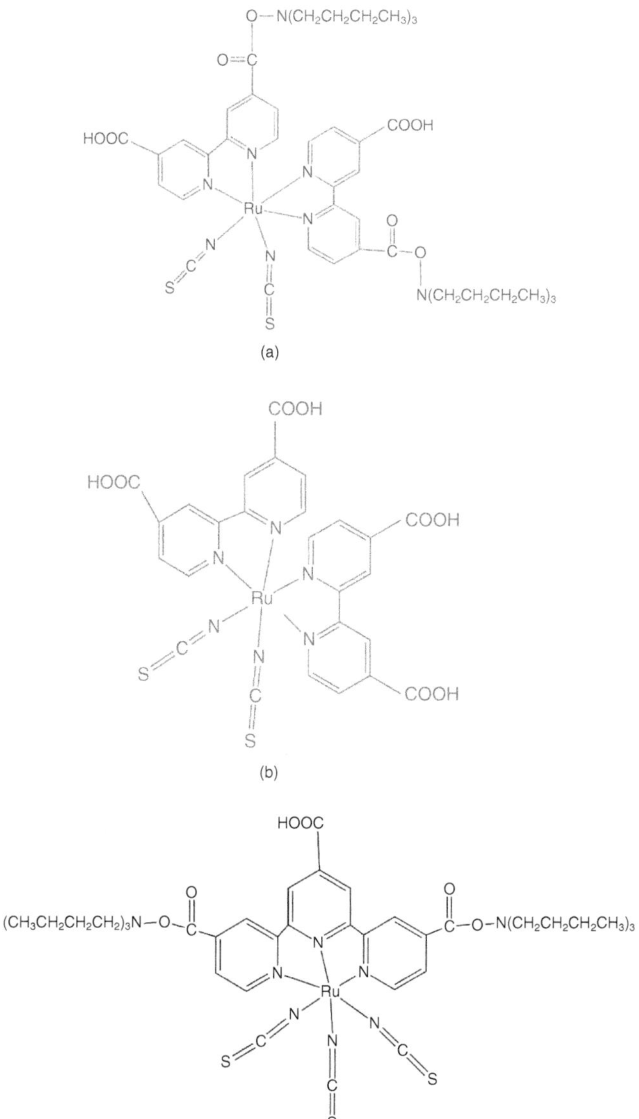

Figure 1.3 Structures of (a) N719, (b) N3, and (c) black dyes used in DSSCs.

and glass provides the back support. Platinum thin-film coated TCO/glass is used as a counter electrode, N719 dye as a sensitizer and iodide/tri-iodide as an electrolyte. Compactness with flexible DSSCs can be achieved by modifying the configuration of DSSCs. The modification may be done in TCO substrate, electrode, dye, electrolyte, or redox mediator. The configuration of DSSCs is attempted using thin metal foil as a substrate replacing both TCO and rigid glass substrate. It can withstand high sintering temperature and when TiO_2 is coated on the metal foil, the interconnectivity of the particle increases and internal resistance was also reduced. Initially, Grätzel's group deposited TiO_2 on the titanium foil [50]. The first solid-state flexible DSSCs were developed by Meyer et al. using titanium and stainless steel foil as a substrate and formed the TiO_2 material over the substrate by rapid sintering method. TiO_2 was adsorbed with the dye followed by the insertion of spiro-OMeTAD a hole-transporting material into the active matrix of TiO_2. This configuration showed very low efficiency of 0.8% under 1 sun illumination [51]. These metal substrates were then replaced by several other metals such as Pt, Co, Ti, Al, and W [52] as flexible substrates for photoanode, but at the interface, they form a corresponding insulating oxides that affect the conductivity. This could be retarded by making the ITO layer as an interface in between metal substrate and TiO_2 because it prevents the direct contact of TiO_2 with the metal that avoids the oxidation of metals. In these configurations, TCO–glass substrates were used as the counter electrode that allows the photons during the back illumination process [50].

1.4.2 Metal Substrates for Counter Electrode and Glass/TCO for Photoanode

Recently, researchers focused on enhancing the efficiency of DSSCs using metal substrates as counter parts of the cells such as platinized nickel, stainless steel, titanium, aluminum, copper. But Al and Cu used as counter electrodes showed the weak stability over iodide/tri-iodide electrolyte. So they remain unfit for making the cell, whereas, Pt–nickel and Pt–stainless steel counter electrodes gave high-electrical conductivity and low-internal resistance. The platinum that is coated over the nickel using the sputtering, electroplating technique gives an efficiency of 5.4%, which is comparatively higher than that of the Pt–TCO glass substrate [53]. Kim and Rhee investigated stainless steel-based counter electrodes (stainless steel–Pt) on the performance of DSSCs. The platinum thin-film sputtered on stainless steel substrate counter electrode showed the highest efficiency of 7.7% when coupled with glass–FTO/titanium oxide nanoparticles (TNPs) as a photoanode [54].

1.4.3 Metal Substrate for Photoanode and Polymer Substrate for Counter Electrode

The flexible DSSCs with TNPs on titanium foil as working electrode and poly(ethylene naphthalate)-indium-doped tin oxide/platinum-thinfilm, PEN-ITO/Pt as a counter electrode were fabricated by Grätzel and coworkers that exhibits an efficiency of 7.2% with back illumination process. Though the efficiency was highest, it is lower than that of the efficiency achieved by glass substrates. In another configuration, stainless steel substrate coated with ITO and SiO_x followed by TiO_2 paste as working electrode and Pt coated polymer was used as a counter electrode showing the efficiency of 4.2% at back illumination [55] which is still lower than DSSCs configured by glass substrate. Grätzel and coworkers build the DSSCs with TiO_2 nanotube coated on the titanium foil as the working electrode and glass–ITO/Pt substrate as the counter electrode. On back illumination at counter electrode, the efficiency was about 3.3%. Same configuration is built using different polymer substrates such as PEDOT, PET, PEN instead of glass (polymer–ITO/Pt) as the counter electrode that exhibited the efficiency of 3.6% at back illumination [56].

1.4.4 Polymer Substrates for Flexible DSSCs

The rigid glass substrate was replaced with flexible transparent polymer flims such as poly(ethylene terephthalate) (PET), poly(ethylene naphthalate) (PEN), and poly(ether sulfone) (PES) coated with TiO_2 as a photoanode. The TiO_2 material works well only at the calcination temperature of $450\,°C$, which is not the favorable for the transparent conducting low melting point polymer films. So the calcination temperature was limited to $200\,°C$. Though this temperature is not optimized for the fabrication of TiO_2, various low temperature methodologies such as microwave heating, hydrothermal crystallization, mechanical compression, electrophoretic deposition, chemical sintering were adopted for the fabrication of flexible solar cells irrespective of fabrication methods. The polymer substrates-based flexible solar cells was found to give a very low efficiency compared to glass substrates because of the poor adhesion of the TiO$_2$ material on polymer substrate and the lack of interparticle contact between TiO$_2$ particles due to the low calcination temperature [50]. But Arakawa and coworkers obtained the highest efficiency of 8.1% by fabricating flexible DSSCs using plastic substrate, coating TiO$_2$ paste on the substrate by mechanical pressing method [57]. This is the highest efficiency of DSSCs using plastic as a substrate.

1.4.5 Glass/TCO-Free Metal Substrates for Flexible DSSCs

The conductive mesh-based flexible DSSCs without TCO substrates were fabricated by Fan et al. using stainless steel mesh coated with porous TiO$_2$ as photoanode and 0.5 mm thick Pt foil as counter electrode. The efficiency was 1.49% with 10 μm thick TiO$_2$ layer [58]. Yoshida et al. configured DSSCs with TCO-free substrate. They used stainless steel mesh-TiO$_2$ as a working electrode and Pt-coated titanium foil as counter electrode that showed the efficiency of 5.56% from front illumination [59]. Quasi solid-state DSSCs using a gel electrolyte and stainless steel mesh as a substrate were fabricated by Huang and coworkers. MgO deposited TiO$_2$ colloidal solution was spray coated on the stainless steel mesh and sintered at high temperature followed by deposition of 1D TiO$_2$ nanofibers by electrospinning technique. Stainless steel substrate coated with polypyrrole or the platinum foil was used as the counter electrode. The efficiency was 2.8% with platinum foil as counter electrode, whereas the polypyrrole-coated stainless steel mesh achieved the efficiency of 2.3% [60]. A p–n junction-based flexible DSSCs using copper mesh as a substrate were fabricated by Heng et al. A p-type cuprous iodide was first deposited on copper mesh followed by the deposition of nanostructured TiO$_2$ by electrospun technique. It was then sensitized with Ru-based dye sensitizer. Pt foil acts as the counter electrode. The device was irradiated at Cu-mesh/CuI layer, the direct contact of Cu layer with TiO$_2$ forms the p–n junction that enabled the separation of electron and hole and transports the charge effectively which enhanced the J_{SC} value and efficiency (4.73%) of the cell [61].

1.4.6 Glass/TCO-Free Metal Wire Substrates for Flexible DSSCs

In 2008, Fan et al. designed twisted wire-shaped DSSCs using stainless steel-wire coated with TNPs as working electrodes and poly(vinylidene fluoride) (PVDF) polymer-coated Pt-wire as counter electrodes [62]. Both the electrodes were twisted together and configured to form the wire-shaped DSSCs. Cylinder-shaped DSSCs were fabricated by Liu and Misra by inserting Ti-wires/TiO$_2$-nanotube as working electrodes into the capillary tube along with Pt wires. The efficiency was found to be 2.78% for 55 μm long titanium dioxide nanotubes (TNTs) [63]. Liu et al. constructed TiO$_2$-coated titanium wires as working electrodes and platinum-coated titanium foil as counter electrodes that exhibited the efficiency of 4.1% by front-side illumination [64].

The following is the list of works done by various scientists all over the world using different metallic and polymeric substrates as the working and counter electrodes for flexible DSSCs (see Table 1.1).

1.5 Working Principle of DSSCs

The working principle of DSSCs involves the following processes:

1) Light absorption
2) Charge separation
3) Charge collection

Table 1.1 Various configuration evolved in dye-sensitized solar cells.

Substrate	Working electrode	Counter electrode	Done by
Metal substrate and TCO – glass substrate	Ti/TiO$_2$	FTO/Pt	Kim et al.
	Ti/ZnO	FTO/Pt	Fan et al.
	SS/TiO$_2$	FTO/Pt	Vijayakumar et al.
	Zn/ZnO	FTO/Pt	Gao et al.
Metal substrate and TCO – polymer substrate	Ti and SS/TiO$_2$	ITO–PET/Pt	Chang et al.
	Ti/ZnO	ITO–PEN/Pt	Lin et al.
	Ti/TiO$_2$	ITO–PEN/Pt-SWCNT	Xiao et al.
	TNT/SrO–TiO$_2$	ITO–PET	Chen et al.
Both as metal substrate	Ti/TiO$_2$	Ti/Pt	Xiao et al.
	SS/TiO$_2$	SS/Pt	Bonilha et al.
	SS/TiO$_2$	Glass paper/Pt	Cha et al.
	Ti/TiO$_2$	Ti/Pt/PEDOT	Wu et al.

Source: Balasingam et al. 2013 [50]. Reproduced with permission of Royal Society of Chemistry.

The mechanism resembles and imitates the natural processes of plants that convert sunlight into chemical energy called photosynthesis. In DSSCs, solar to electrical energy conversion occurs by ruthenium-based dye-sensitized nanocrystalline TiO$_2$ photoanode. In DSSCs, charge separation is carried out by a kinetic approach in the same manner as photosynthesis process that leads to photochemical action. The well-configured DSSCs have its unique and specific mechanism [65].

Prior to the discussion about the working principle of DSSCs, the factors influencing the solar to electrical conversion efficiency of the cell should be considered. The efficiency of DSSCs depends on the three criterial aspects: (i) the frontier molecular orbitals of dye (i.e. highest occupied molecular orbital (HOMO) and the lowest unoccupied molecular orbital, LUMO). (ii) On the Fermi-level of scaffold material and (iii) the redox-potential of the electrolyte shuttle. The energy gap between the HOMO and LUMO of the dye molecules should be as short as possible. So that electrons can be easily excited from ground state to excited state by the absorption of photons. Then the conduction band of titanium dioxide must be lower to the LUMO level of dye molecules to facilitate the excited electrons flow from dye to TiO$_2$ material and then circulate to external circuit via TCO substrate. Working principle of DSSCs is quite interesting due to the flow of electrons from the photoanode to counter electrode, which generates electricity (Figure 1.4). The working mechanism of the DSSCs are as follows.

The heart of the system is a mesoscopic semiconductor oxide film, which is placed in contact with a redox electrolyte or an organic hole conductor. Attached to the surface of the nanocrystalline film is a monolayer of the dye sensitizer. Under sunlight illumination, the dye sensitizer will absorb photons (light) and the energy of the photon is sufficient to excite the electron from HOMO to LUMO level of the dye molecule and become photoexcited. The charge separation is attained across dye and TiO$_2$ interface, where an electron is located in the TiO$_2$ and a hole is located in the oxidized dye molecule. The adsorbed dye molecules inject electrons from the LUMO level of dye into the conduction band of nanocrystalline anatase phase TiO$_2$ which is lower in energy than the excited state of the dye (Figure 1.5). The electrons transferred through the porous network of TiO$_2$ eventually reach the back contact of the working electrode where charge collection occurs. The extracted charge can subsequently perform electrical work in the external circuit and eventually return to the counter electrode where platinum is coated on the

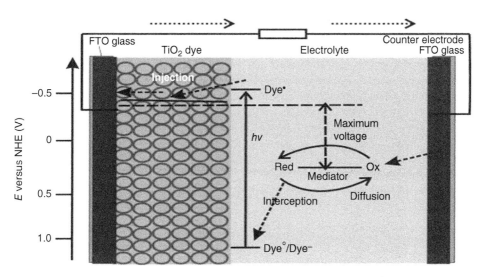

Figure 1.4 Schematic diagram of the working principle of DSSCs.

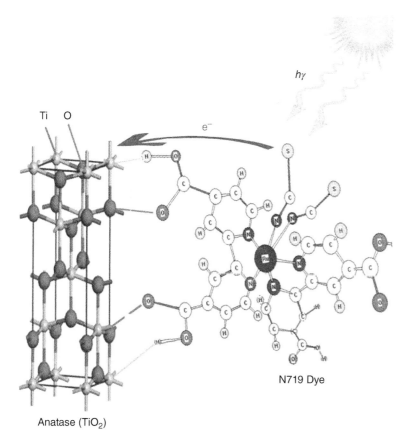

Figure 1.5 Interaction of dye and TiO_2. Reprinted with permission from ref. [66]

FTO glass substrate that acts as a counter part of the cell. This platinum (2+) which already gets oxidized by the electrolyte receives the electron from the external circuit and is reduced to the platinum (0). Now again, the platinum will get oxidized by donating the electron to the electrolyte. The electron enters into the iodine electrolyte that has a potential to carry out the redox reaction and regenerate the dye molecule. In the electrolyte, there is an iodine molecule (I_2) and iodide ion (I^-) that normally is combined as I_3^- (triiodide ion). This ion accepts the electron from the counter electrode and gets reduced and forms three iodide ions ($3I^-$). Again, these iodide ions get oxidized to triiodide ions by donating the electrons to dye. So the liquid redox electrolyte will complete the circuit by reducing the oxidized dye. The voltage generated under illumination corresponds to the difference between the Fermi level of the electron in the solid and the redox potential of the electrolyte [67]. Overall, the device generates electric power from light without suffering any permanent chemical transformation.

1.5.1 Electron Transfer Mechanism in DSSCs

$$TiO_2/S + Photon\ (h\upsilon) \rightarrow S^*\ (\text{Excitation process})$$

$$TiO_2/S^* + TiO_2 \rightarrow TiO_2/S^+ + e_{(CB)}^-\ (\text{Injection process})$$

$$TiO_2/2S^+ + 3I^- \rightarrow TiO_2/2S + I_3^-\ e_{(CE)}^-\ (\text{Regeneration})$$

$$I_3^- + 2e^- \rightarrow 3I^-\ (\text{Reduction})$$

$$I_3^- + 2e^- \rightarrow 3I^-\ (\text{Re} - \text{caption in dark reaction})$$

$$TiO_2/S^+ + e_{(CB)}^- \rightarrow TiO_2/S\ (\text{Recombination in dark reaction})$$

1.5.2 Photoelectric Performance

Under the radiation of solar spectrum with photon flux, the photon-energy-to-electricity conversion efficiency is defined as follows:

$$\eta = \frac{J_{SC} \times V_{OC} \times FF}{J_0}$$

where J_{SC}, short circuit current; V_{OC}, open circuit voltage; FF, fill factor; and J_0, photon flux.

When the cell is exposed to illumination from the direct sun, the photocurrent and voltage drop can be interpreted through $I-V$ characteristics curve. The short circuit current I_{SC} varies with dye concentration. The magnitude of the photocurrent depends upon the nature and concentration of the dye used. When the concentration of the dye is high, the performance of the cell is high. On the other hand, the lower concentration of dyes reduces the efficiency of the cell and the magnitude of the photocurrent [68]. The incident photon-to-current conversion efficiency (IPCE) depends upon the magnitude of the short circuit current, the photoresponse, and the light intensity as given below:

$$IPCE\ (\lambda) = \frac{1240\ (eV\ nm) \times J_{sc}(\mu A/cm^2)}{\lambda\ (nm) \times I\ (\mu W/cm^{-2})}$$

where λ is the wavelength of the adsorbed photon; I, light intensity at wavelength λ.

Acknowledgments

The corresponding author, Dr. S.N. Karthick would like to thank the funding support of University Grant Commission - Basic Scientific Research (UGC-BSR) Start-Up-Grant (2018-2020) from the University Grant Commission, New Delhi, India, the National Research Foundation of Korea (Project No. 2015R1D1A4A01019537) and BK 21 PLUS, creative human resource development programme for IT convergence, Republic of Korea.

References

1 Kalyaanasundram, K. (2010). *Fundamental Sciences in Dye Sensitized Solar Cells*, vol. 1, 543. EPFL Press, Published by CRC Press.
2 Horvath, A. and Rachlew, E. (2016). *Ambio* 45: 38–49.
3 Williams, R. (1960). *J. Chem. Phys.* 32: 1505.
4 Mc Quarrie, D.A. (2016). *Viva Student Edition*. Sausalito, CA: University Science Books.
5 Wong, D., Lee, P., Shenghan, G. et al. (2011). *Eur. J. Phys.* 32: 1059–1064.
6 Chopra, K.L., Paulson, P.D., and Dutta, V. (2004). *Prog. Photovoltaics Res. Appl.* 12: 69–92.
7 Ibn-Mohammed, T., Koh, S.C.L., Reaney, I.M. et al. (2017). *Renew. Sustain. Energy Rev.* 80: 1321–1344.
8 Gerischer, H., Michel-Beyerle, M.E., Rebentrost, F., and Tribijxch, H. (1968). *Electrochim. Acta* 13: 1509–1515.
9 Tributsch, H. and Calvin, M. (1971). *Photochem. Photobiol.* 14: 95–112.
10 Tributsch, H. (1972). *Photochem. Photobiol.* 16: 261–269.
11 Matsumura, M., Matsudaira, S., and Tsubomura, H. (1980). *Ind. Eng. Chem. Prod. Res. Dev.* 19: 415–421.
12 O'Regan, B. and Grätzel, M. (1991). *Nature* 353: 737–740.
13 Wang, P., Zakeeruddin, S.M., Moser, J.E. et al. (2003). *Nat. Mater.* 2: 402–407.
14 Wang, Q., Campbell, W.M., Bonfantani, E.E. et al. (2005). *J. Phys. Chem. B* 109: 15397–15409.
15 Bai, Y., Cao, Y., Zhang, J. et al. (2008). *Nat. Mater.* 7: 626–630.
16 Chen, C.-Y., Wang, M., Li, J.-Y. et al. (2009). *ACS Nano* 3: 3103–3109.
17 Mathew, S., Yella, A., Gao, P. et al. (2014). *Nat. Chem.* 6: 242–247.
18 Kakiage, K., Aoyama, Y., Yano, T. et al. (2015). *Chem. Commun.* 51: 15894–15897.
19 Yang, Z., Gao, S., Li, T. et al. (2012). *ACS Appl. Mater. Interfaces* 4: 4419–4427.
20 Lin, C.-Y., Lai, Y.-H., Chen, H.-W. et al. (2011). *Energy Environ. Sci.* 4: 3448–3455.
21 Choi, J.-W., Kang, H., Lee, M. et al. (2014). *RSC Adv.* 4: 19851–19855.
22 Tan, B., Toman, E., Li, Y., and Wu, Y. (2007). *J. Am. Chem. Soc.* 129: 4162–4163.
23 Sayyed, S.A.A.R., Beedri, N.I., Kadam, V., and Pathan, H.M. (2016). *Mater. Sci.* 39: 1381–1387.
24 Zheng, H., Tachibana, Y., and Kalantar-zadeh, K. (2010). *ACS Langmuir* 26: 19148–19152.
25 Kim, C.W., Suh, S.P., Choi, M.J. et al. (2013). *J. Mater. Chem. A* 1: 11820–11827.
26 Sayama, K., Sugihara, H., and Arakawa, H. (1998). *Chem. Mater.* 10: 3825–3832.
27 Justin Raj, C., Prabakar, K., Karthick, S.N. et al. (2013). *J. Phys. Chem. C* 117: 2600–2607.
28 Haggerty, J.E.S., Schelhas, L.T., Kitchaev, D.A. et al. (2017). *Sci. Rep.* 7: 15232.
29 Wu, J., Lan, Z., Lin, J. et al. (2017). *Chem. Soc. Rev.* 46: 5975–6023.
30 Zhong, C., Hu, W.B., and Cheng, Y.F. (2011). *J. Power Sources* 196: 8064–8072.
31 Wu, M., Chen, C., Chen, Y., and Shih, H. (2016). *Electrochim. Acta* 215: 50–56.
32 Huo, J., Zheng, M., Tu, Y. et al. (2015). *Electrochim. Acta* 159: 166–173.

33 Nam, J., Park, Y., Kim, B., and Lee, J. (2010). *Scr. Mater.* 62: 148–150.

34 Anothumakkool, B., Agrawal, I., Bhange, S.N. et al. (2016). *ACS Appl. Mater. Interfaces* 8: 553–562.

35 Wu, J., Lan, Z., Lin, J. et al. (2015). *Chem. Rev.* 115: 2136–2173.

36 Kalyanasundaram, K., Vlachopoulos, N., Krishnan, V. et al. (1987). *J. Phys. Chem.* 91: 2342–2347.

37 Liu, Y., Hagfeldt, A., Xiao, X.R., and Lindquist, S.E. (1998). *Sol. Energy Mater. Sol. Cells* 55: 267–281.

38 Law, C.H., Pathirana, S.C., Li, X.O. et al. (2010). *Adv. Mater.* 22: 4505–4509.

39 Nazeeruddin, M.K., Liska, P., Moser, J. et al. (1990). *Helv. Chim. Acta* 73: 1788–1803.

40 Papageorgiou, N., Athanassov, Y., Armand, M. et al. (1996). *J. Electrochem. Soc.* 143: 3099.

41 Xi, C., Cao, Y., Cheng, Y. et al. (2008). *J. Phys. Chem. C* 112: 11063.

42 Burschka, J., Dualeh, A., Kessler, F. et al. (2011). *J. Am. Chem. Soc.* 133: 18042.

43 Boschloo, G. and Hagfeldt, A. (2009). *Acc. Chem. Res.* 42: 1819–1826.

44 Oskam, G., Bergeron, B.V., Meyer, G.J., and Searson, P.C. (2001). *J. Phys. Chem. B* 105: 6867–6873.

45 Wang, M., Chamberland, N., Breau, L. et al. (2010). *Nat. Chem.* 2: 385–389.

46 Shalini, S., Balasundaraprabhu, R., Satish Kumar, T. et al. (2016). *Int. J. Energy Res.* 40: 1303–1320.

47 Choi, H., Baik, C., Kang, S.O. et al. (2008). *Angew. Chem. Int. Ed.* 47: 327–330.

48 Wang, Z.-S., Cui, Y., Dan-oh, Y. et al. (2008). *J. Phys. Chem. C* 112: 17011–17017.

49 Hemalatha, K.V., Karthick, S.N., Justin Raj, C. et al. (2012). *Spectrochim. Acta, Part A* 96: 305–309.

50 Balasingam, S.K., Kang, M.G., and Jun, Y. (2013). *Chem. Commun.* 49: 1471–1487.

51 Meyer, T.B., Meyer, A.F., and Ginestoux, D. (2002). *Proc. SPIE Int. Soc. Opt. Eng.* 13: 4465.

52 Man, G.K., Park, N.G., Kwang, S.R. et al. (2005). *Chem. Lett.* 804.

53 Ma, T., Fang, X., Akiyama, M. et al. (2004). *J. Electroanal. Chem.* 574: 77–83.

54 Kim, J.M. and Rhee, S.W. (2012). *J. Electrochem. Soc.* 159: B6–B11.

55 Ito, S., Ha, N.L.C., Rothenberger, G. et al. (2006). *Chem. Commun.*: 4004.

56 Kuang, D., Brillet, J., Chen, P. et al. (2008). *ACS Nano* 2: 1113–1116.

57 Yamaguchi, T., Tobe, N., Matsumoto, D. et al. (2010). *Sol. Energy Mater. Sol. Cells* 94: 812–816.

58 Fan, K., Chen, J., Yang, F., and Peng, T. (2012). *J. Mater. Chem.* 22: 4681–4686.

59 Yoshida, Y., Pandey, S.S., Uzaki, K. et al. (2009). *Appl. Phys. Lett.* 94: 093301.

60 Huang, X., Shen, P., Zhao, B., Feng, X., et al. (2010). *Sol. Energy. Mater. Sol. Cells* 94: 1005–1010.

61 Heng, L., Wang, X., Yang, N. et al. (2010). *Adv. Funct. Mater.* 20: 266–271.

62 Fan, X., Chu, Z., Wang, F. et al. (2008). *Adv. Mater.* 20: 592–595.

63 Liu, Z. and Misra, M. (2010). *ACS Nano* 4: 2196–2200.

64 Liu, Y., Li, M., Wang, H. et al. (2010). *J. Phys. D: Appl. Phys.* 43: 205103.

65 Grätzel, M. (2003). *J. Photochem. Photobiol., C* 4: 145–153.

66 Grätzel, M. (2001). *Nature* 414: 338–344.

67 Zhao, Z. and Liu, Q. (2007). *J. Phys. D: Appl. Phys.*41: 025105.

68 Yanagida, S., Yu, Y., and Manseki, K. (2009). *Acc. Chem. Res.* 42: 1827–1838.

2

Function of Photoanode: Charge Transfer Dynamics, Challenges, and Alternative Strategies

A. Dennyson Savariraj[1,2] and R.V. Mangalaraja[1,3]

[1]*Department of Materials Engineering, Faculty of Engineering, Advanced Ceramics and Nanotechnology Laboratory, University of Concepcion, Concepcion, Chile*
[2]*Department of Chemical Engineering, Khalifa University of Science and Technology, Abu Dhabi, United Arab Emirates*
[3]*Technological Development Unit (UDT), University of Concepcion, Coronel Industrial Park, Coronel, Chile*

2.1 Introduction

The increase in population demanded heavy energy requirements that accelerated the depletion of fossil fuels [1]. Among several renewable resource-based alternatives, dye-sensitized solar cell (DSSC) is one of the cost-effective and potential substitutes to silicon solar cells that emerged in the early 1990s in the field of photovoltaics, as it involves inexpensive components and simple fabrication process. DSSC is entirely different from the conventional p/n junction solar cells in terms of operating principle; however, it adopts the principle of natural photosynthesis process and, therefore, is it is often addressed as artificial photosynthesis. The monolayer of dye molecules on the photoanode functions like chlorophyll in plants and absorbs the incident light to generate both positive and negative charge carriers in the cell.

The light-harvesting efficiency of DSSC relies upon the photoanode that is comprised of wide band gap materials such as TiO_2, ZnO, and dye. The dyes are designed and synthesized based on the absorption range so as to make the injection of excited electrons sensitizers (dye) to TiO_2 matrix. The performance of any dye depends on the thickness of the TiO_2 film, and for enhanced light harvest, a high uptake of dye molecules on to the surface of the electrode is necessary to promote better light absorption. For this purpose, a thick layer of TiO_2 film with a large surface area is required to make better dye adsorption and to have higher light harvesting [2, 3].

When the sunlight is incident on the DSSC, the monolayer of dye molecules on the mesoporous meal oxide (TiO_2) film get excited, generating an electron hole pair. The electron and hole must get separated without getting recombined again, and the electron is injected in to the conduction band of the TiO_2 film. The injected electron then travel through the mesoporous film by diffusion to reach the photoanode and complete the cycle by supplying an external load followed by reaching the counter electrode. The electron collected at the counter electrode is procured by the electrolyte and utilized to regenerate the dye molecule [1].

In this chapter, we discuss the composition of DSSC, the photoanode in particular, and the function of the photoanode. Several processes such as charge generation, charge separation, electron diffusion, charge recombination, and their mechanisms are explained in details which are very much essential to understand the working principle of DSSC. The challenges in the above-said processes and the strategies to overcome are also suggested.

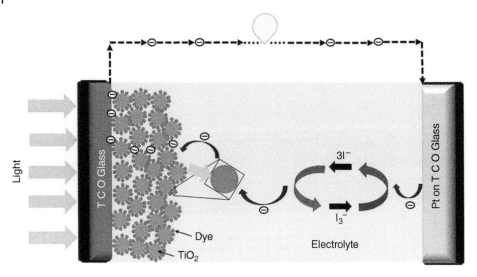

Figure 2.1 Schematic illustration of a typical dye-sensitized solar cell (DSSC).

2.2 The General Composition of DSSC

The typical DSSC consists of five components in total (i) a mechanical support coated with transparent conductive oxides (TCOs); (ii) nanostructured metal oxide semiconducting film, usually TiO_2 or ZnO to transport electrons; (iii) dye sensitizer adsorbed onto the surface of the semiconductor (dye) to harvest solar energy and to produce excitons; (iv) an electrolyte containing a redox mediator or hole-transporting material (iodine/iodide redox couple); (v) a counter electrode capable of regenerating the redox mediator like platinum. A DSSC consists of two electrodes with at least one of them being transparent for light to the visible light. To construct the two electrodes namely photoanode and counter electrode, glass covered with a TCO material either indium tin oxide (ITO) or fluorine doped tin oxide (FTO) is utilized [4, 5]. Figure 2.1 presents the schematic illustration of typical DSSCs.

2.3 Selection of Substrate for DSSCs

As mentioned earlier, the sandwich structure of the DSSCs requires two transparent conducting glass substrates for constructing both photoanode and counter electrode, and they should possess low sheet resistance when subjected to high temperature since the sintering process of the TiO_2 layer takes place at temperatures between 450 and 520 °C. The sheet resistance of the TCO is usually 5–15 Ω/cm^2 and in this regard both indium-doped tin oxide (In:SnO_2, ITO) and fluorine-doped tin oxide (F:SnO_2, FTO) have been used, however, FTO glass is preferred over ITO due to the increase in resistance even at high temperature and low thermal stability of the latter. Flexible substrates can also be employed by depositing conducting materials on the polymer-based substrates, which have the advantages of being flexible, light weight, and easy for roll-to-roll printing when taken for commercialization, although they cannot be subjected to temperature beyond 150 °C. Besides thermal stability and low resistance, it is also necessary that the substrates should possess high transparency to solar radiation in the visible–infrared (IR) window.

2.4   Photoanode

Though several wide-band-gap oxide semiconductors such as TiO_2, ZnO, and SnO_2 have been explored for their potential application as electron acceptors for solar cell, TiO_2 has proven to be superior for its properties such as high chemical stability, nontoxicity, abundance, and high efficiency. Unlike other wide-band-gap semiconductor oxide materials like ZnO, TiO_2 is resistant to photo-degradation when excited [6, 7]. Therefore, the photoanode is made of meso-porous oxide layer of nanometer-sized particles of TiO_2 in anatase phase, which is sintered at high temperature to permit electronic conduction to occur, although alternative materials such as ZnO, SnO_2, and Nb_2O_5 are also used to construct photoanode [8–13].

TiO_2 exists in several crystalline forms, and among them anatase, rutile, and brookite are the important ones [14]. The crystal structure of anatase and rutile are very similar with a tetragonal symmetry of the Ti^{4+} atoms and have sixfold coordination to oxygen atom; however, the position of oxygen atom is different as shown in Figure 2.2. Anatase phase is thermodynamically less stable as the average distance between the Ti^{4+} atoms in anatase phase is shorter than that of rutile phase. At temperature between 700 and 1000 °C, the phase transition from anatase to rutile takes place and that depends on the crystallite size [16] and impurities. Due to the negative shift in the conduction band level by 0.2 eV, anatase has a band gap of 3.2 eV against the indirect band gap of rutile with 3.0 eV [14]. The stoichiometric crystals of TiO_2 are insulating because the TiO_2 consists of partly ionic and partly covalent bond. Therefore, a number of oxygen vacancies-based trap states are introduced during the synthesis procedure, and these vacancies can also be created reversibly either under reduced pressure [17] and/or elevated temperature [18], which can result in a variation of conductivity in several folds. The TiO_2 crystals are doped negatively (n-type) due to the formation of Ti^{3+} state from oxygen vacancies. In DSSC, the photoanode usually consists of double layer of mesoporous metal oxide: an underlying light-scattering layer made of larger size nanoparticles (200–300 nm) to the thickness of 2–4 μm to induce phototrapping effect and over that another layer of thin film layer to the thickness of 2–15 μm consisting of nanoparticles of 10–30 μm.

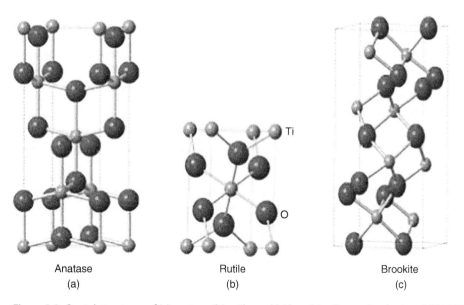

| Anatase | Rutile | Brookite |
| (a) | (b) | (c) |

Figure 2.2 Crystal structures of (a) anatase, (b) rutile, and (c) brookite. *Source:* Etacheri et al. 2015 [15]. Reproduced with permission from Elsevier.

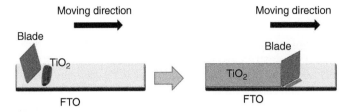

Figure 2.3 Schematic illustration of doctor blade method.

2.4.1 Coating Procedure

The photoanode is made of a wide-band semiconductor metal oxides such as TiO_2, ZnO, SnO_2, and Nb_2O_5. The metal oxides are made into a slurry in a suitable solvent along with a suitable binder so that the paste adheres to the substrate and has rheological properties of a meso-porous film. The paste/slurry thus prepared is deposited on to the TCO substrate wither by doctor-blade method (Figure 2.3) or screen-printing method after which the oxide layers are sintered at 450–520 °C to make the individual particles to come closer to enhance conduction and to improve the charge collection apart from removing the organic matters which would otherwise function as trap states to cater charge recombination. In normal case, the typical thickness of the mesoporous oxide film is between 2 and 15 μm with the particles size of 10–30 nm.

2.4.2 Significance of Using Mesoporous Structure

The high surface area of the mesoporous metal oxide film on the photoanode is a prerequisite to cater to strong absorption of solar irradiation for an efficient device performance by a mono-layer of adsorbed sensitizer dye. Owing to the negligible absorption of light when a monolayer of dye is adsorbed on to the flat interface, mesoporous films are used as they dramatically enhance the interfacial surface over the geometric area to a large figure that is 1000-fold even for a 10 μm thick film, giving way for the absorbance of high visible-light absorbance from several consecutive monolayers of adsorbed dye in the optical path. In addition to the above-stated advantage, usage of a monolayer dye eliminates the requirement for excitation diffusion to the dye/metal oxide interface and avoids the nonradiative quenching of excited states resulting from thicker molecular films. The main disadvantage of mesoporous material is the losses arising from inter-facial recombination [8].

2.5 Sensitizer

Dye functions as the photosensitizer whose fundamental function is absorbing the solar radiation, generate excitons (electron and hole pair) and inject electrons in to the conduction band of TiO_2. To the surface of the nanocrystalline semiconductor film, a monolayer of dye is attached by physical adsorption. A dye should fulfill the following criteria to be an efficient photosensitizer: (i) For enhanced light harvesting, the photosensitizer should have strong light absorption in the visible and near-IR region. (ii) It should be highly soluble in organic solvents so as to have efficient adsorption on to the from the stock solution. (iii) Must possess anchoring ligand groups such as carboxylic or phosphonic acid groups to enable effective interaction with TiO_2 surface followed by coupling of donor and acceptor levels. (iv) For an efficient charge injection in to TiO_2, the lowest unoccupied molecular orbital (LUMO) of the photosensitizer must be sufficiently high in energy and similarly for an efficient regeneration of the oxidized dye molecule by the redox pair (electrolyte), the highest occupied molecular orbital (HOMO) should be sufficiently low in energy. (v) It should have high thermal and chemical stability to yield high efficiency and longevity of the device. (vi) The transfer of electron from the dye to TiO$_2$ should be rapid without undergoing decay to the ground state for recombination.

Over the past two decades, several dyes, based on metals such as Ru [19], Os [20], Re [21], Pt [22], Fe [23], and Cu [24] apart from transition metal complexes such as porphyrin [25] and phtalocyanine dyes [26] and organic molecules such as hemicyanines [27], squaraines [28], coumarins [29], and indolines [30] are also investigated for their potential application as photosensitizes.

Nevertheless, the dyes with superior performance are reported to be the polypyridine complexes of Ru(II) with at least one anchoring group such as carboxylic and phosphonic acid on the surface. Till date, the most commonly used dye is the Ru-bipyridyl dye N719, which is the ditetrabutyl-ammonium salt of RuL$_2$(NCS)$_2$, where L = 4,4'-dicarboxy-2,2'-bipyridine.

When light of energy equal to $h\upsilon$ is incident on the sensitizer dye molecules, they are subjected to photoexcitation, whereby the electrons are injected in to the conduction band of the oxide where the dye molecules oxidized are denoted as S$^+$. The regeneration of the dye is carried out by restoring it to the ground state through oxidation with the help of an electron donated by the electrolyte. The electrolyte usually employed is iodide/triiodide redox couple dissolved in a liquid organic solvent.

In DSSC, the important thing that determines the overall photo conversion efficiency of the cell is the charge separation which is dominated by the materials, design of the device, and parameters. In addition to it, the two important processes that determine photo conversion efficiency are electron injection from the dye into the metal oxide electrode (photoanode) and the regeneration of the dye to the ground state by the redox electrolyte. The electrical circuit is completed through electron migration in the external load, while the redox electrolyte does the job of scavenging the hole to regenerate the dye [8].

2.6 Charge Transfer Mechanism

Charge transfer mechanism takes place in several steps.

In the metal complex dye, the composition is a central metal ion surrounded with chelating ligands with at least one or more anchoring group to adhere to the mesoporous oxide surface. The electron promotion from the metal to ligand takes place when the photosensitizer absorbs light in the visible region, and this exodus of electron is usually from the d orbital of the metal ion to the π^* orbital of the ligand, by d(π)– π^* transition [31].

After excitation, the electron gets injected to the d orbital of the TiO$_2$ layer from the π^* orbital of the ligand on the metal ion. While electron injection to TiO$_2$ takes place after photoexcitation, a temporary interfacial Ti^{3+}–ligand–Ru^{2+} charge transfer complex is formed induced by Ru-complex through back-bonding reaction [32].

2.7 Interfaces

To achieve an efficient electron injection in to the conduction band of the metal oxide, the dye/metal oxide interface is to be designed in such a way that the oxidation potential of the excited dye (LUMO) is sufficiently negative. When light is incident on the photosensitizer (dye molecule), it absorbs energy and attains excited energy state (LUMO) and generates excitons (electron–hole pair) followed by diffusion in to the dye/metal oxide interface developing a built- in energy gradient (ΔE) at the metal oxide interface, caused by the energy difference between LUMO state of excited dye and the conduction band of the metal oxide (E_{CB}). The energy gradient (ΔE) attracts the electrons in the excitons against the binding energy (E_{EX}) and when the energy gradient (ΔE) exceeds the binding energy (E_{EX}) of the excitons, dissociation of the excitons occurs. Forward electron transfer occurs by injecting the electron from the dye to the metal oxide as the energy level of freed electrons is equal to the conduction state of the metal oxide (E_{CB}) as given in the following equation [33]

In order to facilitate electron injection effectively, the LUMO of the excited dye (photosensitizer) has to be in-line with the lower limit of the metal oxide's conduction band [34, 35]. For enhanced light-harvesting charge injection and the number of sensitizer molecules attached on the surface of the metal oxide are very essential, and more important is that those sensitizer molecules must have functionalities attached to them. More number of sensitizer molecules will be accommodated only when they properly orient themselves whereby the coverage area per molecule is reduced to give a close packed arrangement of them. The relative orientation of the donor and acceptor moieties determine the rate constant for the migration of the excited energy and the anchoring group attached to the sensitizer molecule directs and dictates their orientation.

None the less, the above is made possible only if the dye molecule is not adsorbed as aggregates because aggregation of the dye molecule will bring down the transfer of electron from the LUMO of the dye to the conduction band of the metal oxide as aggregation causes inappropriate energetic position of the LUMO level. In addition to this, the binding of the sensitizer molecule directly affects the injection of electrons and eventually the current density.

The relative orientation of the LUMO of sensitizer (donor) and the conduction band of metal oxide (acceptor) and enchantment of electron injection is dependent on the anchoring groups that link the sensitizer and metal oxide surface, while the relative position of the sensitizer's LUMO as well as the anchoring groups governs the electronic coupling strength [36, 37]. For effective electron injection, the distance between the anchoring group and the nearest LUMO should be short, besides the influence of dye metal oxide interface [38].

2.8 Significance of Dye/Metal Oxide Interface

As stated previously, electron injection is highly dictated by the binding between the sensitizer molecule and the metal oxide surface, in other words, the dye/metal oxide interface plays a key role for electron injection process. Figure 2.4 presents a schematic diagram of different materials and interfaces involved in DSSC. This interface is formed by grafting the dye's anchoring group making a coordination linkage with the metal oxide and the two common anchoring groups are carboxylic acid and phosphonic acid. While dyes with phosphonic acid group demonstrated better moisture stability than those with carboxylic acid group [39], the dyes bearing the latter as anchoring group exhibit enhanced electron injection [40] giving rise to better photocurrent in twofolds as compared to the former [41]. This implies that sufficient care should be taken to reduce the distance between the sensitizer and the metal oxide, and employing bidentate ligands will make the electron transfer more rapid as the bonds formed are stronger and shorter [42–44].

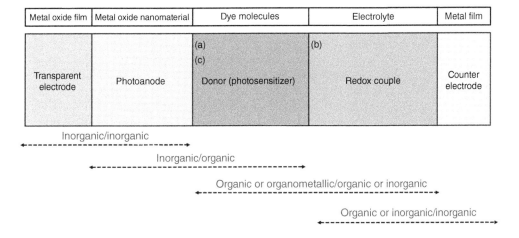

(a) – Metal oxide/dye interface, (b) – Dye/electrolyte interface and (c) – Metal oxide/electrolyte interface

Figure 2.4 Combination of different materials and interfaces involved in DSSC. *Source:* Thavasi et al. 2009 [33]. Reproduced with permission from Elsevier.

Between the cationic metal and oxygen atom of the carboxylic acid, there exits electronic coupling to bind the carboxyl group that will boost transfer of electron when photoexcitation occurs. The efficient electron injection depends on the nature of carboxyl group to the metal oxide group. Normally, the carboxyl groups form two types of linkages with metal oxide: (i) ester like linkage C—O and (ii) carboxylate linkage (C—O—O), and it was found that the latter yielded a decrease in ligand's electron density, bearing low energy shift in bands. When employing N3 dye (*cis*-di(thiocyanato)-*bis*(2,2′-bipyridyl-4,4′-dicarboxylic acid)-ruthenium(II)), the N3/TiO$_2$ interface demonstrates higher conversion efficiency owing to the anchoring by two carboxyl group [44] and due to high rotational freedom of N3 dye molecule, the carboxylic group makes a bidentate linkage for want of stability. Similar conversion efficiency is also exhibited by black dye [(C$_4$H$_9$)$_4$N]$_3$[Ru(Htcterpy)(NCS)$_3$] (tcterpy = 4,4′,4′-tricarboxy-2,2′,2′-terpyridine) only with carboxylic anchoring group [45].

In DSSC, both stability and efficiency are to be taken in to account and a compromise has to be made since efficient electron injection and stability are facilitated by carboxylate group and phosphonate groups, respectively. Considering the strong electron withdrawing nature and stability with a covalent like bonding ester linkage is opted to achieve better electron injection [33]. Therefore, ester linkages are introduced into dyes by reflux treatment, and a comparison study made with the dyes having carboxylate groups revealed that ester linkages do heap in achieving higher conversion efficiency [46].

2.9 Factors That Influence Efficiency in DSSC

2.9.1 Dye Aggregation

Dye aggregation arising due to unfavorable adsorption geometry seems to be a major bottleneck directly affecting the conversion efficiency via poor electron injection [47]. Though a dye may have several anchoring groups, not necessarily all the anchoring groups get attached to the surface of the metal oxide. Both the number and type of anchoring groups attached to the metal oxide surface, influence the difference in the energy level which in turn will determine the efficiency of electron injection. Therefore, the alignment of dye on the metal oxide surface plays a crucial role in the performance of DSSC. Therefore, for constructing a device to yield highest efficiency, zero-aggregated or least aggregated dye anchoring on the metal oxide should be chosen to enable efficient electron transfer. Dye aggregation caused by excessive sonication or stirring, will result in slower electron injection or self-quenching leading to inefficient electron injection and dye regeneration by the redox couple [48].

In the case of N3 dye, it has two bipyridine ligands and four carboxyl groups and adsorption can take place through protonation of one or more of the available. The extent of dye protonation greatly influences the energy conversion performance because increased number of protons on the sensitizer alters the energetics of the metal oxide's conduction band which in turn brings down the driving force between the dye and the redox couple (electrolyte) decreasing open-circuit voltage (V_{OC}) and short-circuit current density (J_{SC}). The influence of protonation has been investigated and reported by Nazeerudin et al. using N3 dye: The monoprotonated N3 dye seemed to have exhibited a photo conversion efficiency of 9.3% against 7.4% for the one with four-protonation [44] and a high of 100% electron injection by monoprotonated N3 in to TiO_2 has been reported [49]. Monoprotonated dye yields better photoconversion efficiency for the DSSCs with high values of V_{OC} and J_{SC}, and in the case of monoprotonated N3 dye, it gives a highest J_{SC} of 19 mA/cm^2, while diprotonated N712 yields 13 mA/cm^2, and zeroprotonated N719 exhibit 17 mA/cm^2.

The second type of aggregation taking place is due to anchoring via intermolecular hydrogen bonding [47]. Black dye is a typical example for this: the long alkyl chain present in the dye molecule forms hydrogen bonding, while anchoring on to the metal oxide surface results in aggregation and eventual poor electron injection. Carboxyl groups while functioning as a better anchoring group and enhancing electron transfer do have uncoordinated binding when the number is very high. The reason attributed for this is the steric hindrance. Therefore, fewer number of carboxyl groups will be preferred for enhanced electron transfer with lesser H bonding resulting in aggregation [50]. Z907 dye has better photoconversion efficiency due to well organized self-assembly when the few COOH group anchor on to the metal oxide avoiding self-quenching at the interface between the dye and the metal oxide, however, dyes with only one COOH group such as cyanine-based organic dyes L3 and L4 suffer from poor electron injection efficiency because of unfavorable orientation or binding of the anchoring group with the metal oxide [51]. Therefore, aggregation of the dye can be overcome by taking in to account the orientation, nature of binding and the structure of the dyes while designing and synthesizing dyes as they determine the electron injection efficiency which in turn reflects the overall photo conversion of the DSSC.

2.9.2 Effect of Metal Oxide on the Performance of Metal Oxide/Dye Interface

Electron injection from dye to the metal oxide normally takes place in ultrafast timescale and slow timescale, and they are libeled fast and sow components based on biphasic (two-state) model as depicted in the Figure 2.5. Unthermalized and relaxed excited states of the dye are responsible for fast and slow components [52]. The electron injection efficiency depending on the rate of electron injection is determined by the unthermalized excited state, which is responsible for the fast component. A comparison of the rate of electron injection from N3 to different metal oxide shows that TiO_2 is a better material for fabricating photoanode. The electron injection from N3 to TiO_2 takes place within 250 fs [53], while it is 1.5 ps in the case of ZnO (fast component), [54] and it is observed to be slower in the case of SnO_2 with 5–10 ps [55–57].

The energetics of metal oxide plays a key role in determining the electron injection speed, although the LUMO level of N3 dye is unchanged. Though TiO_2 and ZnO do possess similar

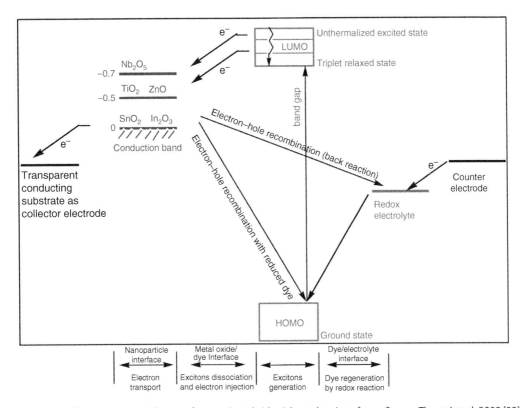

Figure 2.5 The energy-level diagram for metal oxide/dye/electrolyte interfaces. *Source:* Thavasi et al. 2009 [33]. Reproduced with permission from Elsevier.

band gap and conduction band edge value (E_{CB}) [58, 59], TiO$_2$ has better electron injection efficiency than ZnO and SnO$_2$ even with its E_{CB} edge position 0.5 V lower than that of TiO$_2$ and has lower electron injection ability. The difference in the injection rate cannot be explained based on the difference in energetics, which can be understood based on electronic structure. The overall electron injection efficiency of the metal oxides are in the following order: TiO$_2$ > Nb$_2$O$_5$ > SnO$_2$ ~ZnO ~ ZNO [60, 61].

2.9.3 Role of Electronic Structure of Metal Oxides

As proposed by Marcus theory, density of states (DOS) in the conduction band of the metal oxide semiconductor has an impact on the interfacial electron transfer and the rate of electron injection. In any system, s, p, d, and f orbitals make the electronic structure of the condition band. While the conduction band of TiO$_2$ is comprised of empty d orbitals of Ti^{4+}, SnO$_2$ consists of empty s and p orbitals of Sn^{4+}. The DOS of broader s and p orbitals is smaller in magnitude than the narrow d-type conduction bands. The effective mass (m_e) of the conductive electron has to be larger for the d-orbital material because the electron transfer integrals for the d-orbitals is smaller than that of the s-orbitals. The electron injection rate from dye to metal oxide is directly proportional to the electron effective mass. The electron effective mass of TiO$_2$ and Nb$_2$O$_5$ is higher than that of SnO$_2$ and ZnO. Therefore, the electron injection rate at dye/TiO$_2$ and dye/Nb$_2$O$_5$ interfaces is faster than dye/SnO$_2$ and dye/ZnO interfaces [32]. The different properties of various metal oxides that influence the performance of dye/metal oxide interface are presented in Table 2.1.

Table 2.1 The different properties of various metal oxides that influence the performance of dye/metal oxide interface [33].

Material	Isoelectric point	Band gap	$E_{CB} = 0.0$ V versus normal hydrogen electrode (NHE)	Electron affinity (eV)	Density of states (effective electron mass)	Electronic structure
Anatase TiO_2	5–6	3.3	0.5 V	3.9	$5–10\,m_e$	d orbitals of Ti^{4+} density of states is higher than others
ZnO	4–9	3.3	Close to TiO_2 ~0.5 V higher than SnO_2	4.5	$0.3\,m_e$	s, p orbitals of Zn^{2+}
SnO_2	2.5–4.0	3.5	0.5 V lower than TiO_2	4.8	$0.3\,m_e$	s and p orbitals of Sn^{4+}
In_2O_3	7.1	3.6	0.5 V lower than TiO_2	4.45	$0.3\,m_e$	s orbitals of In^{3+}
Nb_2O_5	2.6–4.5	3.4	0.2–0.3 eV higher than TiO_2	2.34	$3\,m_e$	d orbitals of Nb^{5+}

When light is incident on dye, the electron is sent from dye to the ligand orbital, which is in contact with the metal oxide's conduction band. At this point, a temporary interfacial metal oxide–ligand–dye charge transfer complex. The electron transferred to the ligand is further pumped in to the conduction band of the metal oxide. In the case of TiO_2, the above-mentioned interfacial charge transfer complex and the back bonding of the metal oxide and the dye increases both the stability and electron injection. On the other hand, in the case of ZnO back bonding cannot be established due to the absence of vacant d-orbital that make it handicapped with decreased stability and poor electron injection efficiency. Therefore, TiO_2 is more preferred over ZnO to fabricate photoanode [62].

2.10 Kinetics of Operation in DSSCs

1) The mesoporous metal oxide film forming photoanode should have high surface area so as to adsorb solar irradiation strongly by the dye sensitizer molecules [8].
2) The light-harvesting ability of the DSSC can be increased by incorporating tandem structure and by incorporating two individual cells to photoanodes [63, 64].
3) The key to achieve high efficiency is to have maximum charge injection and minimum charge recombination [33].
4) The recombination of electron with electrolyte can be prevented by introducing a blocking layer/compact layer between the FTO substrate and the mesoporous metal oxide that can improve the photoconversion efficiency besides improving the adhesion between the FTO substrate and the mesoporous metal oxide [65, 66].

As depicted in Figure 2.6, in DSSC, the two primary charge separation steps namely electron injection from the excited state of dye to the anode made of metal oxide and regeneration of the dye molecule to the ground state anode, are intervened by several parameters, and there is always a kinetic competition between charge separation and the decay of the excited state to the

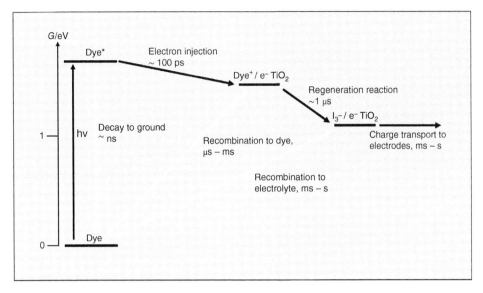

Figure 2.6 State diagram representation of the kinetics of DSSC function. *Source:* Durrant and Gratzel 2008 [67]. Reproduced with permission of The American Chemical Society.

ground state and charge recombination [8]. The overall reaction consists of both forward and loss pathways. In Figure 2.6, the forward process of light absorption, electron injection, dye regeneration, and charge transport are indicated by black arrows, while the loss pathways of excited-state decay to ground and electron recombination with dye cations and oxidized redox couple are shown by white arrows. The vertical scale corresponds to the free energy stored in the charge-separated states [8, 67].

DSSC has processes that are very parallel to natural photosynthesis and the kinetic competitions between the different forward and loss pathways are of paramount importance in deciding the quantum efficiencies of charge separation and collection. Every charge transfer step is accompanied by an increased spatial separation of electrons and holes whereby the lifetime of the charge-separated state is increased at the cost of the free energy stored in the same state which is similar to that of photosynthesis [8].

In DSSC, the efficiency of electron injection depends on the magnitude of injection kinetics related to the decay of excited-state to ground and not upon the absolute kinetics of electron injection. In most cases, the rate of excited-state decay lies between picoseconds and nanoseconds. Dye triplets are usually formed by intersystem crossing from the singlet excited state that has a longer life with less energy in comparison to the singlet state, and these triplet states contribute to the potential required for the electron injection. The quick electron injection dynamics demand very strong electronic coupling of the dye LUMO orbital with the conduction band state of the metal oxide and the large density of states in TiO_2 that are energetically obtainable from the dye excited state. The thermalization taking place in the injected electron after injection gives rise to a loss in the free energy and eventual drop in the device efficiency. For an increased efficiency of the device, it is necessary to have efficient dye regeneration by the redox couple and charge collection by the external electrode against recombination of injected electron. The device efficiency is brought down not only by the material composition but also by the operating conditions all of which cause the energetic losses. An efficient device should have quick electron injection, charge regeneration, and efficient charge collection and less of decay of the excited state and recombination [8].

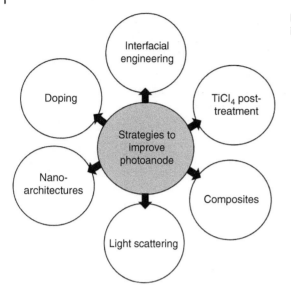

Figure 2.7 Typical TiO$_2$ nanoarchitectures for improving DSSC performance.

2.11 Strategies to Improve the Photoanode Performance

The performance of photoanode can be improved by adopting certain strategies presented in Figure 2.7 that help in enhancing electron injection and reducing electron recombination to achieve high photoconversion efficiency.

2.11.1 TiCl$_4$ Treatment

Among the several approaches to TiCl$_4$, the post treatment of the metal oxide (TiO$_2$) with photoanode has been a familiar method to improve the overall performance of DSSC; however, the exact mechanism of this method is not presented with clarity. In this method, the previously sintered TiO$_2$ nanocrystalline and other nanostructures-coated substrates have been immersed into a TiCl$_4$ solution of moderate-concentration and a temperature of (60–70 °C) for a period of 30 minutes and sintered again to obtain a thin layer of TiCl$_4$ of enhanced surface area. As stated earlier, the effect of TiCl$_4$ treatment in improving the performance has not been completely understood, although several effects such as increase in surface area, enhanced electron transport, TiO$_2$ purity, light scattering, and dye anchoring are attributed [68, 69]. While Lee et al. have presented the effect of treating TiCl$_4$ of different concentrations with the morphology of film, charge carrier dynamics, and the overall performance of the device, transport, and recombination of electron [70], Sommeling et al. have made a systematic study and reported that neither increase in light absorption nor decrease in the light loss during transport is responsible for the increased photocurrent rather the downward shift in the band edge [69]. The effect of TiCl$_4$ treatment to increase the overall efficiency of the DSSC is limited when the untreated photoanode achieves maximum quantum efficiency through high photocurrent density and photoconversion efficiency because the band edge of the TiO$_2$ is at the proper potential [68].

2.11.2 Composites

Photoanodes can also be made of composites consisting of TiO$_2$ or ZnO with carbon-based materials such as graphene and carbon nanotubes, and the plasmonic effect induce noble metals such as gold and silver and other transition metal oxides to enhance the photoconversion efficiency through synergetic and intrinsic properties. Graphene boosts electron mobility, offers high specific area, and enhances mechanical properties besides lowering charge recombination and resistance at the electrolyte–electrode interface as well as facilitating electron transport from the semiconductor film to the conductive substrates. Although carbon nanotubes enhance the device performance, graphene is more preferred as its two-dimensional structure offers better anchoring sites and causes fewer recombination than the former. The composites made of noble metals like Au and Ag with their surface plasmonic effect can enhance the spectral absorption, increase free carrier density, reduce recombination, and increase the overall photoconversion efficiency of the device. Metal oxides such as SnO_2 are also used along with TiO_2 and ZnO to make a bilayer in order to slow down the rate of recombination and to increase the efficiency [68].

2.11.3 Light Scattering

The enhanced optical absorption and light-harvesting properties are achieved by coating light-scattering material on to the photoanode film. The materials making the light-scattering layer comprise of larger particles, and they help to prolong the light path by reflection and/or scattering of light incident on the photoanode, and this is achieved by establishing a "double-layer structure" and "mixture structure" in the photoanode composition. The double-layer structure is achieved by coating a layer of nanoparticles of relatively larger size as compared to the photoanode materials comparable to the wavelength of light (e.g. 200–500 nm) which reduces transmission-assisted light loss by reflecting the incident light. The size of the scattering particles are of paramount importance for light reflection [71–73] and scattering as their size reflect scattering properties [74, 75]. The mixture structure consist of large-sized scattering particles embedded into the nanocrystalline film that absorbs the sensitizer dyes, and these large particles function as light scatterers to cause light scattering in order to stretch out the path of light in the photoanode's vicinity [76]. Both double-layer and mixed-layer nanomaterials of different morphology are employed as scattering materials, however, several yardsticks like film thickness scattering contents have to be kept in mind besides morphology while designing the scattering layer [68].

2.11.4 Nanoarchitectures

Electron transfer within the photoanode occurs by percolation via a network of sites and thermal accessibility to energy sites. So the electron diffusion coefficient and the energy properties of the photoanode materials are greatly affected by different parameters such as particle diameter, pore size, surface area, orientation, and crystallinity of the nanostructures. Nevertheless, the electronic concentration and/or distribution of trap energy are also influenced in turn [68, 77]. Therefore, it is necessary to carefully design the photoanode with metal oxide of proper morphology to tailor desired properties. In this regard, one-dimensional TiO_2 nanostructures, such as nanotubes, nanowires nanofibers, and nanorods with large aspect ratio, have been investigated as alternative morphologies for constructing photoanode. Although these morphologies supply enough energy levels and exhibit higher-level dye absorption with increased surface area, better light absorption, and longer diffusion length, they yield to easy electron recombination through defects and grain boundaries that serve as trap states and "random walk-"assisted electron transfer [68, 78]. In general, the photoconversion efficiency by the photoanode comprising one-dimensional nanomaterial array surpasses the conventional photoanodes with adjustable length, facile electron transfer, and low recombination.

2.11.5 Doping

Even the purest form of TiO_2 does have oxygen deficiencies that can create electron–hole pairs. The oxidizing holes play a dual role of either reacting with the dye or destroying it, besides being scavenged by the redox couple [79] leading to a shorter life time of DSSCs. This can be overcome by replacing the oxygen deficiency with nitrogen by doping TiO_2, and the nitrogen atoms can make TiO_2 active to visible-light and stabilize the photoelectric properties to improve the photoconversion efficiency [79–81]. Doping of TiO_2 with nitrogen can be done both by physical sputtering, implantation [82] and sintering TiO_2 under nitrogen-containing atmosphere [83] and chemical methods (solvothermal and sol-gel) [84, 85]. Ammonia, triethylamine, and urea are used as nitrogen dopants, and ammonia-assisted nitrogen doping outperformed the other two candidates in terms of efficiency [85]. In the case of ZnO, it can be doped using boron to improve the light-harvesting efficiency and photon-to-electron transfer [85]. Boron-doped ZnO nanoporous sheets have been used as photoanodes and as scattering layer materials to boost the device performance [86, 87].

The photoanode materials can also be doped with transition metals such as Ti(III), W, Mg, Sn, Nb, Ni, Ta, and Zn to alter the energy levels and to reduce recombination [88]. When rare earth metals are used as dopants to modify the energy levels and electron density, the energy level of the photoanode is increased by p-type doping as the rare earth metal ions are mostly positive trivalent cations [68].

2.11.6 Interfacial Engineering

Interfacial engineering is adopted to reduce the electron back recombination that brings down J_{SC}, V_{OC}, and eventually reducing the photoconversion efficiency of the device. This electron back recombination takes place while conducting substrate/electrolyte and photoanode/electrolyte interfaces. The electron recombination at the substrate/electrolyte interface can be overcome by introducing a thin compact layer (TiO_2) or depositing a blocking material on to the substrate, and the electron recombination at the photoanode/electrolyte can be reduced by covering the photoanode with a materials whose conduction band is more than that of the photoanode materials or by creating an energy barrier at the interface which can be achieved by introducing an electronically insulating layer. To coat the compact layer (TiO_2) on the conducting substrate, many deposition methods such as atomic layer deposition, spray pyrolysis, DC-Magnetron sputtering, electrochemical deposition and sol-gel can be adopted. The back electron transfer or recombination at the photoanode/electrolyte interface can be minimized by using a thin layer of metal oxides such as MgO, ZrO_2, Nb_2O_5, Al_2O_3, Ta_2O_5, SiO_2, and Ga_2O_3, which can increase VOC and the photoconversion efficiency by curtailing the back electron recombination.

2.12 Conclusion

The overall performance of the DSSCs is determined by three core elements, namely metal oxide, dye sensitizer, and metal oxide/dye/electrolyte interface. Therefore, a lot of attention is devoted to research these elements in order to engineer suitable DSSC composition toward obtaining high photoconversion efficiency. In this chapter, we have elaborately discussed the charge transfer dynamics, the different processes, and parameters influencing the photoconversion efficiency in depth. In particular, the function of the photoanode, materials making up the photoanode, role of dye, electron injection along with mechanism, charge recombination, dye aggregation, and the strategies required to improve the efficiency have been dealt and presented in detail. Dye aggregation and charge recombination are the main causes that bring down the performance of the cells.

Acknowledgments

The corresponding author A. Dennyson Savariraj gratefully acknowledges the FONDECYT Post-doctoral Project No. 3170640, Government of Chile, Santiago, for the financial assistance.

References

1 Gong, J., Sumathy, K., Qiao, Q., and Zhou, Z. (2017). *Renewable Sustainable Energy Rev.* 68: 234–246.
2 Chi, C., Su, S., Liu, I. et al. (2014). *J. Phys. Chem. C* 118 (31): 17446–17451.
3 Huang, C., Hsu, Y., Chen, J. et al. (2006). *Sol. Energy Mater. Sol. Cells* 90 (15): 2391–2397.
4 Nazeeruddin, M.K., Baranoff, E., and Gratzel, M. (2011). *Sol. Energy* 85 (6): 1172–1178.
5 Hug, H., Bader, M., Mair, P., and Glatzel, T. (2014). *Appl. Energy* 115: 216–225.
6 Gerischer, H. and Sorg, N. (1992). *Electrochim. Acta* 37 (5): 827–835.
7 Frank, S.N. and Bard, A.J. (1977). *J. Phys. Chem.* 81 (15): 1484–1488.
8 Listorti, A., O'Regan, B., and Durrant, J.R. (2011). *Chem. Mater.* 23 (15): 3381–3399.
9 Zhang, Q., Dandeneau, C.S., Zhou, X., and Cao, G. (2009). *Adv. Mater.* 21 (41): 4087–4108.
10 Jose, R., Thavasi, V., and Ramakrishna, S. (2009). *J. Am. Ceram. Soc.* 92 (2): 289–301.
11 Chen, S.G., Chappel, S., Diamant, Y., and Zaban, A. (2001). *Chem. Mater.* 13 (12): 4629–4634.
12 Kay, A. and Gratzel, M. (2002). *Chem. Mater.* 14 (7): 2930–2935.
13 Rensmo, H., Keis, K., Lindstrom, H. et al. (1997). *J. Phys. Chem. B.* 101 (14): 2598–2601.
14 Kavan, L., Gratzel, M., Gilbert, S.E. et al. (1996). *J. Am. Chem. Soc.* 118 (28): 6716–6723.
15 Etacheri, V., Di Valentin, C., Schneider, J. et al. (2015). *J. Photochem. Photobiol., C* 25: 1–29.
16 Banfield, J.F., Bischoff, B.L., and Anderson, M.A. (1993). *Chem. Geol.* 110 (1): 211–231.
17 Blumenthal, R.N., Baukus, J., and Hirthe, W.M. (1967). *J. Electrochem. Soc.* 114 (2): 172–176.
18 Von Hippel, A., Kalnajs, J., and Westphal, W.B. (1962). *J. Phys. Chem. Solids* 23 (6): 779–799.
19 Nazeeruddin, M.K., Kay, A., Rodicio, I. et al. (1993). *J. Am. Chem. Soc.* 115 (14): 6382–6390.
20 Kuciauskas, D., Freund, M.S., Gray, H.B. et al. (2001). *J. Phys. Chem. B.* 105 (2): 392–403.
21 Hasselmann, G.M. and Meyer, G.J. (1999). *Z. Phys. Chem.* 212: 39.
22 Geary, E.A.M., Yellowlees, L.J., Jack, L.A. et al. (2005). *Inorg. Chem.* 44 (2): 242–250.
23 Ferrere, S. and Gregg, B.A. (1998). *J. Am. Chem. Soc.* 120 (4): 843–844.
24 Alonso-Vante, N., Nierengarten, J., and Sauvage, J. (1994). *J. Chem. Soc., Dalton Trans.* (11): 1649–1654.
25 Baluschev, S., Jacob, J., Avlasevich, Y.S. et al. (2005). *ChemPhysChem* 6 (7): 1250–1253.
26 Komori, T. and Amao, Y. (2003). *J. Porphyrins Phthalocyanines* 07 (02): 131–136.
27 Chen, Y., Li, C., Zeng, Z. et al. (2005). *J. Mater. Chem.* 15 (16): 1654–1661.
28 Burke, A., Schmidt-Mende, L., Ito, S., and Gratzel, M. (2007). *Chem. Commun.* (3): 234–236.
29 Hara, K., Wang, Z., Sato, T. et al. (2005). *J. Phys. Chem. B.* 109 (32): 15476–15482.
30 Horiuchi, T., Miura, H., Sumioka, K., and Uchida, S. (2004). *J. Am. Chem. Soc.* 126 (39): 12218–12219.
31 Ardo, S. and Meyer, G.J. (2009). *Chem. Soc. Rev.* 38 (1): 115–164.
32 Tributsch, H. (2004). *Coord. Chem. Rev.* 248 (13): 1511–1530.

33 Thavasi, V., Renugopalakrishnan, V., Jose, R., and Ramakrishna, S. (2009). *Mater. Sci. Eng. R Rep.* 63 (3): 81–99.

34 O'Regan, B. and Gratzel, M. (1991). *Nature* 353 (6346): 737–740.

35 Nilsing, M., Persson, P., Lunell, S., and Ojamäe, L. (2007). *J. Phys. Chem. C* 111 (32): 12116–12123.

36 Rochford, J., Chu, D., Hagfeldt, A., and Galoppini, E. (2007). *J. Am. Chem. Soc.* 129 (15): 4655–4665.

37 Meerheim, R., Nitsche, R., and Leo, K. (2008). *Appl. Phys. Lett.* 93 (4): 043310.

38 Jose, R., Kumar, A., Thavasi, V., and Ramakrishna, S. (2008). *Nanotechnology* 19 (42): 424004.

39 Bae, E., Choi, W., Park, J. et al. (2004). *J. Phys. Chem. B* 108 (37): 14093–14101.

40 Nilsing, M., Persson, P., and Ojamäe, L. (2005). *Chem. Phys. Lett.* 415 (4): 375–380.

41 Park, H., Bae, E., Lee, J. et al. (2006). *J. Phys. Chem. B* 110 (17): 8740–8749.

42 Vittadini, A., Selloni, A., Rotzinger, F.P., and Grätzel, M. (2000). *J. Phys. Chem. B* 104 (6): 1300–1306.

43 Finnie, K.S., Bartlett, J.R., and Woolfrey, J.L. (1998). *Langmuir* 14 (10): 2744–2749.

44 Nazeeruddin, M.K., Humphry-Baker, R., Liska, P., and Gratzel, M. (2003). *J. Phys. Chem. B* 107 (34): 8981–8987.

45 Bauer, C., Boschloo, G., Mukhtar, E., and Hagfeldt, A. (2002). *J. Phys. Chem. B* 106 (49): 12693–12704.

46 Murakoshi, K., Kano, G., Wada, Y. et al. (1995). *J. Electroanal. Chem.* 396 (1): 27–34.

47 Shklover, V., Ovchinnikov, Y.E., Braginsky, L.S. et al. (1998). *Chem. Mater.* 10 (9): 2533–2541.

48 Wenger, B., Gratzel, M., and Moser, J. (2005). *J. Am. Chem. Soc.* 127 (35): 12150–12151.

49 Wang, P., Zakeeruddin, S.M., Comte, P. et al. (2003). *J. Phys. Chem. B* 107 (51): 14336–14341.

50 Gratzel, M. (2005). *Inorg. Chem.* 44 (20): 6841–6851.

51 Hagberg, D.P., Marinado, T., Karlsson, K.M. et al. (2007). *J. Organomet. Chem.* 72 (25): 9550–9556.

52 Guo, J., Stockwell, D., Ai, X. et al. (2006). *J. Phys. Chem. B* 110 (11): 5238–5244.

53 Furube, A., Murai, M., Watanabe, S. et al. (2006). *J. Photochem. Photobiol., A* 182 (3): 273–279.

54 Asbury, J.B., Hao, E., Wang, Y. et al. (2001). *J. Phys. Chem. B.* 105 (20): 4545–4557.

55 Iwai, S., Hara, K., Murata, S. et al. (2000). *J. Chem. Phys.* 113 (8): 3366–3373.

56 Bauer, C., Boschloo, G., Mukhtar, E., and Hagfeldt, A. (2002). *Int. J. Photoenergy* 4 (1): 17–20.

57 Benko, G., Myllyperkio, P., Pan, J. et al. (2003). *J. Am. Chem. Soc.* 125 (5): 1118–1119.

58 Kakiuchi, K., Hosono, E., and Fujihara, S. (2006). *J. Photochem. Photobiol., A* 179 (1): 81–86.

59 Keis, K., Magnusson, E., Lindström, H. et al. (2002). *Sol. Energy Mater. Sol. Cells* 73 (1): 51–58.

60 Ai, X., Anderson, N.A., Guo, J., and Lian, T. (2005). *J. Phys. Chem. B.* 109 (15): 7088–7094.

61 Ai, X., Guo, J., Anderson, N.A., and Lian, T. (2004). *J. Phys. Chem. B.* 108 (34): 12795–12803.

62 Tributsch, H. (2001). *Appl. Phys. A* 73 (3): 305–316.

63 Murayama, T.M.M. (2007). *J. Phys. D* 40 (6): 1664.

64 Wenger, S., Seyrling, S., Tiwari, A.N., and Gratzel, M. (2009). *Appl. Phys. Lett.* 94 (17): 173508.

65 Sivakumar, R., Ramkumar, J., Shaji, S., and Paulraj, M. (2016). *Thin Solid Films* 615: 171–176.

66 Choi, H., Nahm, C., Kim, J. et al. (2012). *Curr. Appl Phys.* 12 (3): 737–741.

67 Durrant, J.R., Gratzel, M. (2008), *Nanostructured and Photoelectrochemical Systems for Solar Photon Conversion.* Imperial College Press, 3, 503–536.

68 Fan, K., Yu, J., and Ho, W. (2017). *Mater. Horiz.* 4 (3): 319–344.

69 Sommeling, P.M., O'Regan, B.C., Haswell, R.R. et al. (2006). *J. Phys. Chem. B.* 110 (39): 19191–19197.

70 Lee, S., Ahn, K., Zhu, K. et al. (2012). *J. Phys. Chem. C* 116 (40): 21285–21290.

71 Park, Y., Chang, Y., Kum, B. et al. (2011). *J. Mater. Chem.* 21 (26): 9582–9586.

72 Yu, I.G., Kim, Y.J., Kim, H.J. et al. (2011). *J. Mater. Chem.* 21 (2): 532–538.

73 Sun, X., Zhou, X., Xu, Y. et al. (2015). *Appl. Surf. Sci.* 337: 188–194.

74 Usami, A. (1997). *Chem. Phys. Lett.* 277 (1): 105–108.

75 Ferber, J. and Luther, J. (1998). *Sol. Energy Mater. Sol. Cells* 54 (1): 265–275.

76 Zhang, Q., Myers, D., Lan, J. et al. (2012). *Phys. Chem. Chem. Phys.* 14 (43): 14982–14998.

77 Macaira, J., Andrade, L., and Mendes, A. (2013). *Renewable Sustainable Energy Rev.* 27: 334–349.

78 Ku, C. and Wu, J. (2007). *Appl. Phys. Lett.* 91 (9): 093117.

79 Mrowetz, M., Balcerski, W., Colussi, A.J., and Hoffmann, M.R. (2004). *J. Phys. Chem. B.* 108 (45): 17269–17273.

80 Asahi, R., Morikawa, T., Ohwaki, T. et al. (2001). *Science* 293 (5528): 269.

81 Ma, T., Akiyama, M., Abe, E., and Imai, I. (2005). *Nano Lett.* 5 (12): 2543–2547.

82 Tian, H., Hu, L., Zhang, C. et al. (2010). *J. Phys. Chem. C* 114 (3): 1627–1632.

83 Fu, H., Zhang, L., Zhang, S. et al. (2006). *J. Phys. Chem. B.* 110 (7): 3061–3065.

84 Etacheri, V., Seery, M.K., Hinder, S.J., and Pillai, S.C. (2010). *Chem. Mater.* 22 (13): 3843–3853.

85 Guo, W., Shen, Y., Boschloo, G. et al. (2011). *Electrochim. Acta* 56 (12): 4611–4617.

86 Mahmood, K. and Park, S.B. (2013). *J. Mater. Chem. A* 1 (15): 4826–4835.

87 Mahmood, K. and Sung, H.J. (2014). *J. Mater. Chem. A* 2 (15): 5408–5417.

88 Yang, M., Kim, D., Jha, H. et al. (2011). *Chem. Commun.* 47 (7): 2032–2034.

3

Nanoarchitectures as Photoanodes

Hari Murthy

Department of Electronics and Communication Engineering, CHRIST (Deemed to be) University, Kanminike, Bengaluru, India

3.1 Introduction

Photovoltaic (PV) devices are the proficient way to obtain electrical energy from solar energy to meet the ever-increasing global energy demand. Silicon (Si)-based PV cells have reached an efficiency of 24.7% though at the cost of sophisticated technologies and expensive techniques [1], hampering low-cost production and limiting their widespread utilization [2]. Some of the drawbacks of the Si-PV devices were overcome by second-generation thin-film PV devices that are lightweight, flexible, and low-cost but are less efficient. The thin-film PV devices suffer from complex deposition process, difficulty in controlling stoichiometry and the presence of structural defects that adversely affects their performance [3]. The third-generation PV technology including organic photovoltaics (OPVs), dye-sensitized solar cells (DSSCs), quantum-dot dye-sensitized solar cells (QD-DSSCs), and perovskite solar cells have fulfilled the condition of low-cost simple fabrication process, and the research focuses is on enhancing the efficiency, performance, and stability [2]. OPVs possess low efficiency but consist of toxic materials, while perovskite solar cells suffer from moisture instability and poor reproducibility. DSSCs offer a lot of advantages such as excellent stability, low toxicity, good conversion efficiency [4], simple device design, and low-cost fabrication process that supports large-scale production [2]. Even though the highest certified conversion efficiency of DSSCs (11–13%) [4] is half of the advanced thin film or crystalline (26.4%, 27.6%, respectively), their unique functionalities make them attractive for research [5].

The Global DSSC market valued at US$49.6 million in 2014 is expected to grow at over 12% from 2015–2022 to cross US$130 million by 2023 [2]. Increasing user awareness regarding ecological impact of conventional fuels, increasing reliance on renewable energy resources, and increased energy requirements for residential areas are the impetus to the market growth [3]. The ability of DSSC modules to provide optimum power even in low, diffused, and indoor light conditions makes it suitable for smartphones, camera, and indoor electronics applications [5], with portable charging accounting for 33% of revenue share. The primary objective of this technology would be to build integrated photovoltaics (BIPVs), and automobile integrated photovoltaics (AIPVs) that are expected to witness a significant growth exceeding US$30 million by 2022.

Interfacial Engineering in Functional Materials for Dye-Sensitized Solar Cells, First Edition.
Edited by Alagarsamy Pandikumar, Kandasamy Jothivenkatachalam and Karuppanapillai B. Bhojanaa.
© 2020 John Wiley & Sons, Inc. Published 2020 by John Wiley & Sons, Inc.

3.2 DSSC Operation

Optical excitation of the dye is the basis of DSSC operation, where an electron is excited from the dye molecule into the conduction band (CB) of a wideband metal oxide [6]. The device consists of five main components – transparent conducting oxides (TCO, normally indium tin oxide [ITO]/fluorine-doped tin oxide [FTO]), semiconductor photoanode (metal oxide semiconductor, MOS), sensitizer dye, counter-electrode (Pt, carbon), and electrolyte [3]. The first DSSC was fabricated using trimeric ruthenium complex $RuL_2((\mu\text{-}CN)Ru(CN)L_2)_2$ (L = 2,2′-bipyridine-4,4′-dicarboxylic acid; L′ = 2,2′-bipyridine) as dye, 15 nm nanocomposite titanium oxide (TiO_2) (3D interconnected nanoparticles deposited as a 10 μm thick film on ITO) as photoanode, the I^-/I^{3-} redox couple in an electrolyte with acetonitrile solvent and bare FTO as counter electrode. The device showed a conversion efficiency of 7.12% under the AM1.5 spectral distribution with a 90% internal photocurrent efficiency (IPCE) in the 450–700 nm wavelength [2]. DSSCs are highly interfacial devices making them distinct from other PV devices [5] – the generation of electric charges occurs at the oxide/dye/electrolyte interface, functioning at the molecular and nanoscale [7]. The operation steps in DSSCs are (as shown in Figure 3.1) [3]:

(i) excitation of dye molecules on absorbing light causing a state change from ground state to excited state,
(ii) transfer of electrons to conduction band of the mesoporous MOS,
(iii) electron diffusion in semiconductor photoanode via trapping and de-trapping,
(iv) oxidation–reduction reactions of electrolyte redox couples via electrons traveling through the outer circuit, and
(v) dye regeneration due to reduction of oxidized dye by redox couple, making it ready for photoelectron generation [2, 5, 7].

$$S(adsorbed\ on\ photoanode) + h\nu \rightarrow S^*(adsorbed\ on\ photoanode)$$

$$S^*(adsorbed\ on\ photoanode) \rightarrow S^+(adsorbed\ on\ photoanode) + e^-\ (injected)$$

$$S^*(adsorbed\ on\ photoanode) + 1.5I^- \rightarrow S^*(adsorbed\ on\ photoanode) + 0.5I^{3-}$$

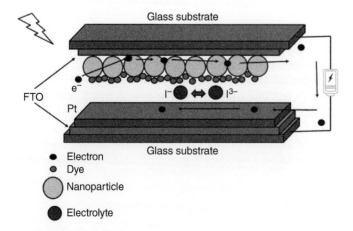

Figure 3.1 Schematic diagram of dye-sensitized solar cell (DSSC) assembly.

The charge separation at the dye during illumination takes place via two mechanisms:

(i) Transportation of photoelectrons because of the higher energy of lowest unoccupied molecular orbital (LUMO) of the dye than the MOS conduction band, and
(ii) Electron density variations between the MOS and dye.

A proper combination of the three components (dye, electrolyte, and metal oxide) separates the light harvesting and carrier charge transport [6]. Ruthenium-based dyes and the I/I^{3-} redox couple are the most commonly used materials achieving a long-standing efficiency of 11.1%, which has been replaced in recent times with porphyrin dyes and cobalt-based redox couple, improving the conversion efficiency to 12.3% [6]. The $J-V$ curves provide information on DSSC device performance from which the open circuit voltage (V_{OC}), short circuit current density (J_{SC}), fill factor (FF), and power conversion efficiency (PCE, %η) under a certain intensity of light (AM1.5 solar spectrum) are calculated [8]. V_{OC} is the maximum voltage that a PV device supplies to an external circuit obtained from charge carrier separation. It is proportional to the difference in the Fermi level of the MOS and redox potential of electrolyte. J_{SC} is the maximum current output of a PV device per unit area under short circuit condition and is affected by several factors as dye molecular structure, amount of dye adsorbed on photoanode, and electrochemical properties of photoanode in the presence of electrolyte [9]. FF is the ratio of the maximum power output (P_{max}) to the product of V_{OC} and I_{SC}, i.e. FF $= \frac{P_{max}}{V_{OC} \times J_{SC}}$ and varies between 0 and 1 (ideality of PV device). PCE of a solar cell is the ratio of the maximum power generated (P_{max}) to the incident power (P_{in}), i.e. η (%) $= \frac{P_{max}}{P_{in}}$ [4]. The device performance depends on the intensity and wavelength of incident light, and high J_{SC}, V_{OC}, and FF at low incident power give high PCE.

Each component of DSSC contributes to the photocurrent density of the device, and a proper DSSC design is always a trade-off and optimization between them [3, 4]. Thus the critical features of DSSCs are [5]

(i) Mesoporous photoanode film to achieve high η,
(ii) High-electron mobility for efficient electron transport within minimum loss through the photoanode, and
(iii) Minimum overpotentials at the various interfaces, which influences V_{OC}.

Recombination of photoelectrons in the FTO/photoanode/dye interface and electrolyte redox couple results in the loss of the photoelectrons [2, 5]. The difference between the redox potential of the electrolyte and highest occupied molecular orbital (HOMO) of the dye causes loss of electron at the dye/electrolyte interface, producing a second overpotential. Thus, choice of appropriate materials with suitable band edge energies to minimize recombination losses is one of the main requirements for improved photovoltaic parameters [9]. Some critical points to be considered while designing a DSSC device are

(i) Substantial charge transfer from dye to the MOS at a rate higher than the rate of recombination,
(ii) The LUMO of the dye must be more negative, i.e. higher than the conduction band of the MOS, which in turn must be more negative, i.e. higher than that of the conducting glass (FTO/ITO/MOS), and
(iii) Redox potential of the electrolyte should be more negative than the HOMO of the dye [2].

Photoanodes participate in the fundamental DSSC operations involving photon absorption, recombination, and electron transport [9], and its properties play a critical role in the device performance [3]. Any modifications of the photoanode have a direct impact on device performance and efficiency [2]. It acts as a site for dye adsorption and a medium for charge transport and collection. The ideal characteristics of a photoanode are [9, 10]

(i) High surface area with appropriate morphological structure for maximized dye loading and effective light absorption,
(ii) High optical transparency to reduce incident photon loss,
(iii) Good electron acceptor with high conductivity and mobility to assist electron transport,
(iv) Stable against redox electrolyte to decrease recombination, and
(v) Contains –OH groups or defects for optimum contact with dye molecules,
(vi) High resistance to photocorrosion.

Also, other parameters that need to be analyzed are the electron collection of electrodes and photogenerated electron properties (generation rate, injection, diffusion, and transferability) [11].

The light-harvesting efficiency (LHE) of photoanode plays an important role for high-efficiency DSSCs, which is dependent on the amount of dye adsorbed on the photoanode, its molar extinction coefficient, and the optical path, given as [7] :

$$\text{LHE}(\lambda) = 1 - 10^{\gamma}\sigma(\lambda)$$

γ is the number of moles of dye molecules per square centimeter of projected film surface area, and σ is the absorption cross section (cm^2/mol). The LHE determines the IPCE, the amount of charge injected (φ_{inj}), and the efficiency of charge collection at the back contact (η_c). Almost all the absorbed photons get converted to conduction band electrons that then travel through the interconnected particles network present in the photoanode without significant loss to be collected at the substrate. Figure 3.2 shows a schematic of the numerous methods investigated to improve DSSC device.

The main challenges are to enhance the performance and stability of the low-cost fabrication and installation technology. Progress in the field of DSSCs is relatively slow, due to several

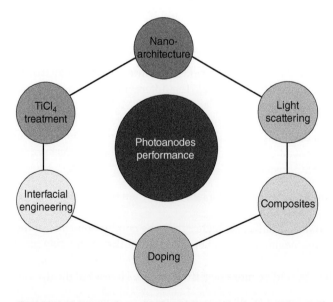

Figure 3.2 Strategies for improving photoanode device performance, ranging from doping with suitable materials, interfacial engineering, and use of nanoarchitectures of various dimensionality.

factors, such as the presence of liquid electrolyte that can expand or freeze at different temperatures and the low efficiency for flexible module limiting large-scale deployment [2]. However, the advantages are also quite a few such as transparency and/or multicolor options, lower production cost, easy integration into BIPVs, less energy payback time, significant reduction of greenhouse gases emission, with a predicted cost of 0.2 US$/W compared to 0.5–1 US$/W offered by the second-generation photovoltaics [3, 6].

Considerable research focus has been on photoanodes to harvest light efficiently along with reduced recombination reaction, increased dye adsorption, light scattering ability, and better charge transportability contributing to high conversion efficiency [2]. Novel semiconductor photoanode nanostructures enhance multiple internal reflections, infrared (IR) absorption, increasing dye loading and optical scattering effects [7], and optimizing the semiconductor material to a preferable morphology in photoanode that can help in improving device performance [4].

Current efficiency of commercially available DSSCs lies within the 12–15% range that is significantly less than their silicon counterparts, and extensive efforts are directed to optimize the functions of the each DSSC component with a focus on nanostructured photoanodes. This chapter shall look into providing detailed information on the state-of-the-art and recent trends on materials and nanoarchitectures for improved photoanode device. The objective of this chapter is to provide a roadmap for researchers toward optimization of photoanodes using advanced material engineering.

3.3 Nanoarchitectures for Improved Device Performance of Photoanodes

Conventional semiconductor materials such as mono- and poly-Si, GaAs, CdS, and InP react with the electrolyte that reduces the device lifetime considerably. TiO_2 and zinc oxide (ZnO) are the most promising materials [11], with TiO_2 reporting better performance due to superior surface area, though their high band gap allows them to be active only in the UV spectrum and not the visible range. Other materials, such as organic compounds, perovskite materials, carbon-metal oxide composite materials, SnO_2, Nb_2O_5, and Zn_2SnO_4 are potential replacement materials for photoanodes [2]. The first DSSC was fabricated using ZnO as the photoanode though TiO_2 has subsequently replaced it as the preferred material for photoanodes due to increased photostability and performance compared to ZnO and SnO_2. Table 3.1 lists the electronic properties of the commonly used photoanode materials.

Efficient photoanodes with improved device performance requires photoanodes with different nanoarchitectures, such as nanoparticles (NPs), nanowires (NWs), nanorods (NRs), nanotubes (NTs), nanosheets (NSs), and mesoporous structures. Asim et al. [14] summarized the different preparation and deposition methods used for photoanode materials, as electrodeposition, hydrothermal, spray, chemical bath deposition, laser pulse deposition, and sol–gel. Their advantages and limitations are emphasized for a researcher to be able to choose and optimize a given method carefully.

3.3.1 TiO_2

Among the various materials investigated as photoanode materials, TiO_2 is the best fit as photoanode material for DSSC fabrication with the highest reported efficiency of 13–14% [15]. Some of its properties which make it an ideal material compared to other transition metal oxides include relatively cost-effective, abundant, biocompatible, nontoxic, stable [2], high

Table 3.1 Materials used as DSSC photoanodes with their properties.

Property	TiO_2	ZnO	SnO_2	Zn_2SnO_4	Graphene	WO_3	CeO_2	Fe_2O_3	$SrTiO_3$	Nb_2O_5
Band gap (eV)	3.2	3.35	3.6	3.35–4.1 [12]		2.6–3.1	3–3.6	2.2	3.2	3.4–4.0
e^- mobility ($cm^2/V\,s$)	0.1–1	100	~100–200	10–15 [13]	2×10^5	6.5		0.2		
Isoelectric point (pH)	6–7	~9	4–5		~1.7 (oxide)	1.5	6.75–8	8.1–8.5	7.8–8.5	4.1
Effective mass (electrons)	$9\,m_e$ $(1–50\,m_e)$	$0.24–0.34\,m_e$	$0.17–0.30\,m_e$	$0.23\,m_e$			$0.42\,m_e$	$1.5–1.6\,m_e$		
Dielectric constant	114	8.5	12.5	15–40	~1.83–3	2.4	23–26		300	33–40

mesoporous nature, high conduction band edge [9], electron affinity, and sufficient surface area for dye loading [8]. The conduction band consists of pure 3d Ti orbitals, while the valence band comprises hybridized O 2p and Ti 3d orbitals. The transition probability for the electrons to return to the valence band decreases due to the indirect band gap resulting in lessened recombination probability and increased electron lifetime. A comparative study on the electron mobility and injection dynamics of mesoporous ZnO, SnO_2, and TiO_2 thin film show that TiO_2 has a faster electron injection due to its higher effective mass ($5–10\,m_e$) resulting in a higher available density of states (DOS). The bulk conductivities of ZnO, SnO_2, and TiO_2 are of the order $TiO_2 < ZnO < SnO_2$. TiO_2 exists in three main crystal structures – (i) rutile, (ii) anatase, and (iii) brookite. Anatase and rutile are the most common and thermodynamically stable TiO_2 polymorphs, while brookite is extremely difficult to obtain due to its metastable crystal structure. Anatase TiO_2 ($E_g \sim 3.2\,eV$) is more preferred as it is chemically more stable than rutile ($E_g \sim 3.0\,eV$) owing to its higher dielectric constant, wider band gap, low packing density, and better PCE [9]. In some cases, both anatase and rutile can coexist which is undesirable, although some studies suggest that some amount of rutile in anatase can enhance device performance.

Nanoporous TiO_2 in DSSCs is a promising alternative to Si-based solar technology due to their relatively higher solar to electric PCE and low cost [16]. Compared to the bulk counterparts, TiO_2 mesoporous nanostructures have higher Hall mobility than ZnO and SnO_2. Electron transportation in bulk TiO_2 occurs by electron–phonon coupling (polarons), while in nanostructured TiO_2, it depends on the characteristics of the individual nanoparticles, modified electronic structure, extent of interparticle connectivity, and geometric shape of the assembly. Also, quantum effects are also observed in TiO_2, though considering that the particle size (2–50 nm) is greater than the Bohr radius (10–25 nm), the effects are minimum. The mesoscopic structure with a percolated links of nanoparticles and irregular secondary particles plays a significant role in increasing cell efficiency by providing greater surface area for dye loading and eventually higher light harvesting [3]. The electron transport in TiO_2 occurs by the process of trapping/detrapping that affects the electron lifetimes by reducing the diffusion length and increases recombination [17]. For considerable electron collection at the anode, the diffusion length must be greater than the semiconductor film thickness, and one way to achieve it is to modify the surface with nanoarchitectures [18]. The photoconversion efficiency is highly dependent on pore size as it influences the intake and diffusion of dye molecules [19]. Addition of oxides as Al_2O_3, SiO_2 within the mesoporous structure opens alternative routes to enhance the LHE, where the thin oxide layer block the charge recombination during the photoelectron generation process. As of now, mesoporous structures due to their desirable features, such as high surface area, low film conductivity, and no built-in electrical field are regarded as the most favorable electrodes and used as a standard to study DSSCs. There are however certain drawbacks that limit further enhancement of PV performance [20] – insufficient connectivity between particles, low charge diffusion, recombination reactions [3], and inferior electron transporting property [2]. A delicate modulation of the shape and size of the materials along with a careful dye selection to improve the light scattering properties of the photoanode are essential to improve the PCE and LHE. Strategies introduced to improve the PV performance include functionalized photoanode electrodes by introducing a blocking and modifier layer at the FTO/TiO_2 interface and TiO_2/dye interface respectively, coating the photoanode with light scattering materials, and developing hierarchical morphologies to name a few.

TiO_2 nanoparticles (10–20 nm) offer excellent catalytic activity properties where the device performance can be tuned by controlling the crystal structure and size of the nanoparticles [3, 8]. Crystalline TiO_2 0D nanoarchitectures (nanoparticles, quantum dots [QDs]) exhibit increased surface area for better dye loading, lower charge recombination, and higher charge separation [2]. The properties of the 0D nanoarchitectures are highly dependent on the specific surface area, morphology, size, composition, crystallinity, and porosity [20]. The specific surface area and band gap energy increase drastically as particle size decreases, which facilitates greater interactions between anchoring groups of the dye and TiO_2 and consequently dye adsorption and charge transport efficiency get enhanced. Capping agents are added to avoid agglomeration and control particle size and efforts have been made to modify the nanoparticle surface to enhance the dye–electrode interaction by doping with suitable materials. The (101) facet dominates majority of the total surface area of anatase TiO_2 crystals (>94%) becoming the primary location for dye adsorption [4]. The (011) and (001) facets have higher reactivity for rutile and anatase TiO_2, respectively. The former possess negative flat band potential, while the latter is the most active among the facets of (101), (001), and (010) [21]. By appropriately controlling the (001) and (100) facets, the surface activity enhances charge transportation because of their high energy, reactivity, and activity in photocatalysis and catalysis. Anatase TiO_2 with 34% exposed (001) facets and TiO_2 nanoparticles has yielded DSSCs with a PCE of 7.06% and 3.47%, respectively [22]. The conversion efficiency increases with increasing percentage of exposed (001) facet in different morphologies, such as nanosheets, nanospheres, and nanorods. Exposing the desired facet influences the morphology and crystal phase, resulting in efficient dye-loading, facile electron transfer, increases electron transfer, reduces recombination rate, increases V_{OC}, and enhances light scattering effect.

The electron transfer in a 0D network occurs through a "random walk" direction [23] requiring 1×10^3 to 1×10^6 interparticle hopping steps to traverse a distance of few microns due to low electron diffusion coefficient compared to bulk (shown in Figure 3.3).

The presence of misaligned crystallites, lattice mismatches, defects, and grain boundaries causes small electron diffusion length (10–35 μm) [23], low electron transport characteristics, reduced electron lifetime, electron recombination, and electron scattering decreasing the DSSC efficiency [4, 8, 9, 16]. 0D nanoarchitecture-based networks have several drawbacks concerning the light scattering, electron transport, charge recombination at the grain boundaries, and defects as discussed above [2].

Polycrystalline and monocrystalline crystal 1D nanoarchitectures (nanowires, NWs; nanorods, NRs; nanotubes, NTs; and nanofibers, NFs) with high aspect ratio provide several benefits over 0D nanoarchitectures by providing smooth directional electron mobility which accelerates the charge transfer from the photoanode and FTO [8]. Other features such as increased electron diffusion length [24], excellent electron transport [2, 8], high diffusion

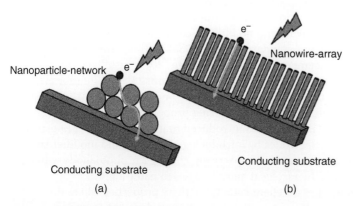

Figure 3.3 Electron movement in a photoanode – (a) 0D nanoarchitecture comprising of nanoparticles, quantum dots have a diffusion-dominated mechanism, while (b) 1D nanoarchitectures have a directional electron movement to the conducting substrate.

coefficient, enhanced light harvesting, tunable length, low recombination [4] makes 1D nanoarchitectures more preferred than 0D. Its geometry allows for a large surface area for increased dye loading improving the light scattering property capacity and the performance of the device. Increasing the length of the 1D nanoarchitectures is a way to increase the scattering effect by enlarging the surface area, though there is an optimized length at which the device provides improved efficiency and can outperform the 0D-based DSSCs, beyond which the V_{OC} reduces due to cross-linking and fast recombination rates.

Nanorods have high electron transportation due to the large surface area, large pores, and well-aligned filaments resulting in an efficiency of 6.2% [2]. Although in the bulk phase, anatase TiO_2 have better photocatalytic activities than rutile TiO_2, rutile TiO_2 NRs has high refractive index, better chemical stability, and low production cost [8]. In contrast to NWs and NRs, vertically aligned NTs can supply both the inner and outer walls as support, resulting in increased surface area for dye intake [4], enhanced optical absorption due to scattering, and reduced recombination loss [8]. The NT length can be tuned to achieve higher light scattering effect and improved performance with the maximum efficiency of 6.9% achieved at NT length ~20 μm [25]. There are complexities involved in the preparation of ultralong NTs as they tend to detach from the substrate due to stress at the interface of nanotube/substrate. Owing to the opacity of metal substrate used to grow and support NTs, the incident light on NT-based DSSCs can only illuminate from the transparent Pt counter-electrode side (back illumination), while for a conventional FTO based DSSC, the incident light illuminates from the front side (front illumination). Back illumination results in loss of light-harvesting capability due to light scattering and absorption at Pt layer and electrolyte before reaching the NT-array. To address the issue of back illumination, NTs are transferred to transparent FTO glass substrate rendering front illumination possible with adhesion of TiO_2 NPs with improved device performance. Front illumination for TiO_2 NT-array DSSCs helps in improving device efficiency.

The main limitation of 1D nanostructures is the low surface area leading to poor dye loading ability [3, 9] resulting in low light harvesting, limiting their application in solar cells [24]. Reducing pore size of the individual NT allows more NTs to be accommodated on the substrate, although the process reduces the individual surface area. Two dimensional (2D) nanoarchitecture materials such as branched NWs, nanosheets, and nanofilms show better performance than bare TiO_2 NPs due to comparable light scattering ability and dye adsorption [26]. Its low-dimensionality characteristics along with unique shape-dependent properties differ from the bulk properties, but its use for DSSCs is yet underutilized due to its low conversion efficiency. The inclusion of nanocomposites improves device performance because of their high porosity and surface area. Although considerable development is taking place at the research level, their commercialization still requires considerable improvements [26].

Hierarchical nanoarchitectures combining 0D and 1D nanoarchitectures, as shown in Figure 3.4, present distinct advantages over its components, regarding higher dye loading due to large surface area [2], good electron transport properties, better adsorption due to scattering of light, and enhanced LHE.

Hierarchical nanostructures yield efficiency of up to 8% due to [2] – (i) improved electron transport properties via the single wall structure, (ii) passivation of surface defects, and (iii) high dye loading capability achieved by the decorated nanoparticles. The three-dimensional (3D) dendrites and dendritic hollow urchins enhance the light-scattering effect and light harvesting. However, the growth and fabrication of hierarchical nanostructures are a difficult task because of lattice mismatch between the FTO and hierarchical photoanode. Hierarchical materials of different morphology at different length scale consists of primary structures (such as, ZnO, TiO_2 nanocrystallites) which is either low dimensional, spherical, or 3D on a secondary structure that is of a higher order [4]. The primary structure provides a higher specific surface

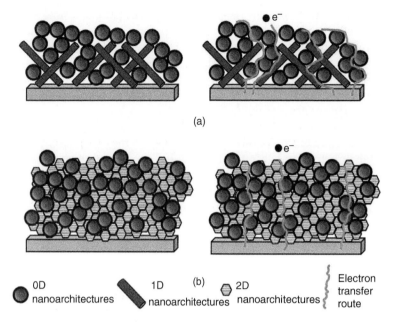

Figure 3.4 Hierarchical nanoarchitectures combining (a) 0D nanoparticles and 1D nanoarchitectures, and (b) 0D and 2D nanoarchitectures. Hierarchical nanoarchitectures offer the advantages of individual components.

area to load dye molecules, while the secondary structure aids in higher LHE, lower electron recombination, and facilitates electrolyte diffusion.

As with mesoporous TiO_2, {001} tuned crystalline facets has recently drawn attention [2] due to increased reactivity and reduced charge recombination as seen with TiO_2 nanosheets (PCE ~ 4.56%), microspheres (7.91%), nanosheet-based hierarchical spheres (7.51%) having different ratios of {001}, and {101} facets. Nanotube hierarchical array (ZnO NP/TiO_2 NTs) exhibit higher efficiency compared to bare TiO_2-NTs due to higher surface area, fast electron transfer, increased dye intake, and improved lifetime [27]. Apart from the use of TiO_2/ZnO nanocomposites for DSSCs, other nanostructures have been investigated such as TiO_2/ZnO hybrid heterostructures (TiO_2 NW/ZnO NRs, TiO_2 NW/ZnO nanosheet) with improved properties for superior quality solar cells. A small amount of ZnO NWs (8.35 wt%) embedded in TiO_2 NP films can significantly enhance the conversion efficiency by 26.9% due to the reduced resistance and enhanced light scattering [28]. Hierarchical nanostructures involving nanowires on nanosheets show better dye loading through increased internal surface area and higher light scattering behavior through increased optical path length within the photoanode. Apart from the NW/NT arrays, hierarchical mesoporous nanoarchitectures comprising of vertically aligned NTs with high internal porosity separated by wide pores exhibit direct conduction path, fast electron injection rate, larger electron diffusivity, and excellent light scattering ability [29]. Integrating high-surface area mesoporous materials and nanocrystalline TiO_2 to achieve high efficiency. TiO_2 spherical aggregates have shown an improved efficiency of up to 10% compared to bare TiO_2 NPs due to high light scattering ability and excellent interaction between the nanoarchitectures [2]. Various hierarchical beads/spheres as alternatives to NPs shows considerable improvement in device performance due to enhanced electron transfer within the electrode [4]. Nanoribbons/nanobelts/wires/rods assemble as sphere-like hierarchical materials can be utilized as the active film in the photoanode, while another nanoarchitecture that has gained significant interest is the hierarchical hollow spheres with

Hierarchical nanoforest morphology

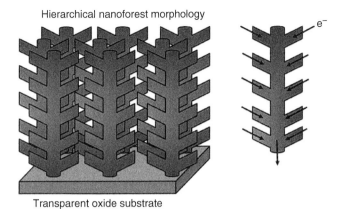

Transparent oxide substrate

Figure 3.5 Schematic representation of nanoforest-like hierarchical morphology which facilitates electron transfer without recombination losses.

single or multiple shells. These nanostructures exhibit facile electron transfer, high surface area, and superior light scattering effect as a result of reflected light from the shells, which can be controlled by the optimizing the number of shells. Nanostructures with a nanoforest morphology comprising of high density [30], long branched tree-like multigeneration hierarchical crystalline nanowires exhibited higher efficiencies than upstanding nanowires (Figure 3.5). The direct conduction pathways along with light harvesting, enhanced surface area, higher dye loading, and reduced charge combination along the nanotree multigeneration branches are responsible for the enhanced efficiency.

Self-organized mesoporous TiO_2 NPs or NWs networks improves the efficiency with a conversion efficiency of 9% [31, p. 2, 19]. Hierarchical (quasi) 1D nanomaterials offer a direct pathway for electron transfer with reduced recombination with the hierarchical structure providing sufficient area for dye loading [4]. These have several branches and along the 1D axis, increasing the surface area to enable increased dye loading, increased LHE, and tunable to match different DSSC requirements. These devices can outperform NP-based DSSC devices with a reported efficiency of 8% [32] which is among the highest obtained for solid-state DSSCs. The 3D nanostructured materials exhibit the properties of porous photoanodes like high dye adsorption, superior light absorbing capacity, and excellent optical transparency along with the added properties of improved electron transport and electron penetration properties [26]. Mali et al. [16] have studied the different 3D nanoarchitectures of TiO_2 from nanocorals, nanorods to nanoflowers thin films and found the nanoflowers to exhibit the highest J_{SC}, and PCE due to increased surface area and reduced recombination.

Electron back recombination occurring at the FTO substrate/electrolyte and photoanode/electrolyte interfaces significantly restricts the PCE [20]. Besides efficient electron transport through the device, mesoporous TiO_2 also cause electron leakage by allowing the liquid electrolyte to interact with substrate. Functionalized photoanodes with an interface modifier assist in electron injection and shielding the undesirable reactions. Depending on the position of the interfacial layer, there are three mechanisms to suppress electron recombination and improve injection (Figure 3.6) – (i) energy barrier mechanism, where a material has a higher conduction band level than that of the electron transport layer to form an energy barrier between the electron transport layer and the electrolyte or dye. The recombination rate is reduced indicating an increased V_{OC}. These include insulating MOS such as Al_2O_3, MgO, ZrO_2, ZnO, Nb_2O_5, Ga_2O_3, Ta_2O_5, SiO_2, In_2O_3, and some polymeric based materials [2], (ii) surface dipole forming mechanism, where the conduction band potential of the

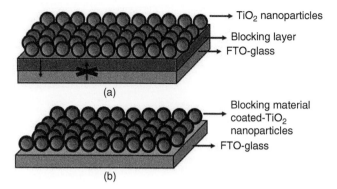

Figure 3.6 Schematic of the function of (a) blocking layer, and (b) surface modification of nanoparticles with blocking materials, to reduce back electron transfer/recombination in photoanode.

photoanode shifts in the negative direction. The J_{SC} values are reduced by the surface dipoles while increasing the V_{OC}, thereby improving the efficiency of the device, (iii) surface states passivating mechanism, where nanosized particles or using a thin TiO_2 compact layer are applied on the electron transport layer to fill the surface states of the photoelectrode effectively.

Among the various materials investigated, a compact $TiCl_4$ layer is widely used as a blocking layer for several metal oxide-based photoanodes. $TiCl_4$ treatment improves the adhesion [2], minimizes electron recombination rate, and improves electron transfer by providing additional electron pathways from photoanode to TCO. There are still several disputable points regarding its contribution to the device performance and the mechanism of posttreatment of $TiCl_4$ in improving the device performance remains unclear due to conflicting results and must be investigated in the future research [4]. The surface modifications (e.g. Eu^{3+}, Tb^{3+}) and codoping help in decreasing the recombination process, improving charge separation at the p–n junction and increasing carrier concentration and improving electron transport process (PCE, NiO–TiO_2 ~8.8%). Deposition of an ultrathin TiO_2 layer (~25 nm) improves the efficiency from 4.15% to 5.16% due to suppressed electron recombination at the FTO/electrolyte interface [33, p. 4] along with retardation of V_{OC} and increased electron lifetime [9, 34]. A SiO_2 insulating layer on dye-loaded TiO_2 architecture demonstrated an efficiency increase of 36% compared to semiconducting architecture. There is a range of emerging nanomaterials that has gained significant research attention, like graphite carbon nitride (g-C_3N_4) modified TiO_2 nanosheets [35] which contribute additional electrons and act as a blocking layer for electron recombination and enhances device performance. The coating layers (depending on the thickness) lead to increased V_{OC} and conversion efficiency by blocking electron recombination. Besides the insulating property of the oxides, other properties such as conduction band position, oxidation state, and isoelectric property should also be taken into consideration in order to suppress back reaction [4]. Compact layers deposited from electrodeposition (~60 nm), atomic layer deposition (ALD) (>5 nm), and thermal oxidation exhibit better blocking properties and provide <1% effective area of pinholes, while those deposited by spray pyrolysis are poor in their properties though they are the preferred route for practical DSSC production. Thus, a careful design of interface engineering of the TiO_2/insulating layer is required to fabricate high-performance DSSCs.

Surface functionalization of photoanode with light scattering materials is another efficient and popular method to improve the device performance by enhancing light absorption [4]. Scattering materials are relatively larger particles which when incident with light, scatter them through the photoanode film resulting in a prolonged light path, light harvesting properties, and enhanced optical absorption. Measures to improve the LHE and in turn device efficiency, include

(i) Increasing the photoanode internal surface area, thus increasing the dye intake over a given film thickness and area,
(ii) Introducing scattering centers and hierarchical structures to increase the optical path length, and
(iii) Enhancing the dye absorption by introducing plasmonic metal semiconductors structures.

F.E. Galvez et al. [36] suggest that when the electrode thickness is less than the electron diffusion length, then light scattering layers can be used as back reflectors to improve light absorption, otherwise the top scattering particles in the nanocrystalline paste yield a better output.

Surface plasmon resonance (SPR) effect is a promising pathway to increase the device performance by enhanced spectral absorption, enhanced photocurrent density, and light harvesting due to excitation of localized SPR along with low recombination. Plasmonic photoanodes are still an emerging field in DSSC spanning a wide range of materials where the paramount challenge is coming up with effective strategies to incorporate suitable plasmonic structures into nanocrystalline and nanostructured electrodes. The use of plasmonic effects aids in increasing light absorption in photoanodes, generated by the resonance between the electric fields of free electrons and electromagnetic waves [4]. Various nanoarchitectures in different configurations such as nanoplates, NWs, spheres, NRs, and nanofibers, nanobelts have been developed [26] and applied as scattering layers of the DSSCs, with the nanoparticles/hollow spheres exhibiting the highest efficiency due to the light scattering effect of the hollow spheres [2]. A mixture or a double-layer structure is used to apply light scattering materials in photoanodes, as shown in Figure 3.7.

In the double-layer structure, the nanocrystalline film embeds an extra film composed of larger particles. Compared to the size of the nanoparticles (~10–50 nm) used to supply sufficient surface area for dye loading, the size of the scattering layer is larger and comparable to the wavelength of light (~200–500 nm) which enables reflection and generates different degrees of light scattering. The optical path length within the photoanode increases, leading to a higher

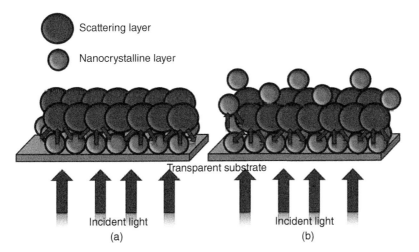

Figure 3.7 A typical structure for photoanodes containing light scattering materials, where they are used in two configurations – (a) double-layer architecture comprising of one layer of nanocrystallites and one layer of scattering materials, and (b) mixed architecture.

probability of photoabsorption by the sensitized nanocrystalline film [4]. Light reflection is influenced by the particle size in the scattering layer – large TiO_2 nanoparticles (100–500 nm) along with other nanoarchitectures such as spheres and sphere-like materials are the most commonly used in the scattering layer with a recorded conversion efficiency of 10% [37, 38]. TiO_2 spheres used for scattering layers are composed of small nanoparticles as primary structures. The large particles in the double-scatter layer hinder charge transport and increase the internal resistance within the solar cells, while in the mixture structure, the large particles will cause an unavoidable loss of photoanode internal surface area. A double-layer structure with rutile NWs containing anatase NPs as under-layers and disordered spherical voids as over-layers exhibited excellent light scattering capability [9] and efficient dye loading resulting in enhanced efficiency ($\eta \sim 4.07\%$) compared to bare 1D TiO_2 NWs ($\eta \sim 1.14\%$) [39]. Higher efficiency of ($\eta \sim 9.40\%$) has been reported using multilayer anatase TiO_2 NWs arrays consisting of three layers – densely packed TiO_2 NWs as bottom layer, hierarchical TiO_2 NWs with short branches as intermediate layers, and loosely packed TiO_2 NWs as upper layer [21] which allows the combination of several desirable features. Nanospindles has been used as bridges to improve the connectivity of the film leading to improved electron diffusion coefficient [40], exhibiting high η ($\sim 8.3\%$) in the double-layer structure, where one layer consists of larger spindles (for enhanced light scattering) and the other layer consists of smaller spindles (for better dye loading) [41]. The introduction of multi-walled carbon nanotubes (MWCNTs) inside the TiO_2 NRs led to a high $\eta \sim 10.24\%$ [42] due to fast electron diffusion, longer electron lifetime, and strong light scattering. Coating with luminescent materials converts UV and near-IR light to visible light, increasing the LHE [4, 43, 44] and incorporating upconverting phosphor nanoparticles [45]. Hybrids of hydrogels [46], CaF_2 [47], SiO_2 [48], SiC [49], Li_2SiO_3 [50], and CdS [51] and perovskite materials have been applied in TiO_2 composite photoanodes resulting in enhanced conversion efficiency. In-depth analysis of the mechanism of the plasmonic resonance embedded in the nanoporous electrode and their influence on the photocurrent conversion efficiency is an unexplored area. It is also important to identify other low cost-effective metallic nanoparticles such as Zn and Cu which could perform better. Thus, DSSC device design using scattering layers require a carefully rational design and factors, such as film thickness, morphology, and scattering contents, must be considered to obtain the optimal synergistic improvement.

Photonic crystals exhibit periodicities in their refractive index on the order of the wavelength of light and can be coupled to the photoanode to increase the LHE by different mechanisms, such as (i) photon localization and enhanced red light absorption, (ii) light reflection within the photonic band gap at various angles, (iii) formation of photon resonance modes within the dye-sensitized layer [7]. The application of photonic crystals can obtain a significant improvement in the LHE via a fine control over the reflected, diffracted, or refracted light passing through the layer. However, the main drawback in the integration of photonic crystals is the incompatibility between fabrication routes for photonic structures and photoanodes. There have been studies aimed to overcome the difficulty, such as developing a seamless photonic crystal-TiO_2 NT-assembly [52] or coupling porous and highly reflecting 1D photonic crystals into a nanocrystalline TiO_2 electrode [53]. The mesoporous structure allows the electrolyte to soak the electrode without interfering with the charge transport, while the photonic crystal layer efficiently localizes the incident light within the photoanode in a targeted wavelength range. Also, planar waveguides have also been investigated to improve the LHE, where the 3D structure effectively helps in increasing the light absorbing area due to internal multiple reflections without increasing the electron path length to the collecting electrode [54]. Plasmonic photoanodes made using Au and Ag NPs with a layer of thin SiO_2 and TiO_2 layer reduce metal corrosion in the electrolyte or undesired electron–hole recombination [37]. Novel $Au@TiO_2$ and $Ag@TiO_2$ core–shell nanostructures blended with TiO_2-NP as photoanode [55–58], result in conversion efficiency of 10%, while Au@Ag core–shell NPs embedded on TiO_2 hollow NPs showed a PCE of 9.7% [59]. Shape-controlled Au@SiO_2 nanocubes embedded in a TiO_2 photoanode show a PCE of 7.8% higher than bare TiO_2 photoanode [60], with significant enhancement in the optical absorption in the 400–600 nm range attributed to the dipole–dipole coupling between the nanocubes and dye molecules. Upon illumination, the nanocubes absorb the incident light at the surface (plasmonic near-field) which gets coupled with the dye molecules, acting as a secondary source of light resulting in enhanced light absorption and increased photocurrent density of the DSSC device. The high conversion efficiency is due to enhanced charge transfer processes [7], increased dye loading and light absorption, and reduced recombination as well as localized SPR enhancement. The absorption and scattering properties of the metal nanostructures also depend on the size and shape of the nanostructures [2], with isotropic and anisotropic nanostructures generating intense fields localized at the edges and corners of nanostructures and acting as "nanosized light concentrators." The shape-controlled metal nanostructures coupled with dyes helps in improving the device performance via scattering and absorption of the incident photons by – (i) enhancement in dye absorption: the incident photons are scattered by the nanostructures and further gets absorbed by the dye molecules, (ii) increased total absorption cross-section: the nanostructures act as antennas that couple the plasmonic near-field to the dye molecules, and (iii) enhancement in photocurrent.

One of the common methods to improve the electronic properties and efficiency of photoanode materials is by doping [20] to maximize the dye–TiO_2 interface and light absorption that can increase the electron lifetime, reduce recombination, and improve optoelectronic response. Introducing metal and nonmetal dopants, such as In, Co, V, Fe, C, S, and B can tune the electronic and optical properties of TiO_2 by providing shallow trap sites for the conduction band electrons [2]. The requirements of a doping ion needs to be:

(i) Ionic radius similar to Ti/O ions to reduce lattice mismatch,
(ii) Reducing the energy gap for excellent visible light absorption,
(iii) Altering the growth rate of TiO_2 particles to enhance the surface area, and
(iv) Tailoring the electronic structure to improve charge transport mechanism.

Doping with metal cations takes place at the Ti sites, while doping with nonmetal anions takes place at the O sites. Oxygen deficiencies in the TiO_2 crystal structure create electron–hole pairs, shorten the electron lifetime, and induce TiO_2 to become visible light active and enhance PCE. N-doping in TiO_2 (N-TiO_2) shows an increased efficiency of about 8% in the IPCE and PCE due to increased absorption in the 370–530 nm spectrum. Doping with B improves the efficiency of ZnO-based DSSC in a similar manner to N-TiO_2 based DSSCs [4]. Co-doping strategy with two or more dopants is also another approach to increase device performance, where each dopant has an individual influence on cell performance, either affecting J_{SC} and/or V_{OC}. Co-doping N-TiO_2 with various nonmetals as graphene, C, S, and Cr [20] has yielded DSSCs with high PCE with the highest PCE of DSSCs with a 9.32% for graphene–N–TiO_2, suggesting that the addition of graphene is beneficial to get higher dye loading, lower internal resistance, and fast carrier transportation for N-TiO_2. Transition metals, such as Ti, W, Mg, Sn, Nb, Ni, Ta, and Zn make the energy levels or electron recombination preferable for DSSCs by shifting the band and facilitating electron transfer thereby increasing V_{OC} and J_{SC} [4]. Doping with suitable transition metals to replace metal cation in TiO_2 results in increased electron DOS, extended band edges, and higher Fermi level. Engineering the electron DOS and energy levels by doping with rare earth metals (e.g. Y, Yb, Er, Lu, and Tm) has also yielded positive results. Most rare earth metals are of the kind M^{3+} (M = Y, Yb, Er, Lu, Tm) causing p-type doping resulting in increased energy level of the photoanode, improved LHE, and enhanced V_{OC}. Using metal oxide composites have several advantages over bare n-type MOS because of the formation of heterojunction structures [26], which reduce recombination to a large extent. Several combinations such as ZnO–TiO$_2$, MgO–TiO$_2$, CNTs–TiO$_2$, Ag/Au–TiO$_2$ have been investigated among which ZnO–TiO$_2$ has emerged as the best composite structure for DSSCs. TiO$_2$ integrated with ZnO takes advantage of the high reactivity of TiO$_2$ and the high electron mobility of ZnO, thereby improving the process of electron–hole transfer. ZnO forms an energy barrier at the electrolyte/electrode interface, leading to reduction in recombination rate and increase in performance. The combination can be further enhanced by various nanoarchitectures such as NRs, NPs, nanocomposites, NTs, nanobelts, nanodots, nanowires, to name a few, where the optical and electrical properties of the material depend on the morphology and phase of the nanoarchitectures (seen in Section 3.3.1). Other parameters that play a vital role includes the dopant source, doping mechanism, and doping level which significantly impacts the electron transport layer structure, including the anatase–rutile transition and accumulation on grain boundaries.

Graphene, a 2D sp^2 bonded carbon sheet and carbon-based materials has attracted a lot of attention in recent times due to their unique structure, and physical properties, as tunable band gap, photon absorption, high electrical conductivity [4], mechanical flexibility, surface area (2.6×10^6 m^2/kg), carrier mobility (20 m^2/V s), chemical stability, and optical transparency [8]. Surface modification of photoanode with graphene increases the surface area to improve the dye loading and suppress photogenerated electron recombination, ultimately enhancing device performance [18]. The photogenerated electrons separated at the TiO$_2$–dye interface are transported through the TiO$_2$ matrix to FTO. The TiO$_2$–FTO interface under the influence of a favorable energy band position enables the final charge collection but is also the primary location for recombination that puts a limit on the PCE. The work function of graphene (~4.42 eV) is between the conduction band of TiO$_2$ (~4.21 eV) and FTO (~4.4 eV) substrate, facilitating charge transfer from TiO$_2$ to FTO without any energy barrier, while suppressing TiO$_2$ agglomeration, recombination, lowering resistance, and enhancing the PCE [18]. It also helps to overcome the requirement of a TiCl$_4$ pretreatment at the FTO/TiO$_2$ interface that creates microstructural defects in FTO. Incorporation of carbon materials in the TiO$_2$ matrix improves TiO$_2$-NP dispersion, yielding a highly porous nanocomposite [2]. The incorporation of graphene into the TiO$_2$ photoanode results in a 45% improvement in the I_{SC} without sacrificing the V_{OC} with a total efficiency of 6.97% [61] which is a 39% improvement compared to nanocrystalline TiO$_2$ photoanode and also higher than that of a 1D CNT-TiO$_2$ photoanode. The graphene–TiO$_2$ nanocomposites also have great LHE with the efficiency improving to 7.68% from 4.78% with the addition of graphene [62]. Graphene increases the surface area and photoconversion [63] [64] of the hybrid photoanode thereby increasing the amount of dye loading while acting as a support for TiO$_2$ when deposited on flexible substrates [64]. The rapid transport of the photogenerated electrons in the 3D graphene mesh reduces charge transfer resistance and recombination rate at the photoanode/electrolyte interface while providing a pathway for the electrons transported from the oxide layer to the flexible substrate. Graphene performs better than carbon nanotubes (CNTs) because of its 2D structure, where the TiO$_2$ can anchor in, and the graphene network can capture and transfer the photoinduced electrons. In contrast, 1D CNTs composites have fewer intermolecular forces leading to weak TiO$_2$ anchoring, transfer barrier and easy recombination. The amount of graphene and CNTs in the photoanode must be less than 1 wt% in the film, as higher loadings of graphene or CNTs, will shield the visible light from being absorbed by the dyes and decreasing the photogenerated electrons [4]. The quality and the nanostructures of graphene incorporated into the TiO$_2$ also influence device performance. Since obtaining high-quality graphene is a tedious task, there is considerable work that is required before graphene–TiO$_2$ photoanodes becomes commercially viable. Proper contact is essential between each TiO$_2$ nanoparticle and graphene network for improved photocurrent, which is possible only when a good graphene sheet dispersion in the inorganic framework is achieved.

3.3.2 ZnO

The large band gap of TiO_2 (~3.2 eV) limits its application under visible light and many dyes fail to inject electrons to the conduction band owing to insufficient electron injection [65]. In this regards, ZnO has a band gap similar to TiO_2 (~3.37 eV) and has been proposed to improve the device performance due to its higher excitonic electron mobility (~115 cm^2/V s) compared to TiO_2 (~10 cm^2/V s).

There is considerable interest in the use of ZnO as an alternate to TiO_2 in DSSCs and similar technologies such as inverted polymer solar cells, QD solar cells, and hybrid light emitting diodes (LEDs) [6] owing to its electronic band position, large exciton binding energy (60 meV at room temperature) [66], high electron mobility [20], excellent bulk electron conductivity (an order higher than anatase TiO_2) [6], and the richest family of well-aligned nanostructures (NPs, NWs, nanosheets, etc.).

However, the highest PCE achieved is still 7.5% [67] which is far from that achieved for TiO_2 due to low PCE and recombination processes. Analogous to TiO_2, a trap-limited transport mechanism is characteristic of randomly packed ZnO NP network [6]. Without considering the defects, traps, and grain boundaries, ZnO is superior to TiO_2 as regards to bulk conductivities, confirmed by the works of Solbrand et al. [68], Noack et al. [69], Quintana et al. [70], Tiwana et al. [71], though in the long term both TiO_2 and ZnO perform the same. The electron diffusion coefficient of ZnO cells fabricated with different dyes and electrolytes have yielded values from 10^{-4} to 10^{-6} cm^2/s and a multiple-trapping behavior. A direct comparison between TiO_2 and ZnO-based photoanodes is difficult due to the variety of fabrication methods and process conditions (dye, electrolyte, particle size), which can influence the chemical nature, and the level of doping affecting the reproducibility of the literature results.

The collection efficiency of a photoanode is analyzed in terms of diffusion length (L_n), which is dependent on the electron diffusion coefficient (D_n) and electron lifetime (τ), given as [6]:

$$L_n = \sqrt{(D_n \tau_n)}$$

For ZnO-based devices, 100% collection efficiency is possible, which suggests that there is significant current loss due to poor injection or poor dye regeneration [72–74] resulting in lower J_{SC} and FF with a tendency for lower V_{OC}. ZnO is more basic making it less stable in acidic dyes with carboxyl groups, resulting in the dissolution of ZnO to Zn^{2+} by the adsorbed dye molecules, and blocking the transport of the injected electrons from the dye to photoanode [20]. The reasons for the slower injection kinetics has been attributed to intermediate states for ZnO NPs functionalized with the dyes, which occur at the semiconductor/dye interface. The higher dielectric constant of ZnO ($\varepsilon \sim 30$–170) compared to TiO_2 ($\varepsilon \sim 8$) makes the exciton Columbic bound energy relatively large, hence preventing fast photogeneration. Besides dielectric constant, other factors such as high density of intra-band defect states and low DOS in the conduction band are also responsible for the lower performance. Good light absorption and an efficient electron injection determine the efficacy of photoanode sensitization with organic and metal-organic dyes [6]. The immersion time of ZnO electrode in Ru-based dye solution is critical for good device performance – long immersion times diminishes the photocurrent due to the dye molecules aggregation in the semiconductor surface which inhibits efficient electron ejection. The optimum immersion time depends on dye concentration – higher concentrations lead to a quicker sensitization, but also to a rapid photocurrent degradation. Xanthene dyes like eosin, indoline do not show this behavior due to lower acidity and lack of complexing agents, while the presence of co-adsorbents as chenodeoxycholic acid prevent dye aggregation. ZnO-based solar cells sensitized with xanthene dyes exhibited similar quantum efficiency to that of the Ru-based dyes with panchromatic absorption spectrum, high extinction coefficient and good LHE (>5%). However, the lower dye–ZnO interaction, while avoiding complex formation, provokes a weak and unstable sensitization effect. This causes its detachment from the surface when the electrolyte acts as a good solvent or when there are adsorbing species competing for the adsorption sites at the semiconductor surface.

In order to overcome the challenges of dye uptake, charge collection, and light absorption, a number of techniques have been proposed such as developing photoanodes with novel morphologies, using nanoarchitectures, hybrid core–shell structures [9], engineering interfacial modifier layers, designing new dyes, doping with metallic cations/nonmetallic anions, and so on. As seen earlier, nanoarchitectures provide high surface area to volume ratio, which will, in turn, facilitate high LHE, though interfacial recombination at grain boundaries and defects cause significant losses [66]. Zero-dimensional (0D) nanoarchitectures (NPs) of 40 nm size show the best conversion efficiency among the particles with sizes ranging from 20 nm to 2 μm. ZnO 1D and 2D nanoarchitectures (NWs, nanobelts, nanoclusters, nanoflowers, and nanocolloids) offer distinct advantages over in terms of electron transport by allowing unidirectional movement for photoexcited electrons in photoanode which facilitates rapid electron injection (about 1–2 orders faster than in 0D nanostructures) from dye to the photoanode [6] leading to high scattering effects. ZnO NRs exposed with {1010} facets are widely used in DSSCs, with studies showing that {1011} facets have superior performance due to the strength of binding between the dye molecules and reactive O-terminated {1011}'s [75], thereby preventing dye aggregation. ZnO NWs are found to be superior to randomly packed NP-networks due to directional charge transport and minimized recombination, improving electron collection due to the existence of a grain boundary-free direct pathway towards the external circuit. The electron density in NWs is several hundred times faster with additional properties such as antireflective and light trapping properties, enhanced surface area, and solution processability. Hierarchical branched NW configuration can achieve higher efficiency than conventional upstanding ZnO NWs due to the modified surface area which allows higher dye loading and reduced recombination rate. Despite being more efficient, NW-arrays do not have high PCE due to the stronger recombination, where the presence of donor states below the conduction band and higher DOS in the ZnO NWs create additional recombination pathways affecting the collection efficiency. Another reason for the low performance of the NW-array is the preparation methods which impacts the properties.

ZnO instability at acidic pH can be overcome by surface passivation as ZnO–TiO$_2$ core–shell NW arrays (PCE ∼ 7%) [6] or by surface modification with Al$_2$O$_3$, TiO$_2$, ZnO, and SiO$_2$, which prevents the formation of Zn^{2+}/dye aggregates reducing recombination [66, 76–78]. Other ways to achieve NW surface modification include coating the NW with a thin shell of nanoparticles. As a compromise between the surface area and electron transfer rate, 2D nanoarchitectures (nanosheets) have achieved a high conversion efficiency of 5.41%. The far superior performance of mesoporous nanosheets is attributed to the higher effective electron diffusion coefficient and dye adsorption (3.59×10^{-3} cm^2/s, 2.66×10^{-7} mol/cm^2, respectively) than nanoparticles (1.12×10^{-3} cm^2/s, 1.99×10^{-7} mol/cm^2, respectively). The perpendicular structure of nanosheets with elongated pores enables rapid dye adsorption and reduces formation of Zn^{2+}/dye aggregates. Such architecture is compatible for diffusion of I$^-$/I^{3-} through the semiconductor matrix and enhanced light absorption by the dye molecules. In the absence of the elongated pores, the aggregates occupy the sites on the adsorbed dye molecules blocking them from light exposure. Hierarchical ZnO NW–nanosheet structure overcomes the issue by filling the gaps with NW-branches offering larger internal surface area and direct electron transport from the branched nanowires to the nanosheet [9]. ZnO tetrapod nanoarchitecture has attracted much attention due to their hierarchical nature with an interconnected network [2]. It allows a low surface defect, high porosity, mechanical strength, high electron mobility, long diffusion length, long carrier lifetimes, and scattering facets that lead to an enhanced LHE. The PCE of ZnO tetrapod is dependent on the arm length: diameter ratio (packing density) since this determines the interface carrier separation and transportation in the device. SnO_2 NPs/ZnO tetrapods exhibit good charge collection, large roughness factors, customizable light scattering properties while repressing recombination [20]. The device has shown an efficiency of 6.31% indicating that ZnO tetrapod based DSSCs can achieve higher efficiency than devices based on ZnO NW arrays and spherical particles.

A double-layer film consisting of a lower layer of ZnO NPs and an upper layer of ZnO nanosheets shows a PCE of 4.65% [79]. Various morphologies for the first layer has been attempted such as NRs [80], microflower [81] that showed an improved device performance. Smooth and mesoporous microspheres and another hybrid structure comprising of ZnO aggregates and 3D upconverting materials improve the scattering properties while broadened the light absorbing range to near IR. Metal-organic frameworks in ZnO-based DSSCs resulted in the formation of Zn parallelepipeds which act as an active light scattering layer leading to significantly enhanced performance. It has been observed that doping ZnO with rare earth metals (La, Ce, Nd, Sm, and Gd) form an energy barrier and maintain a lower charge recombination probability [66]. Such rare earth ion modifications can passivate the surface states of the ZnO electrode, with modifications using Nd, Sm, and Gd enhancing the V_{OC} and FF, while those with La, Ce, and Nd have led to a decrease in the J_{SC}. Materials as I, CNTs, reduced graphene oxide, B, Ga, Li, Sn, Mg, Y, and polyoxometalates has revealed several desirable features as longer electron lifetime, stronger peak current, lower charge transfer resistance, reduced interfacial recombination, and extended visible light activation, which can all promote the performance of the cell.

3.3.3 SnO$_2$

SnO_2 makes an attractive material as photoanodes owing to its high visible light transparency, high electron mobility ($\mu \sim 250 \, cm^2/V \, s$ for single crystals, $150 \, cm^2/V \, s$ for nanocrystals), high band gap ($E_g \sim 3.6 \, eV$), superior photostability under UV light compared to TiO_2 and ZnO, and reduced degradation of dye molecules on solar irradiation [5]. As an oxygen-deficient n-type semiconductor, SnO_2 is promising for applications that require optical transparency and high conductivity [20]. While TiO_2 and ZnO remain the most widely researched material, there are limitations with regards to low electron mobility ($\mu < 1 \, cm^2/V \, s$ for single crystals, $10^{-5} \, cm^2/V \, s$ for nanocrystals) and slow electron diffusion ($10^{-5} \, cm^2/s$). SnO_2 characterized by its lower effective mass ($m_e \sim 0.17–0.30 \, m_o$) has better electron-hole separation ability and superior charge transport properties that improves the device stability.

Despite high μ, SnO_2-based DSSCs still suffers from low efficiency (<5%) due to its lower isoelectric point (pH 4–5) compared to TiO_2 (pH-7) and ZnO (pH 9) and the lowest conduction band edge among all MOS employed as photoanodes. This eventually results in inferior dye loading, lower J_{SC}, larger over-potential at the dye/MOS interface, and higher recombination. The unfavorable band alignment of SnO_2 to the LUMO of commercial Ru-based dyes such as N3, N719, and black dye results in a much-reduced V_{OC}. Enhancing the surface area alone is not capable of improving the efficiency and two approaches to enhance V_{OC} in SnO_2-based photoanodes. One, by doping of transition metals of similar ionic radii to raise its Fermi level, and the other is employing insulating as MgO and Al_2O_3 layers as a shell, to reduce recombination.

Cojocaru et al. [82] showed a morphology-dependent performance of SnO_2-based DSSCs, employing NPs of various sizes (15–20 nm and 50–150 nm, surface area ~44–57 m²/g) as the bottom and top layer, to produce a bilayer photoanode with an efficiency ~4%. Wang et al. [83] employed hollow SnO_2 nanospheres with a Brunauer–Emmett–Teller (BET) surface area of ~64.2 m²/g and reported an efficiency ~0.86% which improved to 6.02% post-$TiCl_4$ treatment. Li et al. [84] in a comparative study showed that SnO_2 macroporous particles of larger size and lower surface area resulted in a 10% improvement in the efficiency compared to smaller NPs due to enhanced light scattering and higher collector efficiency in photoanodes employing macroparticles. Introduction of ZnO NRs on the SnO_2 nanoparticle layer significantly increases electron lifetime and reduces recombination, enhancing the electron transport compared to bare SnO_2 film [4]. Nanostructured TiO_2/ZnO–SnO_2 composites exert a positive impact on the electron transport and charge recombination. Photoanodes made from bare NPs or 1D nanostructure show inferior efficiency since the materials are characterized by only one of the two desired parameters – high electron transport and high surface area. Hybrid nanostructures are widely employed to utilize both key features simultaneously in a photoanode, such as mixing the SnO_2 nanostructures with TiO_2 or ZnO which exhibited a remarkable efficiency of ~6.5% [5]. Among the various dopants, Zn^{2+} and Mg^{2+} having similar ionic radii as Sn^{4+} show exceptional performance. Li et al. [85] compared the performance of Zn-doped SnO_2 based and pure SnO_2-photoanodes, where the former showed improved performance with an efficiency almost three times. The Zn-doped SnO_2 demonstrated faster electron transport and reduced electron recombination, as shown by enhanced FF. A posttreatment with $TiCl_4$ further increases the efficiency, as shown by Pang et al. [86] and Dou et al. [87]. Similar work on Li-doped ZnO NFs mixed with SnO_2 NPs exhibited an FF of 69% [88]. Improved efficiencies can be achieved using core–shell composite nanostructures, porous nanofibers, nanotubes, and doping, though it is still nowhere close to that of TiO_2-based devices. One reason for the reduced efficiency is the lattice strains in doped SnO_2 which can be overcome by suitable doping with transition metals of similar ionic radii such as Nb^{5+}, W^{4+}, Ti^{4+}, In^{3+}, and Ga^{3+} to enhance performance. One strategy to overcome the conduction band disadvantage of SnO_2 is to use sensitizers (perylene dye, PbS QDs) with lower energy LUMO and an extended absorption coefficient in the near-IR region, in bifacial dye-solar cells or tandem configuration to enhance absorption cross-section. The low ionization potential in SnO_2 is another crucial issue that restricts the dye uptake, which can be overcome using a core–shell composite [5]. Here, a MOS with higher CB minimum such as ZnO, TiO_2, and Nb_2O_5 form the shell and SnO_2 forms the core. The shell builds an energy barrier at the SnO_2/electrolyte interface and enhances dye absorption due to the higher isoelectric point. There are two approaches to fabricating the core–shell photoanodes – (i) synthesizing SnO_2 NPs as core material which is subsequently coated with a layer prior to use in the working electrode, (ii) developing SnO_2 NP-based film at the working electrode and then coated with another layer of the shell material. In the first approach, the barrier is formed between each NP and the electrolyte, while in the second approach, the barrier is formed between the shell and electrolyte. Thus electron transport takes place through the interconnected NPs, enabling faster electron transport with reduced electron trapping and recombination. As the architecture comprises of two different materials, it is crucial to have a downhill band alignment for favorable electron injection from dye–shell–core photoanode. The conduction band of the shell must be lower than the LUMO of the dye and higher than the conduction band of the core material (SnO_2). The core–shell architecture shows a considerable improvement in efficiency as shown by Pang et al. [86]. They reported on a SnO_2–ZnO core–shell where the photoexcited electrons can be directly tunneled to SnO_2 or injected into the ZnO conduction band, followed by a downhill transition to SnO_2. The thin layer of ZnO reduces recombination resulting in enhanced efficiency of ~7.3%, which is further improved by creating multiple shells around a single core to enhance dye loading. Dong et al. [89] reported similar structures using quintuple shelled structures with an efficiency of ~7.18% compared to ~5.21% in a single shell structure. A TiO_2 shell increases the dye uptake, and the electrons are injected into the SnO_2 and collected immediately on the working electrode due to the favorable band alignment and higher electron mobility of SnO_2. The concerns of low surface area and unguided electron transport are overcome by having single tubular and multichannel tubular nanostructures. Multiporous nanofibers have shown an increased charge collection, electron lifetime, and electron diffusion length due to the improved contact between the nanoparticles because of the broadened contact interface, which makes the electrons traverse more easily and reduces recombination.

One of the critical challenges reported on hierarchical and composite SnO_2 photoanodes is the poor choice of material combination employed [5]. Employing hierarchical nanostructures (NPs or a composite of NPs and submicron particles) as the composite results in higher surface area and light scattering, but at the expense of electron transfer.

3.3.4 Nb_2O_5

Nb_2O_5 is n-type transition metal oxide and considered a more suitable photoanode material due to its higher band gap ($E_g \sim 3.4$–4.0 eV) [23], high conduction band [24], high dielectric constant ($\varepsilon = 33$–40) [20], and electronic injection efficiency thus having the potential to reach higher V_{OC}. Also, they also possess superior properties such as high refraction index, excellent chemical resistance, thermal stability, corrosion resistance and similar optical band gap, and electronic properties as TiO_2 [90]. The charge carriers exhibit much longer effective lifetimes, smaller diffusion rates, and fewer recombination in comparison to those in TiO_2 NT-arrays [24]. While Nb_2O_5 nanoarchitecture networks provide continuous and directional electron transfer pathways, their photocurrents and photoconversion efficiencies have not reached the desired level due to reduced dye-loading sites leading to smaller pore diameter and low PCE [24]. Increased dye adsorption and lesser dye agglomerates on the oxide surfaces leads to improvement in J_{SC}. The large unit cell dimension of orthorhombic Nb_2O_5 in comparison to anatase TiO_2 makes it a challenging task to obtain the optimum Nb_2O_5 morphology for DSSC applications. Nb_2O_5 is currently employed as a bi-layer on TiO_2 or as a blocking layer at the FTO/TiO_2 interface due to its chemical stability [20].

3.3.5 Graphene

The unique properties of graphene such as fast electron transfer, good optoelectronic properties [8], exceptional high electron mobility (~20 000 cm^2/V s), large theoretical surface area (2630 m^2/g), excellent thermal conductivity (5000 W/m K), and light weight make it an interesting candidate for solar cell applications. Although graphene and CNTs have similar properties, the latter have slightly different chemical and electronic properties due to the high aspect ratio and strain effects. The monoatomic 2D sp^2 hybridized graphene layer has a much larger and thinner surface, leading to higher reactivity and transparency [91]. Graphene-based photoanodes have appropriate properties to strengthen each part of the photoanode – high visible light transparency, high Young's modulus, ultrafast electron mobility, semiconducting properties, tunable band gap, and photon absorption [11]. Majority of the work on graphene-based materials focuses on graphene as a counterelectrode or to replace FTO as the TCO layer due to the ease of fabrication, easy chemistry, along with excellent mechanical, electrical, and optical properties. Its applications for photoanodes are relatively unexplored [92–96].

The photoanode performance is closely related to the properties of the graphene layer which in turn is influenced by the thickness, size, concentration, and surface conditions and can be controlled by the process conditions [11]. There is a concentration limit beyond which graphene forms aggregates which act as trapping sites obstructing the fast charge collection at the electrodes [91]. The preparation method introduces defects (holes, grain boundaries) in the materials that impact the properties in several ways, from chemical to electronic to mechanical and optical properties which influences the overall performance of the device. The addition of graphene to SnO_2 was shown to enhance the performance of the device by 91.5% [97]. To overcome the issue of producing high-quality graphene at low cost, a small quantity of graphene sheets is added as additives to the photoanode materials to enhance the electron transfer process. Graphene nanosheets provide a much larger surface area to anchor TiO_2 NPs and the photo-induced electrons are captured and transferred in a more efficient way.

The mechanisms for improved performance using graphene are yet to be understood. One of the challenges on using graphene is that most methods produce graphene with some defects, and it is difficult to obtain pure monolayer on an industrial scale, and mixed with multilayer graphene having lower transparency and high sheet resistance due to surface defects and oxidation. Also, the lowered mechanical strength and inherent defects of graphene sheets lead to easy cracking during transfer process. Unprotected graphene surface, when exposed to electrolyte, may lead to a high ratio of electron waste. The depth and width of research on employing graphene in photoanode need to be significantly extended presenting opportunities for researchers.

3.3.6 Other Photoanode Materials

Some of the other metal oxides that have been considered as candidates for photoanodes in DSSCs are [20].

1) *Tungsten trioxide (WO_3):* A transition metal oxide, WO_3 has several unique properties, such as high carrier mobility, though its efficiency is lower than other counterparts due to its positive conduction band edge and reduced adsorption of dye molecules leading to a small J_{SC} and unfavorable charge combination. The photoconversion efficiency of WO_3 improves by introducing a blocking layer to suppress charge recombination and fabricating the photoanode with a high surface area for higher dye loading and light harvesting. WO_3 has the potential to be used as the photoanode material in DSSCs with acceptable efficiency.
2) *Cerium(IV) oxide (CeO_2):* It has extensive photo-response property in the visible light region owing to its molecular structure and configuration. It has similar properties with TiO_2, as strong UV light absorption, nontoxicity, and high stability to chemical and photocorrosion. It can absorb a large fraction of the solar spectrum and has better redox nature that supports charge carrier transfer to the catalyst surface and creates oxygen vacancies. CeO_2 is a better photocatalyst material though its photocatalytic properties are not of practical use with solar light due to low light absorption efficiency and high recombination rate of photogenerated charge carriers. Instead of a single CeO_2 photoanode, the use of a bilayered photoanode with CeO_2 NPs as a mirror-like scattering layer on top of TiO_2 layer has exhibited an increase in the LHE.
3) *Iron oxide (α-Fe_2O_3)* – Hematite has many desirable features, such as superior long-term stability, chemical stability, low processing cost, environmental inertness, and high resistance to corrosion. Generally used as counterelectrode, it has the potential to be used as the photoanode due to its low band gap. It, however, suffers from poor PCE due to low charge carrier mobility, small optical absorption coefficient, short carrier lifetime, bulk recombination, and a short charge collection length.

4) *Zinc stannate (Zn_2SnO_4)* – Zn_2SnO_4 have better photostability against UV light than TiO_2 cells because of its large band gap (~3.6 eV) and weak work function, along with faster charge injection and electron diffusion efficiencies than TiO_2-based devices. The most attractive attribute for Zn_2SnO_4 is its high chemical stability against acid and polar groups, although its short electron diffusion length limits its application. Improvements in its crystal structure and morphology can result in improved electron transport, dye adsorption, and light scattering leading to better device performance, while the development of novel sensitizers with high molar absorbability and are sufficiently compatible with the energy levels of Zn_2SnO_4 is also essential for enhanced PV performance.

5) *Perovskite-double oxide materials*, such as $SrTiO_3$, $CaTiO_3$, $BaSnO_3$ (BSO), and $BaTiO_3$ have reported higher V_{OC} (up to 0.650 V) since they are n-type materials with comparable band gap with TiO_2. Their ionization potential, electronic structure, band gap, and electron affinity can be modified by altering their atomic composition [2]. However, their lower J_{SC} restricts its practical application. The electronic, optical, and electrical properties can be tuned by atomic substitution and doping. BSO perovskite materials have exhibited an efficiency of 6.2% where the formation of an ultrathin TiO_2 layer on the BSO surface leads to increased charge collection efficiency by improving the charge transport properties and suppressing the recombination reaction. The presence of a scattering layer of TiO_2 or ZnO supports dye adsorption, increases light absorption, and controls the recombination reaction resulting in high efficiency and stability. $SrTiO_3$ is well known for its superconductivity, ferroelectricity, and thermoelectricity [98]. It has a slightly higher conduction edge that is beneficial for collecting photogenerated electrons. The room temperature electron mobility of bulk $SrTiO_3$ (5–8 cm^2/V s) is higher than that of TiO_2 (0.1–4 cm^2/V s) along with the higher dielectric constant provides a good electrostatic shielding of the injected electrons. The charge recombination is reduced [20] and subsequently device performance improves. Although $SrTiO_3$ has attracted significant attention, there is not much work on its use as nanostructured photoanode that leaves scope for future research.

Table 3.2 provides the performance characteristics of commonly used materials as photoanodes, while Table 3.3 provides the performance characteristics of the various nanoarchitectures using different nanomaterials.

Table 3.2 Performance of solar cells from different materials measured under AM1.5 spectrum (1000 W/m^2).

Material	V_{OC} (mV)	J_{SC} (mA/cm^2)	FF (%)	Efficiency η (%)	References
Mono silicon	743.8	42.25	83.8	26.3	[8]
Poly silicon	667.8	39.80	80	21.3	
Amorphous silicon	896	16.36	69.8	10.2	
III–V (e.g. GaN)	1122	29.68	86.5	28.8	
CdTe	875.9	30.25	79.4	21	
CIGS	757	35.70	77.6	21	
DSSC	744	22.47	71.2	11.9	
Organic	780	19.30	74.2	11.2	
Perovskite	1104	24.67	72.3	19.7	

Table 3.3 Nanoarchitectures from different materials and the device performance.

Material	Morphology	Dye	J_{SC} (μA/cm²)	V_{OC} (mV)	FF (%)	PCE (% η)	References
0D nanoarchitecture							
CsPbI$_3$	Quantum dots	—	15.246	1162.6	76.63	13.43	[99]
TiO$_2$	Microspheres	—	531	507	55	0.35	[100]
SiO$_2$@Au@TiO$_2$	Microsphere	N719	16.2	706	60	7.14	[101]
Mixed-phase TiO$_2$	Nanoparticles	N719	6.776	821	70	3.82	[102]
Multishaped AgNPs	Spherical NPs	N719	16.58	760	69	8.69	[103]
	Multishaped AuNPs		16.73	750	71	8.91	
	Spherical Ag and AuNPs		17.58	750	69	9.10	
	Multishaped Ag and AuNPs		19.41	750	68	9.90	
			19.76	750	69	10.22	
TiO$_2$–Sr (50 ppm)	Nanoparticles	N719	17.43	728	62	7.88	[104]
ZnO	Nanoparticles	Organic fruit dyes (dark prune, mango, lemon)	0.006	124	3.95	3.08	[65]
TiO$_2$			0.008	145	2.48	2.88 (dark prune)	
CdSe–TiO$_2$/ CdSeTe–TiO$_2$	Quantum dots	—	19 386–20.98	645–662	60–61	8.5 (PCE)	[105]
TiO$_2$	Nanoparticles	N719	10.27	830	62.5	5.88	[106]
TiO$_2$–Sn			12.76	860	57.7	6.24	
TiO$_2$–Fe			1.14	710	73.1	0.67	
TiO$_2$–Cu			10.68	760	58.5	5.24	
Reduced graphene–N–TiO$_2$	Nanoparticles	—	19.65	744	64.70	9.32 (PCE)	[107]
TiO$_2$	Nanoparticles	N719	6.27	700	59	2.57	[108]
N,S-TiO$_2$			9.78	690	50	3.35	
AgNPs@N,S-TiO$_2$ (Ag: 20%)			29.05	770	37	8.22	

Material	Morphology	Dye	J_{sc}	V_{oc}	FF	η	Ref.
TiO_2–ZnO	Nanoparticles	N719	—	581.1	56.8	0.5	[109]
Au, Ag@TiO_2 (75%, 25%)	Nanocomposites	N719	16.33	760	42	7.33	[110]
MgLa–TiO_2	Nanoparticles	N719	14.2	743	68.7	8	[111]
Mg–TiO_2			12.1	744	68.6	6.8	
La–TiO_2			13.3	751	69	7.7	
CeO_2	Nanoparticles	N719	4.83	654	38	1.201	[112]
SnO_2	Nanoparticles	N719	9.43	530	57	2.85	[113]
$Ga_{0.25}Te_{0.75}ZnO$	Nanoparticles	N719	6.75	562	56	7.08 69.94 (IPCE)	[114]
In–ZnO (0.2%)	Nanoparticles	N719	12.58	421	51	2.7 (PCE)	[115]
In–TiO_2	Nanoparticles	N719	16.97	716	61.4	7.48	[116]
C–TiO_2	Hollow spheres and NPs	N719	14.69	770	62.2	7.02	[117]
Nb_2O_5–nanoporous TiO_2	Nanocrystals blocking layer	N719	11.94	780	74.25	6.94	[118]
Ag@Nb_2O_5	Blocking layer	N719	17.33	780	69.11	9.24	[119]
Ag@Nb_2O_5–SnO_2	Blocking layer	N719	23.12	541	51.73	6.47	[120]
Ag@Fe_2O_3–TiO_2	Mesoporous	N719	9.0177	725	68.91	4.5	[121]
Pt@Fe_2O_3–TiO_2			7.7419	823	61.22	3.9	
Pd@Fe_2O_3–TiO_2			7.0177	692	72.16	3.9	
Fe_2O_3–TiO_2			5.7677	684	76.14	3	
1D nanoarchitecture							
TiO_2–Ag	Nanotube	N719	14.42	643	54	5.01	[122]
ZnO–TiO_2	Nanowire	N719	3.51	686	38	0.91	[123]
	Nanoneedle		5.70	617	42	1.47	
ZnO	Brush-like nanorods	N719	7.97	567	43.3	1.95	[124]
ZnO NWs–mesoporous TiO_2	Nanowires	N719	14.70	814	59.57	7.13 (PCE)	[125]

(Continued)

Table 3.3 (Continued)

Material	Morphology	Dye	J_{SC} (μA/cm²)	V_{OC} (mV)	FF (%)	PCE (% η)	References
$C_3H_9BO_3$–ZnO	Nanotubes	N719	1.9	500	0.31	0.222	[126]
SnO_2	Nanobelts	Z907	1.87	463	55	0.48	[127]
Ni–SnO_2	Hierarchical nanostructures		18.57	414	48	3.69	
MgO/NiO–SnO_2			13.34	552	52	4.14	
TiO_2	Nanosheets	N719	11.9	670	61	4.84	[128]
TiO_2	Nanofibers	N719	13.32	698	55	5.39 (PCE)	[129]
TiO_2–HfO_2	Core-shell–NWs	N719	12.17	713	55.8	4.83	[130]
K–ZnO	NRs	N719,	8.26	910	41	3.1	[131]
Na–ZnO			5.96	900	58	3.09	
NH–ZnO			9.79	650	57	3.62	
HMT–ZnO			7.74	900	59	4.11	
ZnO	Nanoplates	N719	6.58	840	52	2.85	[132]
	NRs		5.35	850	49	2.21	
	NPs		4.23	640	54	1.46	
TiO_2	NRs array	–	13.5	720	54.5	5.3	[133]
CaF_2–TiO_2 (0.5 wt%)	Nanocrystallites	N719	14.5	710	74.2	7.66	[47]
CNT–graphene–TiO_2 (0.1 wt%)	NTs array	N719	14.31	650	56	6.17	[134]
2D nanoarchitectures							
Eu^{3+},Tb^{3+}:NiO–TiO_2	Films	N719	17.40	780	65	8.80	[135]
B–NiO	Films	N719	1.51	144	28.8	0.063	[136]
g-C_3N_4–TiO_2	Nanosheets	N719	15.77	646	72	7.34	[35]
ZnO@SnO_2	Films	N719	14.8	600	56	4.96 (PCE)	[137]
TiO_2	NPs–NPs	–	17.3	580	72	7.22	[138]
	NPs–NS–NP		18.8	580	69	7.54	

Material	Structure	Dye	Jsc	Voc	FF	η	Ref.
Reduced graphene oxide–ZnO	Nanocomposite	N719	6.77	629	60.6	2.58	[139]
ZnO	Long wavelength absorber dyes	D149/D131/OA SQ2/CA+ D149/D131/OA	6.7 8.6	579 539	63 55	2.42 2.55	[140]
ZnO	NW-arrays	N719	1.3	490	45.2	–	[141]
TiO$_2$ Anatase Rutile	Films	N719	14 10.6	730		7.1 5.6	[142]
TiO$_2$–silicate (1 wt%)	NPs-microsheet	N719	18.12	744	68.4	9.22	[143]
Macroporous SnO$_2$	–	–	–	540	41.42	1.53	[144]
TiO$_2$	NWs NTs	N719	5.15 8.84	680 620	46 62	1.60 3.4	[145]
TiO$_2$	Mesoporous NRs–NPs	N719	15.56	751	55	6.88 86 (IPCE)	[146]
TiO$_2$	Mesoporous NSs–NPs	N719	17.16	680	64	7.51	[147]
Nb$_2$O$_5$–TiO$_2$	NWs–NSs arrays	N719	10.5	750	58	4.55	[148]
TiO$_2$	Hierarchical NTs–NPs	N719	16.49	710	62	7.24	[149]
TiO$_2$	Hexagonal mesoporous	N719	11.96	680	62	4.93 (73 IPCE)	[150]
Nc TiO$_2$–VO$_2$	Blocking layer	N719	16.91	832	68.1	9.56	[151]
Nb$_2$O$_5$	Nanoporous layer		10	701.3	58.5	4.1	[24]
S–TiO$_2$	Nanofibers		10.66	683	59	4.27	[152]
TiO$_2$–ZnO (15%)	Heterostructures	N719	8.4	450	75	2.8	[153]
TiO$_2$–graphene	Nanocomposite films	N3	21.4	749	49	7.70	[154]

(Continued)

Table 3.3 (Continued)

Material	Morphology	Dye	J_{SC} (μA/cm²)	V_{OC} (mV)	FF (%)	PCE (% η)	References
TiO$_2$	Nanocomposite films	N3	0.138	290	46	0.019	[155]
SnO$_2$			0.097	360	33	0.014	
TiO$_2$–SnO$_2$			0.3	360	38	0.041	
3D Hierarchical nanoarchitectures							
TiO$_2$	Doughnut	N719	15	817	71	8.8	[156]
	Spherical		15.7	820	72	9.3	
	Disk		17.5	825	72	10.4	
	Biconcave		14.7	805	71	8.4	
BaTiO$_3$	Microflowers	N719	12.34	772	54	5.13 (PCE)	[157]
TiO$_2$ NP			14.09	791	53	57 (IPCE) 5.90 (PCE), 74 (IPCE)	
F–TiO$_2$	Nanocubes	N719	17.621	603	69.84	7.4 (PCE)	[158]
ZnO–TiO$_2$ NTs/ZnO nanoflake	Heterostructure	N719	$I_{SC} \sim 4$ mA	30	—	—	[159]
BaSnO$_3$–TiO$_2$/Zn	Cubic perovskite	N719	16.677 10.020	776 686	44 62	5.68 4.28	[160]
ZnO	Hemispherical shell and NRs	D719	6.70	630	36	1.53	[161]
N-TiO$_2$–Cu$_x$O	Core–shell architecture	N719	13.24	660	52.26	4.57	[162]
TiO$_2$	NPs–NTs hybrid	N719	22.9	747	65.8	11.3 (PCE)	[163]
TiO$_2$	Nanoflower	DN350	14.70	650	62	5.92 (PCE)	[164]
TiO$_2$–B	NTs–NPs	C106	12.98	781.4	70.3	7.13	[165]

Material	Morphology	Dye	J_{sc}	V_{oc}	FF	η	Ref
TiO_2	SWNTs–NPs/DWNTs–NPs	N719	16.01 13.65	750 740	65.12 54.75	7.82 5.53	[166]
Au@Ag–TiO_2	Core–shell	N719	17.3	750	72	9.4	[59]
TiO_2	NRs–branches hierarchical nanostructures	N719	6.72	720	49	2.4	[167]
TiO_2–ZnO	NPs–NRs hybrid	N719	8	930	—	—	[168]
TiO_2–ZnO	NWs–NRs, NWs–NSs	—	12.49 14.23	493 504	52 50	3.20 3.57	[169]
TiO_2/ZnO	Sponge-like porous donuts	N719	16.70	780	69	9	[170]
TiO_2	Meso/macroporous structures hierarchical	N719	12	755	60	5.4	[171]
TiO_2	Microspheres–NPs	N719, C106	22.92	720	69	11.43	[172]
ZnO	Multishelled microspheres	N719	14.64	754	69.4	7.66 (PCE)	[173]
Reduced graphene–TiO_2–CdS	Mesoporous hierarchical	N719	13.27	660	75	6.5	[174]
ZnO	Rectangular prisms	N719	8.91	725	51	3.3	[175]
ZnO	Olive shaped (NPs – olive-like aggregates)	N719	10.54	610	69	4.43	[176]
ZnO	Nanoburger	N719	8.71	700	66	4.03	[177]
ZnO	Pomegranate, hollow spheres	N719	8.80 7.80	686 680	72 62	4.35 (56 IPCE) 3.28 (50 IPCE)	[178]
CNT–TiO_2 (0.025 mol%) Zr–TiO_2 (0.025 mol%)	NPs–NRs	N719	15.5 19.4	712 699	62.6 60.2	6.81 8.19	[179]
Nb–MWCNT–TiO_2 Nb–TiO_2 (3 wt%)	NPs–NRs	N719	18.83 16.52	763 759	62.70 62.61	9.02 7.87	[180]

(Continued)

Table 3.3 (Continued)

Material	Morphology	Dye	J_{SC} (μA/cm^2)	V_{OC} (mV)	FF (%)	PCE (% η)	References
Bilayer TiO$_2$	NWs–NPs–spherical voids	N719	9.79	644	64.5	4.07	[39]
TiO$_2$	NPs	N719	140		62	4.1	[181]
	NTs		100		56	3.8	
CeO$_2$–TiO$_2$	Porous flower-like thin film	N719	18.654	764	69.2	9.86	[182]
TiO$_2$–ZnO	Branched core–shell	N719	2.841	712	36	0.72	[183]
TiO$_2$	Microspheres–pyramids	N719	15.96	755	69.3	8.35	[184]
TiO$_2$	Macro-mesoporous	N719	16.10	610	68	6.7	[185]
TiO$_2$	NTs-hollow microspheres	N719	19.46	707	61	8.38	[186]
TiO$_2$	a-TiO$_2$	N719	15.98	760	69	8.38	[187]
	P25		15.67	760	69	8.22	
	Star		17.97	760	70	9.56	
	Rice		17.09	770	70	9.21	
	Flower		16.73	760	68	8.65	
	a-TiO$_2$/P25–TiO$_2$		16.47	750	69	8.52	
TiO$_2$	Bilayer spindle/hollow NPs	N719	19.07	675	67	8.65	[188]
TiO$_2$–Au	NWs–NPs	N719	17.38	791	71	9.73	[189]
Au–β-NaYF$_4$:Er/Yb@SiO$_2$	Core-shell microprisms	N719	7.29	—	—	3.65	[190]
Zn$_2$SO$_4$	Microspheres	N719	7.82	716	71	4	[13]
Zn2SnO4@TiO$_2$	Core-shell		8.96	740		4.72	
SnO$_2$	Cauliflower-like hollow microspheres, quantum dot sensitized	N719	9	709	55.6	3.6%	[191]

3.4 Future Outlook and Challenges

Enormous research efforts have been undertaken to modify the photoanode to achieve better conversion efficiencies for DSSCs. The urgent requirement in today's time is the efficiency breakthrough of DSSCs in order to compete with emerging types of solar cells, such as perovskite cells with conversion efficiency >20% [192]. The device performance can be improved by (i) increasing the adsorption of the sensitizing dye molecules within the semiconducting film, (ii) increased charge separation and transportation and (iii) reducing photoelectron recombination. Controlling the photoanode thickness is one way to improve device performance though there is a saturation point beyond which the efficiency decreases due to reduced light transmission. This calls for novel materials and nanoarchitectures to improve the photoanode performance, which is still in an infant stage till date and contain significant possibilities for improvement [3].

There are significant challenges which need to be accounted to improve the device performance specifically with regards to its poor efficiency, poor performance, charge recombination, low LHE, stability, and optimization of each component (photoanode, electrolytes, counter-electrode, dye) [26]. Some of the points that need consideration for improved performance are [3, 4]:

(a) Ongoing research in the field of DSSC technology is based on hit and trial which is time-consuming and less effective. In many reports, two or more materials are combined to enhance the device performance. This calls for effective protocols to establish material selection process and evaluate the materials on the basis of their performance indices. A database to guide the research efforts effectively would be handy.

(b) An important area that is yet to be sorted out is a standardized process for a fair comparison between the various types of DSSCs. Several reports compare the performance of elaborate nanoarchitectures with TiO_2 P25 NPs, which in itself is not an ideal material for DSSC, resulting in inaccurate and unreliable inference.

(c) Low energy, low cost, and low-temperature fabrication processes have to be developed, particularly in the field of flexible DSSCs which is still at its lowest efficiency (3–4%). Development of DSSCs on flexible susbtrates with various colors has shown great commercial potential for various applications.

(d) The commonly used dyes and liquid electrolytes are thermodynamically unstable and suffer from inefficient solar absorption with solvent leakage being a common issue. Alternate photosensitizers and dyes have to be developed to replace Pt- and Ru-based metal complexes toward improving the performance of the device.

(e) Research has to be focused on new and cost-effective photoanode materials other than TiO_2 with high oxygen vacancies, high electron mobility, high dye adsorption, good light absorption ability, and excellent corrosion resistance. Doping with different materials and use of nanoarchitectures is a viable approach though there is considerable work that is yet to be made regarding materials and composition that are efficient [26].

(f) Compared to n-type DSSCs, p-type counterparts offer the possibility of higher V_{OC} but show limited overall efficiencies (~1–2%) [17]. Another possible direction for device improvement is the development of efficient photoanodes of p-type DSSCs and fabricating them with the n-type DSSCs to construct a tandem configuration which is predicted to capture a wide range of solar spectrum and improve the performance.

3.5 Conclusion

The chapter casts some light on the performance of various photoanode materials and nanostructures, such as TiO_2, ZnO, SnO_2, Nb_2O_5, Al_2O_3, ZrO_2, CeO_2, $SrTiO_3$, Zn_2SnO_4, and carbon in DSSCs. A change of focus from morphology to surface control and investigating into new dyes specially designed for the specific materials is required which can lead to improvements in device performance. All components of the DSSCs (photoanode, electrolyte, counter-electrode, and sensitizer) play a significant role in controlling the device performance. The efficiency of the photoanode to generate photoexcited electrons and to reduce recombination, a suitable sensitizer to absorb the solar spectrum efficiently and create more photoexcited electrons is all crucial. Nanoarchitectures exhibit multifunctionality to enhance the device performance, by presenting a high surface area, increased amount of dye sensitizers anchored, facile electron transfer, and effective light scattering. The dimensionality, ranging from 0D (QDs, NPs), 1D (NWs, NTs, NRs), 2D (nanosheets, films), and 3D (nanoflowers, core@shell, nanoforest, microspheres), affects the properties. Zero-dimensional (0D) nanoparticles offer a much higher surface area for increased dye loading but offer lower light scattering ability resulting in poor LHE. Presence of defects and grain boundaries cause charge recombination and low power output. The 1D nanoarchitectures (NRs, NWs, NTs, nanofibers) show promising PCE due to direct electron transport and reduced recombination but have lower surface area, which can be improved by $TiCl_4$ treatment. Hierarchical nanoarchitectures take advantage of 0D with 1D nanomaterials, with increased surface area and directional electron transfer. Novel nanoarchitectures as nanoforest has also been investigated by vertically aligned bundles of oxide nanocrystals. A structure combining the high surface area associated with nanoparticles with the electron transport directionality of nanorods/nanotubes would be optimal for DSSCs. Mesoporous hierarchical MOS-based materials from different materials, such as ZnO NWs with TiO_2 NPs with a crystalline framework, high specific surface area, and tunable pore size have received significant attention for energy conversion applications. Another approach to exploit to obtain better performance is to utilize optical structures within the photoanodes for improving photon absorption. However, there are issues that are related to patterning over large areas with considerable stabilization of the fabrication process that needs to be achieved, limiting the practical application of photonic structures to the DSSCs. Material modification with metallic and nonmetallic dopants modulates the band gap and control the injection of the photogenerated electrons from the dye to the photoanode. Application of an interface modifier/blocking layer in the photoanode material also resists the back electron transfer. Currently, most of the research is based on TiO_2 and ZnO photoanodes and focused on conjugating the photoanodes with existing dyes, electrolytes, and cathode materials. It is important to find alternate materials and compatible dyes–electrolyte to improve the light harvesting and current density. Besides the optimization of the photocathode and refinement of the dye-photoanode design, there should be considerable systematic research on the selection and definition of new redox mediators in order to improve the device performance.

References

1 Liu, X., Fang, J., Liu, Y., and Lin, T. (2016). Progress in nanostructured photoanodes for dye-sensitized solar cells. *Front. Mater. Sci.* 10 (3): 225–237.

2 Shaikh, J.S., Shaikh, N.S., Mali, S.S. et al. (2018). Nanoarchitectures in dye-sensitized solar cells: metal oxides, oxide perovskites and carbon-based materials. *Nanoscale* 10 (11): 4987–5034.

3 Shakeel Ahmad, M., Pandey, A.K., and Abd Rahim, N. (2017). Advancements in the development of TiO_2 photoanodes and its fabrication methods for dye sensitized solar cell (DSSC) applications. A review. *Renewable Sustainable Energy Rev.* 77: 89–108.

4 Fan, K., Yu, J., and Ho, W. (2017). Improving photoanodes to obtain highly efficient dye-sensitized solar cells: a brief review. *Mater. Horiz.* 4 (3): 319–344.

5 Wali, Q., Fakharuddin, A., and Jose, R. (2015). Tin oxide as a photoanode for dye-sensitised solar cells: current progress and future challenges. *J. Power Sources* 293: 1039–1052.

6 Anta, J.A., Guillén, E., and Tena-Zaera, R. (2012). ZnO-based dye-sensitized solar cells. *J. Phys. Chem. C* 116 (21): 11413–11425.

7 Gao, X.-D., Li, X.-M., and Gan, X.-Y. (2013). Enhancing the light harvesting capacity of the photoanode films in dye-sensitized solar cells. In: *Solar Cells – Research and Application Perspectives* (ed. A. Morales-Acevedo), 169–202. InTech.

8 Low, F.W. and Lai, C.W. (2018). Recent developments of graphene-TiO_2 composite nanomaterials as efficient photoelectrodes in dye-sensitized solar cells: a review. *Renewable Sustainable Energy Rev.* 82: 103–125.

9 Yeoh, M.-E. and Chan, K.-Y. (2017). Recent advances in photo-anode for dye-sensitized solar cells: a review: recent advances in photo-anode for DSSCs: a review. *Int. J. Energy Res.* 41 (15): 2446–2467.

10 Sengupta, D., Das, P., Mondal, B., and Mukherjee, K. (2016). Effects of doping, morphology and film-thickness of photo-anode materials for dye sensitized solar cell application – a review. *Renewable Sustainable Energy Rev.* 60: 356–376.

11 Guo, X., Lu, G., and Chen, J. (2015). Graphene-based materials for photoanodes in dye-sensitized solar cells. *Front. Energy Res.* 3: 1–15.

12 Alpuche-Aviles, M.A. and Wu, Y. (2009). Photoelectrochemical study of the band structure of Zn_2SnO_4 prepared by the hydrothermal method. *J. Am. Chem. Soc.* 131 (9): 3216–3224.

13 Wang, X., Wang, Y.-F., Luo, Q.-P. et al. (2017). Highly uniform hierarchical Zn_2SnO_4 microspheres for the construction of high performance dye-sensitized solar cells. *RSC Adv.* 7 (69): 43403–43409.

14 Asim, N., Ahmadi, S., Alghoul, M.A. et al. (2014). Research and development aspects on chemical preparation techniques of photoanodes for dye sensitized solar cells. *Int. J. Photoenergy* 2014: 1–21.

15 Archana, J., Harish, S., Sabarinathan, M. et al. (2016). Highly efficient dye-sensitized solar cell performance from template derived high surface area mesoporous TiO_2 nanospheres. *RSC Adv.* 6 (72): 68092–68099.

16 Mali, S.S., Betty, C.A., Bhosale, P.N. et al. (2014). From nanocorals to nanorods to nanoflowers nanoarchitecture for efficient dye-sensitized solar cells at relatively low film thickness: all hydrothermal process. *Sci. Rep.* 4 (5451): 1–8.

17 Cavallo, C., Di Pascasio, F., Latini, A. et al. (2017). Nanostructured semiconductor materials for dye-sensitized solar cells. *J. Nanomater.* 2017: 1–31.

18 Ge, C., Rahman, M.M., Deb Nath, N.C. et al. (2015). Graphene-incorporated photoelectrodes for dye-sensitized solar cells. *Bull. Korean Chem. Soc.* 36: 762–771.

19 Adachi, M., Murata, Y., Takao, J. et al. (2004). Highly efficient dye-sensitized solar cells with a Titania thin-film electrode composed of a network structure of single-crystal-like TiO_2 nanowires made by the 'oriented attachment' mechanism. *J. Am. Chem. Soc.* 126 (45): 14943–14949.

20 Akin, S. and Sonmezoglu, S. (2018). Metal oxide nanoparticles as electron transport layer for highly efficient dye-sensitized solar cells. In: *Emerging Materials for Energy Conversion and Storage*, 39–79. Elsevier.

21 Yang, W., Xu, Y., Tang, Y. et al. (2014). Three-dimensional self-branching anatase TiO$_2$ nanorods: morphology control, growth mechanism and dye-sensitized solar cell application. *J. Mater. Chem. A* 2 (38): 16030–16038.

22 Chu, L., Qin, Z., Yang, J., and Li, X. (2015). Anatase TiO$_2$ nanoparticles with exposed {001} facets for efficient dye-sensitized solar cells. *Sci. Rep.* 5 (1): 12143.

23 Ghosh, R., Brennaman, M.K., Uher, T. et al. (2011). Nanoforest Nb$_2$O$_5$ photoanodes for dye-sensitized solar cells by pulsed laser deposition. *ACS Appl. Mater. Interfaces* 3 (10): 3929–3935.

24 Ou, J.Z., Rani, R.A., Ham, M.-H. et al. (2012). Elevated temperature anodized Nb$_2$O$_5$: a photoanode material with exceptionally large photoconversion efficiencies. *ACS Nano* 6 (5): 4045–4053.

25 Shankar, K., Mor, G.K., Prakasam, H.E. et al. (2007). Highly-ordered TiO$_2$ nanotube arrays up to 220 μm in length: use in water photoelectrolysis and dye-sensitized solar cells. *Nanotechnology* 18 (6): 065707.

26 Boro, B., Gogoi, B., Rajbongshi, B.M., and Ramchiary, A. (2018). Nano-structured TiO$_2$/ZnO nanocomposite for dye-sensitized solar cells application: a review. *Renewable Sustainable Energy Rev.* 81: 2264–2270.

27 Zhuge, F., Qiu, J., Li, X. et al. (2011). Toward hierarchical TiO$_2$ nanotube arrays for efficient dye-sensitized solar cells. *Adv. Mater.* 23 (11): 1330–1334.

28 Bai, Y., Yu, H., Li, Z. et al. (2012). In situ growth of a ZnO nanowire network within a TiO$_2$ nanoparticle film for enhanced dye-sensitized solar cell performance. *Adv. Mater.* 24 (43): 5850–5856.

29 Cho, C.-Y. and Moon, J.H. (2011). Hierarchically porous TiO$_2$ electrodes fabricated by dual templating methods for dye-sensitized solar cells. *Adv. Mater.* 23 (26): 2971–2975.

30 Ko, S.H., Lee, D., Kang, H.W. et al. (2011). Nanoforest of hydrothermally grown hierarchical ZnO nanowires for a high efficiency dye-sensitized solar cell. *Nano Lett.* 11 (2): 666–671.

31 Hu, B., Tang, Q., He, B. et al. (2014). Mesoporous TiO$_2$ anodes for efficient dye-sensitized solar cells: an efficiency of 9.86% under one sun illumination. *J. Power Sources* 267: 445–451.

32 Roh, D.K., Chi, W.S., Jeon, H. et al. (2014). High efficiency solid-state dye-sensitized solar cells assembled with hierarchical Anatase pine tree-like TiO$_2$ nanotubes. *Adv. Funct. Mater.* 24 (3): 379–386.

33 Choi, H., Nahm, C., Kim, J. et al. (2012). The effect of TiCl$_4$-treated TiO$_2$ compact layer on the performance of dye-sensitized solar cell. *Curr. Appl Phys.* 12 (3): 737–741.

34 Kim, D.H., Woodroof, M., Lee, K., and Parsons, G.N. (2013). Atomic layer deposition of high performance ultrathin TiO$_2$ blocking layers for dye-sensitized solar cells. *ChemSusChem* 6 (6): 1014–1020.

35 Xu, J., Wang, G., Fan, J. et al. (2015). g-C$_3$N$_4$ modified TiO$_2$ nanosheets with enhanced photoelectric conversion efficiency in dye-sensitized solar cells. *J. Power Sources* 274: 77–84.

36 Gálvez, F.E., Kemppainen, E., Míguez, H., and Halme, J. (2012). Effect of diffuse light scattering designs on the efficiency of dye solar cells: an integral optical and electrical description. *J. Phys. Chem. C* 116 (21): 11426–11433.

37 Wang, Z.-S., Kawauchi, H., Kashima, T., and Arakawa, H. (2004). Significant influence of TiO$_2$ photoelectrode morphology on the energy conversion efficiency of N719 dye-sensitized solar cell. *Coord. Chem. Rev.* 248 (13–14): 1381–1389.

38 Miao, Q., Wu, L., Cui, J. et al. (2011). A new type of dye-sensitized solar cell with a multi-layered photoanode prepared by a film-transfer technique. *Adv. Mater.* 23 (24): 2764–2768.

39 Sun, P., Zhang, X., Wang, L. et al. (2015). Bilayer TiO$_2$ photoanode consisting of a nanowire–nanoparticle bottom layer and a spherical voids scattering layer for dye-sensitized solar cells. *New J. Chem.* 39 (6): 4845–4851.

40 Wu, D., Wang, Y., Dong, H. et al. (2013). Hierarchical TiO$_2$ microspheres comprised of anatase nanospindles for improved electron transport in dye-sensitized solar cells. *Nanoscale* 5 (1): 324–330.

41 Qiu, Y., Chen, W., and Yang, S. (2010). Double-layered photoanodes from variable-size anatase TiO$_2$ nanospindles: a candidate for high-efficiency dye-sensitized solar cells. *Angew. Chem.* 122 (21): 3757–3761.

42 Yang, L. and Leung, W.W.-F. (2013). Electrospun TiO$_2$ nanorods with carbon nanotubes for efficient electron collection in dye-sensitized solar cells. *Adv. Mater.* 25 (12): 1792–1795.

43 Bella, F., Griffini, G., Gerosa, M. et al. (2015). Performance and stability improvements for dye-sensitized solar cells in the presence of luminescent coatings. *J. Power Sources* 283: 195–203.

44 Griffini, G., Bella, F., Nisic, F. et al. (2015). Multifunctional luminescent down-shifting fluoropolymer coatings: a straightforward strategy to improve the UV-light harvesting ability and long-term outdoor stability of organic dye-sensitized solar cells. *Adv. Energy Mater.* 5 (3): 1401312.

45 Chander, N., Khan, A.F., Komarala, V.K. et al. (2016). Enhancement of dye sensitized solar cell efficiency via incorporation of upconverting phosphor nanoparticles as spectral converters: enhancement of dye sensitized solar cell efficiency. *Prog. Photovoltaics Res. Appl.* 24 (5): 692–703.

46 Das, S., Chakraborty, P., Shit, A. et al. (2016). Robust hybrid hydrogels with good rectification properties and their application as active materials for dye-sensitized solar cells: insights from AC impedance spectroscopy. *J. Mater. Chem. A* 4 (11): 4194–4210.

47 Wang, Z., Tang, Q., He, B. et al. (2015). Titanium dioxide/calcium fluoride nanocrystallite for efficient dye-sensitized solar cell. A strategy of enhancing light harvest. *J. Power Sources* 275: 175–180.

48 Yuan, S., Tang, Q., He, B. et al. (2014). Transmission enhanced photoanodes for efficient dye-sensitized solar cells. *Electrochim. Acta* 125: 646–651.

49 Gondal, M.A., Ilyas, A.M., and Baig, U. (2016). Facile synthesis of silicon carbide-titanium dioxide semiconducting nanocomposite using pulsed laser ablation technique and its performance in photovoltaic dye sensitized solar cell and photocatalytic water purification. *Appl. Surf. Sci.* 378: 8–14.

50 Kim, J.T., Lee, S.H., and Han, Y.S. (2015). Enhanced power conversion efficiency of dye-sensitized solar cells with Li$_2$SiO$_3$-modified photoelectrode. *Appl. Surf. Sci.* 333: 134–140.

51 Sabet, M., Salavati-Niasari, M., and Amiri, O. (2014). Using different chemical methods for deposition of CdS on TiO$_2$ surface and investigation of their influences on the dye-sensitized solar cell performance. *Electrochim. Acta* 117: 504–520.

52 Yip, C.T., Huang, H., Zhou, L. et al. (2011). Direct and seamless coupling of TiO$_2$ nanotube photonic crystal to dye-sensitized solar cell: a single-step approach. *Adv. Mater.* 23 (47): 5624–5628.

53 Colodrero, S., Mihi, A., Häggman, L. et al. (2009). Porous one-dimensional photonic crystals improve the power-conversion efficiency of dye-sensitized solar cells. *Adv. Mater.* 21 (7): 764–770.

54 Wei, Y., Xu, C., Xu, S. et al. (2010). Planar waveguide–nanowire integrated three-dimensional dye-sensitized solar cells. *Nano Lett.* 10 (6): 2092–2096.

55 Du, J., Qi, J., Wang, D., and Tang, Z. (2012). Facile synthesis of Au@TiO$_2$ core–shell hollow spheres for dye-sensitized solar cells with remarkably improved efficiency. *Energy Environ. Sci.* 5 (5): 6914.

56 Dang, X., Qi, J., Klug, M.T. et al. (2013). Tunable localized surface plasmon-enabled broadband light-harvesting enhancement for high-efficiency panchromatic dye-sensitized solar cells. *Nano Lett.* 13 (2): 637–642.

57 Choi, H., Chen, W.T., and Kamat, P.V. (2012). Know thy nano neighbor. Plasmonic versus electron charging effects of metal nanoparticles in dye-sensitized solar cells. *ACS Nano* 6 (5): 4418–4427.

58 Qi, J., Dang, X., Hammond, P.T., and Belcher, A.M. (2011). Highly efficient plasmon-enhanced dye-sensitized solar cells through metal@oxide core–shell nanostructure. *ACS Nano* 5 (9): 7108–7116.

59 Yun, J., Hwang, S.H., and Jang, J. (2015). Fabrication of Au@Ag core/shell nanoparticles decorated TiO$_2$ hollow structure for efficient light-harvesting in dye-sensitized solar cells. *ACS Appl. Mater. Interfaces* 7 (3): 2055–2063.

60 Zarick, H.F., Hurd, O., Webb, J.A. et al. (2014). Enhanced efficiency in dye-sensitized solar cells with shape-controlled plasmonic nanostructure. *ACS Photonics* 1 (9): 806–811.

61 Bavir, M., Fattah, A., and Nazari, A.A. (2015). An investigation of electrochemical impedance of TiO$_2$-ZnO composite and TiO$_2$-graphene composite in dye-sensitized solar cells, as photoanode. *30th International Power System Conference (PSC)*, vol. 2015, IEEE, pp. 328–332.

62 Cheng, G., Akhtar, M.S., Yang, O.-B., and Stadler, F.J. (2013). Novel preparation of anatase TiO$_2$@reduced graphene oxide hybrids for high-performance dye-sensitized solar cells. *ACS Appl. Mater. Interfaces* 5 (14): 6635–6642.

63 Mehmood, U., Ahmed, S., Hussein, I.A., and Harrabi, K. (2015). Improving the efficiency of dye sensitized solar cells by TiO$_2$-graphene nanocomposite photoanode. *Photonics Nanostruct. Fundam. Appl.* 16: 34–42.

64 Zhi, J., Cui, H., Chen, A. et al. (2015). Efficient highly flexible dye sensitized solar cells of three dimensional graphene decorated titanium dioxide nanoparticles on plastic substrate. *J. Power Sources* 281: 404–410.

65 Mohamad, I.S., Ismail, S.S., Norizan, M.N. et al. ZnO photoanode effect on the efficiency performance of organic based dye sensitized solar cell. *IOP Conf. Ser.: Mater. Sci. Eng.* 209.

66 Vittal, R. and Ho, K.-C. (2017). Zinc oxide based dye-sensitized solar cells: a review. *Renewable Sustainable Energy Rev.* 70: 920–935.

67 Memarian, N., Concina, I., Braga, A. et al. (2011). Hierarchically assembled ZnO nanocrystallites for high-efficiency dye-sensitized solar cells. *Angew. Chem. Int. Ed.* 50 (51): 12321–12325.

68 Solbrand, A., Keis, K., Södergren, S. et al. (2000). Charge transport properties in the nanostructured ZnO thin film electrode – electrolyte system studied with time resolved photocurrents. *Sol. Energy Mater. Sol. Cells* 60 (2): 181–193.

69 Noack, V., Weller, H., and Eychmüller, A. (2002). Electron transport in particulate ZnO electrodes: a simple approach. *J. Phys. Chem. B* 106 (34): 8514–8523.

70 Quintana, M., Edvinsson, T., Hagfeldt, A., and Boschloo, G. (2007). Comparison of dye-sensitized ZnO and TiO$_2$ solar cells: studies of charge transport and carrier lifetime. *J. Phys. Chem. C* 111 (2): 1035–1041.

71 Tiwana, P., Docampo, P., Johnston, M.B. et al. (2011). Electron mobility and injection dynamics in mesoporous ZnO, SnO$_2$, and TiO$_2$ films used in dye-sensitized solar cells. *ACS Nano* 5 (6): 5158–5166.

72 Jennings, J.R., Liu, Y., and Wang, Q. (2011). Efficiency limitations in dye-sensitized solar cells caused by inefficient sensitizer regeneration. *J. Phys. Chem. C* 115 (30): 15109–15120.

73 Guillén, E., Azaceta, E., Peter, L.M. et al. (2011). ZnO solar cells with an indoline sensitizer: a comparison between nanoparticulate films and electrodeposited nanowire arrays. *Energy Environ. Sci.* 4 (9): 3400.

74 Guillén, E., Peter, L.M., and Anta, J.A. (2011). Electron transport and recombination in ZnO-based dye-sensitized solar cells. *J. Phys. Chem. C* 115 (45): 22622–22632.

75 Chang, J., Ahmed, R., Wang, H. et al. (2013). ZnO Nanocones with high-index $\{10\bar{1}1\}$ facets for enhanced energy conversion efficiency of dye-sensitized solar cells. *J. Phys. Chem. C* 117 (27): 13836–13844.

76 Qin, Z., Huang, Y., Liao, Q. et al. (2012). Stability improvement of the ZnO nanowire array electrode modified with Al_2O_3 and SiO_2 for dye-sensitized solar cells. *Mater. Lett.* 70: 177–180.

77 Goh, G.K.L., Le, H.Q., Huang, T.J., and Hui, B.T.T. (2014). Low temperature grown $ZnO@TiO_2$ core shell nanorod arrays for dye sensitized solar cell application. *J. Solid State Chem.* 214: 17–23.

78 Prabakar, K., Son, M., Kim, W.-Y., and Kim, H. (2011). TiO_2 thin film encapsulated ZnO nanorod and nanoflower dye sensitized solar cells. *Mater. Chem. Phys.* 125 (1–2): 12–14.

79 Kung, C.-W., Chen, H.-W., Lin, C.-Y. et al. (2014). Electrochemical synthesis of a double-layer film of ZnO nanosheets/nanoparticles and its application for dye-sensitized solar cells: electrochemical synthesis of ZnO nanosheets/nanoparticles. *Prog. Photovoltaics Res. Appl.* 22 (4): 440–451.

80 Lin, L.-Y., Yeh, M.-H., Lee, C.-P. et al. (2012). Enhanced performance of a flexible dye-sensitized solar cell with a composite semiconductor film of ZnO nanorods and ZnO nanoparticles. *Electrochim. Acta* 62: 341–347.

81 Xu, J., Fan, K., Shi, W. et al. (2014). Application of ZnO micro-flowers as scattering layer for ZnO-based dye-sensitized solar cells with enhanced conversion efficiency. *Sol. Energy* 101: 150–159.

82 Cojocaru, L., Olivier, C., Toupance, T. et al. (2013). Size and shape fine-tuning of SnO_2 nanoparticles for highly efficient and stable dye-sensitized solar cells. *J. Mater. Chem. A* 1 (44): 13789.

83 Wang, H., Gao, J., Tang, M. et al. (2012). SnO_2 hollow nanospheres enclosed by single crystalline nanoparticles for highly efficient dye-sensitized solar cells. *CrystEngComm* 14: 5177–5181.

84 Li, K.-N., Wang, Y.-F., Xu, Y.-F. et al. (2013). Macroporous SnO_2 synthesized via a template-assisted reflux process for efficient dye-sensitized solar cells. *ACS Appl. Mater. Interfaces* 5 (11): 5105–5111.

85 Li, Z., Zhou, Y., Yu, T. et al. (2012). Unique Zn-doped SnO_2 nano-echinus with excellent electron transport and light harvesting properties as photoanode materials for high performance dye-sensitized solar cell. *CrystEngComm* 14 (20): 6462.

86 Pang, H., Yang, H., Guo, C.X., and Li, C.M. (2012). Functionalization of SnO_2 photoanode through mg-doping and TiO_2 -coating to synergically boost dye-sensitized solar cell performance. *ACS Appl. Mater. Interfaces* 4 (11): 6261–6265.

87 Dou, X., Prabhakar, R.R., Mathews, N. et al. (2012). Zn-doped SnO_2 nanocrystals as efficient DSSC photoanode material and remarkable photocurrent enhancement by Interface modification. *J. Electrochem. Soc.* 159 (9): H735–H739.

88 Bhattacharjee, R. and Hung, I.-M. (2013). A SnO_2 and ZnO nanocomposite photoanodes in dye-sensitized solar cells. *ECS Solid State Lett.* 2 (11): Q101–Q104.

89 Dong, Z., Ren, H., Hessel, C.M. et al. (2014). Quintuple-shelled SnO$_2$ hollow microspheres with superior light scattering for high-performance dye-sensitized solar cells. *Adv. Mater.* 26 (6): 905–909.

90 Ling, X., Yuan, J., Liu, D. et al. (2017). Room-temperature processed Nb$_2$O$_5$ as the electron-transporting layer for efficient planar perovskite solar cells. *ACS Appl. Mater. Interfaces* 9 (27): 23181–23188.

91 Szostak, R., Morais, A., Carminati, S.A. et al. (2018). Application of graphene and graphene derivatives/oxide nanomaterials for solar cells. In: *The Future of Semiconductor Oxides in Next-Generation Solar Cells*, 395–437. Elsevier.

92 Brennan, L.J., Byrne, M.T., Bari, M., and Gun'ko, Y.K. (2011). Carbon nanomaterials for dye-sensitized solar cell applications: a bright future. *Adv. Energy Mater.* 1 (4): 472–485.

93 Chen, D., Zhang, H., Liu, Y., and Li, J. (2013). Graphene and its derivatives for the development of solar cells, photoelectrochemical, and photocatalytic applications. *Energy Environ. Sci.* 6 (5): 1362.

94 Kavan, L., Yum, J.-H., and Graetzel, M. (2013). Application of graphene-based nanostructures in dye-sensitized solar cells: graphene-based nanostructures in dye-sensitized solar cells. *Phys. Status Solidi B* 250 (12): 2643–2648.

95 Maçaira, J., Andrade, L., and Mendes, A. (2013). Review on nanostructured photoelectrodes for next generation dye-sensitized solar cells. *Renewable Sustainable Energy Rev.* 27: 334–349.

96 Roy-Mayhew, J.D. and Aksay, I.A. (2014). Graphene materials and their use in dye-sensitized solar cells. *Chem. Rev.* 114 (12): 6323–6348.

97 Batmunkh, M., Dadkhah, M., Shearer, C.J. et al. (2016). Incorporation of graphene into SnO$_2$ photoanodes for dye-sensitized solar cells. *Appl. Surf. Sci.* 387: 690–697.

98 Bera, A., Wu, K., Sheikh, A. et al. (2014). Perovskite oxide SrTiO$_3$ as an efficient electron transporter for hybrid perovskite solar cells. *J. Phys. Chem. C* 118 (49): 28494–28501.

99 Sanehira, E.M., Marshall, A.R., Christians, J.A. et al. (2017). Enhanced mobility CsPbI$_3$ quantum dot arrays for record-efficiency, high-voltage photovoltaic cells. *Sci. Adv.* 3 (10): eaao4204.

100 Bhat, T.S., Mali, S.S., Korade, S.D. et al. (2017). Mesoporous architecture of TiO$_2$ microspheres via controlled template assisted route and their photoelectrochemical properties. *J. Mater. Sci. - Mater. Electron.* 28 (1): 304–316.

101 Li, M., Li, M., Zhu, Y. et al. (2017). Scattering and plasmonic synergetic enhancement of the performance of dye-sensitized solar cells by double-shell SiO$_2$@Au@TiO$_2$ microspheres. *Nanotechnology* 28 (26): 265202.

102 Fan, Y.-H., Ho, C.-Y., and Chang, Y.-J. (2017). Enhancement of dye-sensitized solar cells efficiency using mixed-phase TiO$_2$ nanoparticles as photoanode. *Scanning* 2017: 1–7.

103 Song, D.H., Kim, H.-S., Suh, J.S. et al. (2017). Multi-shaped Ag nanoparticles in the plasmonic layer of dye-sensitized solar cells for increased power conversion efficiency. *Nanomaterials* 7 (6): 136.

104 Mehnane, H.F., Wang, C., Kondamareddy, K.K. et al. (2017). Hydrothermal synthesis of TiO$_2$ nanoparticles doped with trace amounts of strontium, and their application as working electrodes for dye sensitized solar cells: tunable electrical properties & enhanced photo-conversion performance. *RSC Adv.* 7 (4): 2358–2364.

105 Feng, W., Li, Y., Du, J. et al. (2016). Highly efficient and stable quasi-solid-state quantum dot-sensitized solar cells based on a superabsorbent polyelectrolyte. *J. Mater. Chem. A* 4 (4): 1461–1468.

106 Ako, R.T., Ekanayake, P., Young, D.J. et al. (2015). Evaluation of surface energy state distribution and bulk defect concentration in DSSC photoanodes based on Sn, Fe, and Cu doped TiO_2. *Appl. Surf. Sci.* 351: 950–961.

107 Kim, S.-B., Park, J.-Y., Kim, C.-S. et al. (2015). Effects of graphene in dye-sensitized solar cells based on nitrogen-doped TiO_2 composite. *J. Phys. Chem. C* 119 (29): 16552–16559.

108 Lim, S.P., Pandikumar, A., Lim, H.N. et al. (2015). Boosting photovoltaic performance of dye-sensitized solar cells using silver nanoparticle-decorated N,S-Co-doped-TiO_2 photoanode. *Sci. Rep.* 5 (1).

109 Boro, B., Rajbongshi, B.M., and Samdarshi, S.K. (2016). Synthesis and fabrication of TiO_2–ZnO nanocomposite based solid state dye sensitized solar cell. *J. Mater. Sci. – Mater. Electron.* 27 (9): 9929–9940.

110 Lim, S.P., Lim, Y.S., Pandikumar, A. et al. (2017). Gold–silver@TiO_2 nanocomposite-modified plasmonic photoanodes for higher efficiency dye-sensitized solar cells. *Phys. Chem. Chem. Phys.* 19 (2): 1395–1407.

111 Tanyi, A.R., Rafieh, A.I., Ekaneyaka, P. et al. (2015). Enhanced efficiency of dye-sensitized solar cells based on Mg and La co-doped TiO_2 photoanodes. *Electrochim. Acta* 178: 240–248.

112 Rajendran, A. and Kandasamy, S. (2017). Synthesis and photovoltaic property characterization of CeO_2 film deposited on ITO substrate for dye sensitized solar cell. *Mater. Res. Innovations*: 1–7.

113 Basu, K., Benetti, D., Zhao, H. et al. (2016). Enhanced photovoltaic properties in dye sensitized solar cells by surface treatment of SnO_2 photoanodes. *Sci. Rep.* 6 (23312): 1–10.

114 Akin, S., Erol, E., and Sonmezoglu, S. (2017). Enhancing the electron transfer and band potential tuning with long-term stability of ZnO based dye-sensitized solar cells by gallium and tellurium as dual-doping. *Electrochim. Acta* 225: 243–254.

115 Chava, R.K. and Kang, M. (2017). Improving the photovoltaic conversion efficiency of ZnO based dye sensitized solar cells by indium doping. *J. Alloys Compd.* 692: 67–76.

116 Bakhshayesh, A.M. and Farajisafiloo, N. (2015). Efficient dye-sensitised solar cell based on uniform in-doped TiO_2 spherical particles. *Appl. Phys. A* 120 (1): 199–206.

117 Tabari-Saadi, Y. and Mohammadi, M.R. (2015). Efficient dye-sensitized solar cells based on carbon-doped TiO_2 hollow spheres and nanoparticles. *J. Mater. Sci. – Mater. Electron.* 26 (11): 8863–8876.

118 Suresh, S., Deepak, T.G., Ni, C. et al. (2016). The role of crystallinity of the Nb_2O_5 blocking layer on the performance of dye-sensitized solar cells. *New J. Chem.* 40 (7): 6228–6237.

119 Suresh, S., Unni, G.E., Satyanarayana, M. et al. (2018). Ag@Nb_2O_5 plasmonic blocking layer for higher efficiency dye-sensitized solar cells. *Dalton Trans.* 47 (13): 4685–4700.

120 Suresh, S., Unni, G.E., Nair, A.S., and Pillai Mahadevan, V.P. (2018). Plasmonic Ag@Nb_2O_5 surface passivation layer on quantum confined SnO_2 films for high current dye-sensitized solar cell applications. *Electrochim. Acta* 289: 1–12.

121 Sanad, M.M.S., Shalan, A.E., Rashad, M.M., and Mahmoud, M.H.H. (2015). Plasmonic enhancement of low cost mesoporous Fe_2O_3-TiO_2 loaded with palladium, platinum or silver for dye sensitized solar cells (DSSCs). *Appl. Surf. Sci.* 359: 315–322.

122 Wei, X., Nbelayim, P.S., Kawamura, G. et al. (2017). Ag nanoparticle-filled TiO_2 nanotube arrays prepared by anodization and electrophoretic deposition for dye-sensitized solar cells. *Nanotechnology* 28 (13): 135207.

123 Marimuthu, T., Anandhan, N., Thangamuthu, R., and Surya, S. (2017). Facile growth of ZnO nanowire arrays and nanoneedle arrays with flower structure on ZnO–TiO_2 seed layer for DSSC applications. *J. Alloys Compd.* 693: 1011–1019.

124 Pace, S., Resmini, A., Tredici, I.G. et al. (2018). Optimization of 3D ZnO brush-like nanorods for dye-sensitized solar cells. *RSC Adv.* 8 (18): 9775–9782.

125 Yang, Y., Zhao, J., Cui, C. et al. (2016). Hydrothermal growth of ZnO nanowires scaffolds within mesoporous TiO_2 photoanodes for dye-sensitized solar cells with enhanced efficiency. *Electrochim. Acta* 196: 348–356.

126 Rahman, M.Y.A., Roza, L., Umar, A.A., and Salleh, M.M. (2016). Effect of dimethyl borate composition on the performance of boron doped ZnO dye-sensitized solar cell (DSSC). *J. Mater. Sci. – Mater. Electron.* 27 (3): 2228–2234.

127 Abd-Ellah, M., Bazargan, S., Thomas, J.P. et al. (2015). Hierarchical tin oxide nanostructures for dye-sensitized solar cell application. *Adv. Electron. Mater.* 1 (9): 1500032.

128 Chen, C., Ikeuchi, Y., Xu, L. et al. (2015). Synthesis of [111]- and {010}-faceted anatase TiO_2 nanocrystals from tri-titanate nanosheets and their photocatalytic and DSSC performances. *Nanoscale* 7 (17): 7980–7991.

129 Mali, S.S., Shim, C.S., Kim, H. et al. (2015). Evaluation of various diameters of titanium oxide nanofibers for efficient dye sensitized solar cells synthesized by electrospinning technique: a systematic study and their application. *Electrochim. Acta* 166: 356–366.

130 Li, L., Xu, C., Zhao, Y. et al. (2015). Improving performance via blocking layers in dye-sensitized solar cells based on nanowire photoanodes. *ACS Appl. Mater. Interfaces* 7 (23): 12824–12831.

131 Çakar, S. and Özacar, M. (2017). Fe-quercetin coupled different shaped ZnO rods based dye sensitized solar cell applications. *Sol. Energy* 155: 233–245.

132 Çakar, S. and Özacar, M. (2016). Fe–tannic acid complex dye as photo sensitizer for different morphological ZnO based DSSCs. *Spectrochim. Acta, Part A* 163: 79–88.

133 Chen, X., Tang, Q., Zhao, Z. et al. (2015). One-step growth of well-aligned TiO_2 nanorod arrays for flexible dye-sensitized solar cells. *Chem. Commun.* 51 (10): 1945–1948.

134 Zhao, Y.L., Yao, D.S., Song, C.B. et al. (2015). CNT–G–TiO_2 layer as a bridge linking TiO_2 nanotube arrays and substrates for efficient dye-sensitized solar cells. *RSC Adv.* 5 (54): 43805–43809.

135 Yao, N., Huang, J., Fu, K. et al. (2016). Reduced interfacial recombination in dye-sensitized solar cells assisted with $NiO:Eu^{3+},Tb^{3+}$ coated TiO_2 film. *Sci. Rep.* 6 (1).

136 Flynn, C.J., McCullough, S.M., Li, L. et al. (2016). Passivation of nickel vacancy defects in nickel oxide solar cells by targeted atomic deposition of boron. *J. Phys. Chem. C* 120 (30): 16568–16576.

137 Milan, R., Selopal, G.S., Epifani, M. et al. (2015). $ZnO@SnO_2$ engineered composite photoanodes for dye sensitized solar cells. *Sci. Rep.* 5 (1).

138 Wu, C., Qi, L., Chen, Y. et al. (2016). Dye-sensitized solar cells based on two-dimensional TiO_2 nanosheets as the scattering layers. *Res. Chem. Intermed.* 42 (6): 5653–5664.

139 Song, J.-L. and Wang, X. (2016). Effect of incorporation of reduced graphene oxide on ZnO-based dye-sensitized solar cells. *Physica E* 81: 14–18.

140 Rudolph, M., Yoshida, T., Miura, H., and Schlettwein, D. (2015). Improvement of light harvesting by addition of a long-wavelength absorber in dye-sensitized solar cells based on ZnO and indoline dyes. *J. Phys. Chem. C* 119 (3): 1298–1311.

141 Tao, P., Guo, W., Du, J. et al. (2016). Continuous wet-process growth of ZnO nanoarrays for wire-shaped photoanode of dye-sensitized solar cell. *J. Colloid Interface Sci.* 478: 172–180.

142 Park, N.G., van de Lagemaat, J., and Frank, A.J. (2000). Comparison of dye-sensitized rutile- and anatase-based TiO_2 solar cells. *J. Phys. Chem. B* 104 (38): 8989–8994.

143 Wang, Z., Tang, Q., He, B. et al. (2015). Efficient dye-sensitized solar cells from curved silicate microsheet caged TiO_2 photoanodes. An avenue of enhancing light harvesting. *Electrochim. Acta* 178: 18–24.

144 Lee, T.H., Seong, Y.B., Kim, Y.S. et al. (2015). Synthesis of macro-porous SnO_2 for dye-sensitized solar cells. *Mol. Cryst. Liq. Cryst.* 617 (1): 204–210.

145 Chai, Z., Gu, J., Qiang, P. et al. (2015). Facile conversion of rutile titanium dioxide nanowires to nanotubes for enhancing the performance of dye-sensitized solar cells. *CrystEngComm* 17 (5): 1115–1120.

146 Ahmed, I., Fakharuddin, A., Wali, Q. et al. (2015). Mesoporous titania–vertical nanorod films with interfacial engineering for high performance dye-sensitized solar cells. *Nanotechnology* 26 (10): 105401.

147 Tao, X., Ruan, P., Zhang, X. et al. (2015). Microsphere assembly of TiO_2 mesoporous nanosheets with highly exposed (101) facets and application in a light-trapping quasi-solid-state dye-sensitized solar cell. *Nanoscale* 7 (8): 3539–3547.

148 Liu, W., Hong, C., Wang, H. et al. (2016). Enhanced photovoltaic performance of fully flexible dye-sensitized solar cells based on the $Nb_2 O_5$ coated hierarchical TiO_2 nanowire-nanosheet arrays. *Appl. Surf. Sci.* 364: 676–685.

149 Chai, Z., Gu, J., Yuan, Y. et al. (2015). Fabrication and integration of quasi-one-dimensional hierarchical TiO_2 nanotubes for dye-sensitized solar cells. *CrystEngComm* 17 (43): 8327–8331.

150 Xiong, Y., He, D., Jin, Y. et al. (2015). Ordered mesoporous particles in titania films with hierarchical structure as scattering layers in dye-sensitized solar cells. *J. Phys. Chem. C* 119 (39): 22552–22559.

151 Elbohy, H., Thapa, A., Poudel, P. et al. (2015). Vanadium oxide as new charge recombination blocking layer for high efficiency dye-sensitized solar cells. *Nano Energy* 13: 368–375.

152 Mahmoud, M.S., Akhtar, M.S., Mohamed, I.M.A. et al. (2018). Demonstrated photons to electron activity of S-doped TiO_2 nanofibers as photoanode in the DSSC. *Mater. Lett.* 225: 77–81.

153 Noor, S., Sajjad, S., Leghari, S.A.K. et al. (2018). ZnO/TiO_2 nanocomposite photoanode as an effective UV–Vis responsive dye sensitized solar cell. *Mater. Res. Express* 5 (9): 095905.

154 Mehmood, U. (2017). Efficient and economical dye-sensitized solar cells based on graphene/TiO_2 nanocomposite as a photoanode and graphene as a Pt-free catalyst for counter electrode. *Org. Electron.* 42: 187–193.

155 Musyaro'ah, Huda, I., Indayani, W., et al. (2017). Fabrication and characterization dye sensitized solar cell (DSSC) based on TiO_2/SnO_2 composite. Paper presented at the International Conference On Engineering, Science And Nanotechnology 2016 (ICESNANO 2016), Solo, Indonesia, p. 030062.

156 Hwang, D.-K., Kim, J.-H., Kim, K.-P., and Sung, S.-J. (2017). Mesoporous TiO_2 hierarchical structures: preparation and efficacy in solar cells. *RSC Adv.* 7 (77): 49057–49065.

157 Gireesh Baiju, K., Murali, B., and Kumaresan, D. (2018). Synthesis of hierarchical barium titanate micro flowers with superior light-harvesting characteristics for dye sensitized solar cells. *Mater. Res. Express* 5 (7): 075503.

158 Subalakshmi, K. and Senthilselvan, J. (2018). Effect of fluorine-doped TiO_2 photoanode on electron transport, recombination dynamics and improved DSSC efficiency. *Sol. Energy* 171: 914–928.

159 John, K.A., Naduvath, J., Mallick, S. et al. (2016). Electrochemical synthesis of novel Zn-doped TiO_2 nanotube/ZnO nanoflake heterostructure with enhanced DSSC efficiency. *Nano-Micro Lett.* 8 (4): 381–387.

160 Rajamanickam, N., Soundarrajan, P., Vendra, V.K. et al. (2016). Efficiency enhancement of cubic perovskite BaSnO$_3$ nanostructures based dye sensitized solar cells. *Phys. Chem. Chem. Phys.* 18 (12): 8468–8478.

161 Hsieh, M.-Y., Lai, F.-I., Chen, W.-C. et al. (2016). Realizing omnidirectional light harvesting by employing hierarchical architecture for dye sensitized solar cells. *Nanoscale* 8 (10): 5478–5487.

162 Guo, E. and Yin, L. (2015). Nitrogen doped TiO$_2$–Cu$_x$ O core–shell mesoporous spherical hybrids for high-performance dye-sensitized solar cells. *Phys. Chem. Chem. Phys.* 17 (1): 563–574.

163 Choi, J., Kang, G., and Park, T. (2015). A competitive electron transport mechanism in hierarchical homogeneous hybrid structures composed of TiO$_2$ nanoparticles and nanotubes. *Chem. Mater.* 27 (4): 1359–1366.

164 Shim, C.S., Mali, S.S., Aokie, R. et al. (2015). Evaluation of a metal free dye for efficient dye sensitized solar cells based on hydrothermally synthesized TiO$_2$ nanoflowers. *RSC Adv.* 5 (111): 91708–91715.

165 Zhao, W., Fu, W., Chen, J. et al. (2015). Preparation of TiO$_2$-based nanotubes/nanoparticles composite thin film electrodes for their electron transport properties. *Thin Solid Films* 577: 49–55.

166 So, S., Hwang, I., and Schmuki, P. (2015). Hierarchical DSSC structures based on "single walled" TiO$_2$ nanotube arrays reach a back-side illumination solar light conversion efficiency of 8%. *Energy Environ. Sci.* 8 (3): 849–854.

167 Liang, J., Zhang, G., Yang, J. et al. (2015). TiO$_2$ hierarchical nanostructures: hydrothermal fabrication and application in dye-sensitized solar cells. *AIP Adv.* 5 (1): 017141.

168 Yao, J., Lin, C.-M., and Yin, S. (2015). Density-controlled ZnO/TiO$_2$ nanocomposite photoanode for improving dye-sensitized solar cells performance. Paper presented at the SPIE OPTO, San Francisco, California, United States, p. 935819.

169 Feng, H.-L., Wu, W.-Q., Rao, H.-S. et al. (2015). Three-dimensional TiO$_2$/ZnO hybrid array as a heterostructured anode for efficient quantum-dot-sensitized solar cells. *ACS Appl. Mater. Interfaces* 7 (9): 5199–5205.

170 Li, F., Jiao, Y., Xie, S., and Li, J. (2015). Sponge-like porous TiO$_2$/ZnO nanodonuts for high efficiency dye-sensitized solar cells. *J. Power Sources* 280: 373–378.

171 Qi, D., Wang, L., and Zhang, J. (2015). Hierarchically mesoporous/macroporous structured TiO$_2$ for dye-sensitized solar cells. *RSC Adv.* 5 (91): 74557–74561.

172 Sheng, J., Hu, L., Mo, L. et al. (2016). TiO$_2$ hierarchical sub-wavelength microspheres for high efficiency dye-sensitized solar cells. *Phys. Chem. Chem. Phys.* 18 (47): 32293–32301.

173 Xia, W., Mei, C., Zeng, X. et al. (2016). Mesoporous multi-shelled ZnO microspheres for the scattering layer of dye sensitized solar cell with a high efficiency. *Appl. Phys. Lett.* 108 (11): 113902.

174 Zhang, Y., Wang, C., Yuan, Z. et al. (2017). Reduced graphene oxide wrapped mesoporous hierarchical TiO$_2$–CdS as a photoanode for high-performance dye-sensitized solar cells: reduced graphene oxide wrapped mesoporous hierarchical TiO$_2$–CdS as a photoanode for high-performance dye-sensitized solar cells. *Eur. J. Inorg. Chem.* 2017 (16): 2281–2288.

175 Al-Agel, F.A., Shaheer Akhtar, M., Alshammari, H. et al. (2015). Solution processed ZnO rectangular prism as an effective photoanode material for dye sensitized solar cells. *Mater. Lett.* 147: 119–122.

176 Chen, H.-S., Yu, W.-C., Chang, W.-C., and Lu, Y.-W. (2016). Olive-shaped ZnO nanocrystallite aggregates as bifunctional light scattering materials in double-layer photoanodes for dye-sensitized solar cells. *Electrochim. Acta* 187: 655–661.

177 Chang, W.-C., Lin, L.-Y., and Yu, W.-C. (2015). Bifunctional zinc oxide nanoburger aggregates as the dye-adsorption and light-scattering layer for dye-sensitized solar cells. *Electrochim. Acta* 169: 456–461.

178 Chauhan, R., Shinde, M., Kumar, A. et al. (2016). Hierarchical zinc oxide pomegranate and hollow sphere structures as efficient photoanodes for dye-sensitized solar cells. *Microporous Mesoporous Mater.* 226: 201–208.

179 Moradzaman, M., Mohammadi, M.R., and Nourizadeh, H. (2015). Efficient dye-sensitized solar cells based on CNTs and Zr-doped TiO_2 nanoparticles. *Mater. Sci. Semicond. Process.* 40: 383–390.

180 Ghartavol, H.M., Mohammadi, M.R., Afshar, A. et al. (2016). Efficient dye-sensitized solar cells based on CNT-derived TiO_2 nanotubes and Nb-doped TiO_2 nanoparticles. *RSC Adv.* 6 (103): 101737–101744.

181 Serikov, T.M., Ibrayev, N.K., Smagulov, Z.K., and Kuterbekov, K.A. (2017). Influence of annealing on optical and photovoltaic properties of nanostructured TiO_2 films. *IOP Conf. Ser.: Mater. Sci. Eng.* 168: 012054.

182 Song, W., Gong, Y., Tian, J. et al. (2016). Novel photoanode for dye-sensitized solar cells with enhanced light-harvesting and electron-collection efficiency. *ACS Appl. Mater. Interfaces* 8 (21): 13418–13425.

183 Zhang, Z., Hu, Y., Qin, F., and Ding, Y. (2016). DC sputtering assisted nano-branched core–shell TiO_2/ZnO electrodes for application in dye-sensitized solar cells. *Appl. Surf. Sci.* 376: 10–15.

184 Huang, N., Chen, L., Huang, H. et al. (2015). Bilayer TiO_2 photoanode consisting of microspheres and pyramids with reinforced interface connection and light utilization for dye-sensitized solar cells. *Electrochim. Acta* 180: 280–286.

185 Wang, P., Wang, J., Yu, H. et al. (2016). Hierarchically macro–mesoporous TiO_2 film via self-assembled strategy for enhanced efficiency of dye sensitized solar cells. *Mater. Res. Bull.* 74: 380–386.

186 Gu, J., Khan, J., Chai, Z. et al. (2016). Rational design of anatase TiO_2 architecture with hierarchical nanotubes and hollow microspheres for high-performance dye-sensitized solar cells. *J. Power Sources* 303: 57–64.

187 Lekphet, W., Ke, T.-C., Su, C. et al. (2016). Morphology control studies of TiO_2 microstructures via surfactant-assisted hydrothermal process for dye-sensitized solar cell applications. *Appl. Surf. Sci.* 382: 15–26.

188 Wang, G., Zhu, X., and Yu, J. (2015). Bilayer hollow/spindle-like anatase TiO_2 photoanode for high efficiency dye-sensitized solar cells. *J. Power Sources* 278: 344–351.

189 Yen, Y.-C., Chen, P.-H., Chen, J.-Z. et al. (2015). Plasmon-induced efficiency enhancement on dye-sensitized solar cell by a 3D TNW-AuNP layer. *ACS Appl. Mater. Interfaces* 7 (3): 1892–1898.

190 Liu, Y., Xia, Y., Jiang, Y. et al. (2015). Coupling effects of Au-decorated core-shell β-NaYF$_4$:Er/Yb@SiO$_2$ microprisms in dye-sensitized solar cells: plasmon resonance versus upconversion. *Electrochim. Acta* 180: 394–400.

191 Ganapathy, V., Kong, E.-H., Park, Y.-C. et al. (2014). Cauliflower-like SnO_2 hollow microspheres as anode and carbon fiber as cathode for high performance quantum dot and dye-sensitized solar cells. *Nanoscale* 6 (6): 3296.

192 Yang, W.S., Noh, J.H., Jeon, N.J. et al. (2015). High-performance photovoltaic perovskite layers fabricated through intramolecular exchange. *Science* 348 (6240): 1234–1237.

4

Light Scattering Materials as Photoanodes

Rajkumar C[1] and A. Arulraj[2]

[1] *Department of Electronics and Communication, University of Allahabad, Allahabad, India*
[2] *Department of Physics, University College of Engineering – Bharathidasan Institute of Technology (BIT) Campus, Anna University, Tiruchirappalli, India*

4.1 Introduction

Dye-sensitized solar cells (DSSCs) are existing to be a promising alternative to conventional silicon solar cells owing to its ease of fabrication, low cost, along with the additional features like flexible devices and building integrated photovoltaic system [1]. In general, DSSC architecture comprised of semiconducting photoanode, counter electrode (catalyst), and an electrolyte [2, 3]. In DSSC, the quantity of light caught by the dye-sensitized photoanode has a strong influence on the overall performance because the electrons are generated directly from the excited dye molecules by the irradiated sunlight. One of the primary key tasks is to boost the performance of the device by enriching the photon absorption in DSSCs such as the introduction of a tandem structure and the introduction of a light-scattering effect. Among these developments, the light-scattering effect had more attention than others because of an easy method for enhancing light absorption [4–7]. The simple theory of light scattering effect was introduced by Usami [8] in 1997, stating that a new photoelectrode structure consisting of top layer made up of a large sized particle film on the below layer made up of a small sized particle film; the light scattered by the top layer triggered an increase in the light absorption. In initial days, several studies had been focused on the analysis of the light-scattering effect in the DSSCs by using various simulation methods based on the Monte Carlo model, Mie theory, and the many-flux model. Subsequently, the light-scattering effect had been widely used as an important technique to develop the performance of DSSCs [9]. So, the introduction of light scattering and light scattering materials are explained in the following sections.

4.2 Introduction to Light Scattering

Light scattering is defined as the redirection of electromagnetic waves (ultraviolet radiation, microwaves, radio waves, X-rays, and γ-radiation, heat radiation) when it strikes an object (a particle or molecules). Figure 4.1a shows that after the light strikes an object, the light should be redirected according to their laws of reflections and refractions. Figure 4.1b shows that the object that creates or causes light scattering is called *scatterer* or *scattering center*. Light scattering is of two types: one is elastic light scattering, and another one is inelastic light scattering. The scattered light that has the same frequency (same energy) as compared to incident light

Interfacial Engineering in Functional Materials for Dye-Sensitized Solar Cells, First Edition.
Edited by Alagarsamy Pandikumar, Kandasamy Jothivenkatachalam and Karuppanapillai B. Bhojanaa.

is called elastic scattering, while the scattered light that has a different energy as compared to incident light is called inelastic scattering. So the inelastic scattering has either shorter wavelengths or longer wavelengths as compared to the incident light. Elastic scattering includes Rayleigh scattering and Mie scattering, while inelastic scattering contains Raman scattering, Brillouin scattering, Compton scattering, and inelastic X-ray scattering. In DSSCs, the elastic scattering is taken for analysis to improve the efficiency. In the elastic scattering, the Mie scattering applies to all sizes of most spherical particles, but the Rayleigh scattering applies to the small dielectric (nonabsorbing) spherical particles. So the Rayleigh scattering is followed by the same condition that is $\alpha \ll 1$ and $|m|\alpha \ll 1$, where $\alpha = \frac{2\pi a}{\lambda}$, $a =$ the radius of spherically shaped particles, and $\lambda =$ the wavelength of the relative scattering. The wavelength of the relative scattering is defined as $\lambda = \frac{\lambda_0}{m}$, where $\lambda_0 =$ incident wavelength at vacuum medium and $m =$ refractive index of the medium. The refractive index (m) of the scattering medium is defined as $m = n - ik$, where $n =$ the refraction of light and the complex part represents to absorption. The complex term from the above equation represents the absorption coefficient which is defined by absorption coefficient $=4\pi k/\lambda$. The particles that have small dimensions compared to incident wavelength is only applicable to Rayleigh scattering method. The intensity of Rayleigh scattered light by single particles is defined as $I = I_0 \frac{1+\cos^2 \theta}{2d^2} \left(\frac{2\pi}{\lambda}\right)^4 \left|\frac{m^2-1}{m^2+1}\right| r^2$, where $I=$ intensity of the incident light, $d=$ distance to the particle, and $\theta=$ scattering angle. The cross-section (σ) of the Rayleigh scattering is defined by the equation, $\sigma_{scat} = \frac{2\pi^5}{3} \frac{(2r^2)}{\lambda^4} \left|\frac{m^2-1}{m^2+1}\right|^2$. The Mie theory is used to describe the scattering from the spherical particles whether the material is absorbing the light or not. The efficiency of the Mie scattering is determined by the equation, $Q_{Scat,\ Mie} = \frac{\sigma_{Scat,Mie}}{\pi r^2}$, where $\sigma_{Scat,\ Mie}=$ the cross-section of the Mie scattering that is defined by this equation, $Q_{Scat,\ Mie} = \sum_{n=0}^{\infty}(2n + 1)(|a_n|^2 + |b_n|^2)$, where a_n and b_n are specified through the Riccati Bessel functions ψ and ζ. The scattering efficiency depends on the diameter of the particles and wavelengths of the incident light that is shown in Figure 4.1 [10, 11]. Figure 4.1c shows that the incident wavelength is set to the fixed wavelength of 532 nm, the scattering efficiency of the TiO_2 scatterer starts increasing from larger than 200 nm of the scatterer diameter. It is specified that the diameter of the particle must be approximately half of the wavelength of incident light. Figure 4.1d shows that the particle diameter is set to 200 nm, and here the scattering efficiency varies according to the different incident wavelength, the scattering efficiency occurs in between ~400 and ~650 nm. From these rough calculations, it is indicating that the diameter of the particle and wavelength of the incident light depends on each other for effective light scattering in DSSCs.

4.3 Materials for Light Scattering in DSSCs

From the literature, the colloidal science tells about the preparation of small particles and how to control the aggregation as well as coagulation of small particles. Mostly, the colloidal particles have the size approximately below the 100 nm, for the quantum effect, the particles size must be less than the first exciton Bohr radius. Optical properties are changing according to the size of the particles. The thin films have large inner surface area compared to their nanometer-sized materials. The semiconductor that absorbs the light below a particular wavelength is called fundamental absorption edge.

$$E_g\ (eV) = \frac{1240}{\lambda\ (nm)}$$

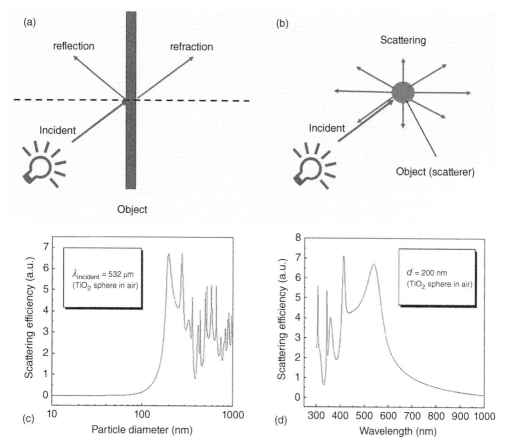

Figure 4.1 (a) Reflection, refraction and (b) scattering of light, (c) scattering efficiencies versus particle diameter, and (d) scattering efficiencies versus wavelength. *Source:* Zhang et al. 2012 [10]. Reproduced with permission of Royal Society of Chemistry.

The extinction of light depends on the exponential value of penetration length of the light (l) and reciprocal absorption length (α).

$$I = I_0 \exp(-\alpha l)$$

The particles sizes are playing the major role for providing absorbance band edge due to quantum effects. In nanotechnology-assisted applications, TiO_2 are widely used including solar cells. In DSSCs, TiO_2 film is deposited on conducting glass substrate fluorine-doped tin oxide (FTO) to create the photoanode. The conductivity of TiO_2 film was increased due to the absorbance of UV illumination, whereas the conductivity of the TiO_2 film was decreased in the dark at room temperature. The semiconductor was immersed into the electrolyte to form the junction of semiconductor/electrolyte in each nanocrystal of the semiconductor. The electron-hole pairs are created during illumination in which the holes are transferring fast to the electrolyte than the recombination of electrons. This process created an electrochemical gradient. The electrons are transported through the interconnected particles to the back contact. Finally, it is concluded that the charge carriers are transported through the method of diffusion in the semiconductor films. The semiconductor films thickness was also playing an important role in the process of recombination. The recombination probabilities are increased due to an increase in the thickness of semiconductor films; here the electrons are traveled through the large number of nanoparticles (NPs) and grain boundaries. In addition to the recombination process, the resistance loss is also increased due to the increased thickness of the semiconductor films. So it is recommended that the film thickness of semiconductors must be as thin as possible to increase the performance of the DSSCs. J. Ferber and J. Luther [12] reported that the light scattering is also based on the dyes in DSSCs. The dye has poor light absorption, which reduces the efficiency of the DSSCs. In DSSCs, the nanoporous TiO_2 layer is coated with a monolayer of the suitable charge transfer dye. The solar energy was absorbed by the dye, but the TiO_2 was responsible for a charge carrier transport in DSSCs. The particle with 10–25 nm size increased the effective surface area up to 1000. Therefore, the 1000 monolayers of the dye are also available in the DSSC cell. If the TiO_2 particles are very small, the pores are not enough to enter the dye and the electrolyte. So the large TiO_2 particles create the pores to enter the dye and the electrolyte, but there is not enough internal surface. Even though the effective surface area adsorbs the dye, in light the dye absorbs is very poor in the certain wavelength range ($\lambda = 600$–800 nm), as shown in Figure 4.2a. The incident path length of the light, as well as the optical absorption, must be increased by increasing the light scattering in the TiO_2 electrode. This effective light scattering can be achieved by the mixture of some large-sized particles into small-sized particles, but the too large-sized particles reduced the internal surface. So the optimum large-sized particles must be mixed with small-size particles to increase the efficiency of the DSSCs. The optical modeling can be done through electrical modeling because the photoconductivity of the TiO_2 particles depends on the size of the particle size. The dark current and the series resistance must be minimized and suppressed. First, the electrical model must be implemented, which was reported earlier [14]. From the optical point of view, the optical absorption must be increased with the help of thicker layers. As per the earlier reports and experience, the optimal thickness of the electrodes must be in the range of 10–15 μm. The path length of the electrons and the iodide/iodine particles in the electrolyte must be too long, which shows the increase in series resistance due to the thickness of the electrodes. In DSSCs, the photoanodes normally contains the ~12–20 nm-sized TiO_2 particles which were transparent in the visible light region and increased the traveling distance of the incident light in the photoelectrode film [15].

4.4 Early Theoretical Predictions of Light Scattering in DSSCs

This part explains the few theoretical studies that are explaining the design and optimization of the structures of DSSCs photoelectrode with the integration of light scattering function. Ferber and Luther [12] investigated the optical absorption of photoelectrode via the computer simulation and explained the influence of size and concentration of scatters, and the structure of the photoelectrode films. He explained that the scattering theory has a single scattering for individual TiO_2 particles and multiple scattering for within the ensemble. Mie theory was applied to determine the single scattering from coated spheres. The dye molecules are having the dimensions of about 1 nm, which is very small compared to the wavelength of the incident light. So there is no requirement to consider dye molecules as small spheres, but the scattering of the small sphere-coated large spheres (TiO_2) must be considered for calculation. For assumption, the TiO_2 spheres have been covered by the homogenous dye layer with an effective absorption coefficient. The scattering from TiO_2 particles which is having the size of 10–25 nm is negligible. The scatterers or particles that are having the size range of the light wavelength is considered as an effective Mie scatterer. Large particles are only scattered in the forward direction. The simulation was based on small (20 nm) TiO_2 nanoparticles (TNPs) mixed with large TNPs to form a binary particle mixture of photoelectrode film, or the small

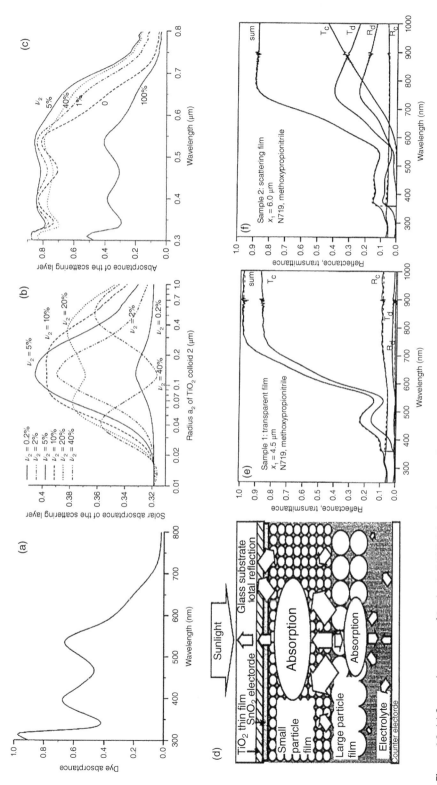

Figure 4.2 (a) Spectral response of Ru dye in DSSC. Solar absorption of scattering layer consists of binary mixtures based on (b) radius of TiO_2 colloid and (c) wavelengths, respectively. (d) Akira Usami proposed DSSC structure. (e, f) Collimated transmittance (T_c), diffused transmittance (T_d), collimated reflectance (R_c), and diffused reflectance (R_d). Transmittance and reflectance spectra of (e) sample 1 and (f) sample 2, respectively. *Source:* (b, c) Ferber and Luther 1998 [12]. Reproduced with permission of Elsevier. (d) Usami 1997 [8]. Reproduced with permission of Elsevier. (e,f) Rothenberger et al. 1999 [13]. Reproduced with permission of Elsevier.

(20 nm) TNPs layer combined with large TiO$_2$ particles layer to form a photoelectrode film for backward scattering. These particles were assumed to be covered with a monolayer of dye molecules. The numerical solution of the radiative transport equation was used to calculate the multiple scattering of the coated sphere scatterers using Mie theory. The calculated results revealed that the scattering of the particles or spheres which is having the size of 10–25 nm are negligible, whereas the particles which are having the size range of light wavelengths are generating the light scattering. The simulated results predicted that the optical absorbance of the photoelectrode increased due to light scattering. The optimal structure of the photoelectrode was proposed as a mixture of large particles (containing 5 wt% with the radius of 125–150 nm) and small nanocrystalline particles (containing 95 wt% with the size of 20 nm) that increase the optical absorbance in DSSC photoelectrodes, as shown in Figure 4.2. The simulation results also predict that the mixture of too many large particles into nanocrystalline film decreased the internal surface area as well as increased the backscattering (reflectance) as compared to absorbance of the photoelectrode film but increased the absorbance in the infrared (IR) region ($\lambda = 550–750$ nm), as shown in Figure 4.2c. From the simulation results, the double-layer structured photoelectrode that is combined with the nanocrystalline film and back-scattering layer with large size particles was not mostly recommended for DSSCs. From the simulation results, it was revealed that the mixture structure photoelectrodes has more optical absorbance than that of double-layer structure photoelectrodes. Even though from the manufacturing point of view, the double-layer structure photoelectrodes required extra work than the mixture structure photoelectrodes. The simulation results may not completely reflect the practical results of light scattering of the system containing the mixture of a nanocrystalline film with large size particles. For example, the large-sized particles that are incorporated into the nanocrystalline film may decrease the internal surface area of the photoelectrode film. This may decrease the adsorption of dye, and it may increase the light absorption due to light scattering in photoelectrode film. So the counterbalance that happened in some degree increased the dye adsorption and increased the light absorption. However, the simulation results show that the optical absorbance enhancement of binary mixture structure photoelectrodes is the same as the double-layer structure photoelectrodes. Nevertheless, the qualitative prediction of higher efficiency in certain sized large particles (250–300 nm in diameter) with a certain weight percentage (5 wt%) mixed with the nanocrystalline film remains valid. Another theoretical study of Vargas and Niklasson [16], using four-flux radiative transfer calculation, reveals that the maximum value of effective scattering was correlated to the volume fraction of the scatterers in addition to the size of the scatterers. From the calculation, the maximum effective scattering coefficient was reached due to the volume fraction of TiO$_2$ particle, which is having a diameter of 250 nm with 25 wt%.

The theoretical study that A. Usami [8] proposed show (Figure 4.2d) a new structure of the dye-sensitized nanocrystalline photoelectrochemical cell, which permits the effective incident solar energy absorption using thinner dye-sensitized film. The proposed photoelectrode film consists of a three-layer structure that includes a layer of conventional TiO$_2$ nanocrystalline film, a layer of large-sized TiO$_2$ particles deposited on top of the nanocrystalline TiO$_2$ film layer for light scattering, and the rutile-structured TiO$_2$ film was inserted between the nanocrystalline TiO$_2$ film and glass substrate for a total reflection of scattered light. The proposed cell (shown in Figure 4.2d) effectively confined the light in the thinner dye-sensitized film by multiple scattering with the help of dispersed TiO$_2$ particles at the bottom and total reflection between inserted TiO$_2$ film and surface of the glass substrate. From the boundary-element method (BEM), A. Usami [8] reveals that the optimal diameter ($1.3 \sim 1.4 \times \pi/k$) of the TiO$_2$ particles are used for effective scattering. From this condition, the backscattering angle has the largest intensity compared to the critical angle at the interface between TiO$_2$ inserted rutile layer and the glass substrate. The optical confinement is not only for short wavelength light, but it is also for long wavelengths, which is having a small absorption coefficient. Finally, the simulation results concluded that the multilayer structure photoelectrode film was having higher optical absorption than the mixed structure photoelectrode film with a certain condition. From these theoretical studies, it is indicating that the structure of photoelectrode film plays a critical role for optical absorption and light propagation in DSSCs to improve the performance of the DSSCs. So a careful design of photoelectrode film will give a good optical absorption for improving the performance as well as the efficiency of the DSSCs. From the theoretical modeling of the optical characterization in DSSCs, Rothenberger et al. [13] reported that the incorporation of large particles or voids could enhance the absorption as a result of scattering. Two kinds of samples were prepared for optical studies: sample 1 consists of anatase TiO_2 with an average diameter of 15 nm to make 4.5 μm thick film and the sample 2 consists of 6.0 μm thick film of anatase particles with large sizes up to about 150 nm. The specific surface area (Brunauer−Emmett−Teller BET) of sample 2 is reduced to 35 m^2/g from the sample 1 surface area of 101 m^2/g because of increasing the particle size. Figure 4.2e,f shows that sample 1 has low diffused transmittance as compared to sample 2. Moreover, sample 2 has high diffused reflectance and low collimated transmittance as a result of increased particle size. The major contribution of the collinear reflectance is from the interface of the glass substrate. So the glass substrate has anti-reflecting materials for improving the efficiency of the solar cell. The absorption of light in solar cells is limited because of backscattering and diffused reflection. The relation between the electron transfer yield and the optical thickness in DSSCs had been indirectly studied by Y. Tachibana et al. [17]. The quantitative analysis of light harvesting efficiency (LHE) and electron transfer yield had been done in DSSCs with the series of light scattering magnitudes (Figure 4.3a,b). The photocurrent of the solar cell is calculated with the help of electron transfer yield and LHE are given by

$$J_{SC} = \phi_{ET} \int qF(\lambda)[1 - r(\lambda)]LHE(\lambda)\,d\lambda$$

where q = electron charge, $F(\lambda)$ = incident photon flux density at the wavelength of λ, $r(\lambda)$ = loss of the incident light and φ_{ET} = average electron transfer yield. The addition of large particles into the transparent film will increase the LHE, specifically in near-infrared wavelengths. The addition of more large particles will reduce the LHE over the whole visible wavelengths due to more light reflection in conducting glass/TiO_2 interface. From the calculations of LHE and short-circuit photocurrent, they demonstrate that the electron transfer yield decreases due to increases of optical thickness (film light scattering magnitude) by a maximum of ≈60%.

4.5 Different Light Scattering Materials

4.5.1 Mixing of Large Particles into Small Particles

J. Zhao et al. [18] prepared TiO_2 hollow spheres (THSs) with different diameters using the hard-template method and mixed with TNP to form a photoanode film. The 600 nm THS (Shown in Figure 4.3c)-embedded photoanode showed 5.84% power conversion efficiency (PCE), which was 68.8% higher than the P25 TiO_2 photoanode due to efficient light scattering in DSSC. Similarly, J. Lin et al. [20] fabricated high scattering hybrid TiO_2 photoanode composed of TNT (shown in Figure 4.3d) and TNP. The nanotube (NT) with 50 wt% based photoanode showed 6.0% PCE due to improvement in light scattering by TNT because NT acted as a light scattering component.

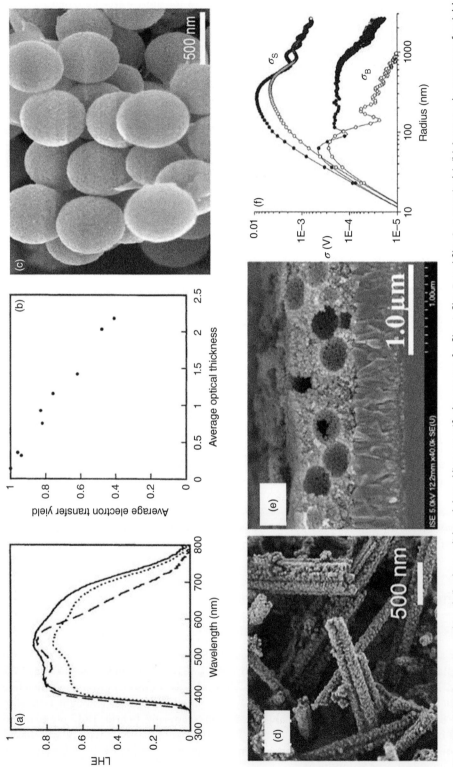

Figure 4.3 (a) LHE versus wavelength (broken, solid, and dotted lines specify the spectra for film-1, film-5, and film-8, respectively). (b) Average electron-transfer yield versus average optical thickness. (c) SEM image of THS having 600 nm size. (d) SEM images of NT powder samples annealed at 650 °C. (e) Spherical voids of TiO_2 film when the polystyrene spheres are burnt out during heating or sintering at 450 °C. (f) Backward (σ_B/v) and total (σ_S/v) scattering coefficients per unit particle volume due to the presence of spherical electrolyte-filled voids (white) and larger TiO_2 solid particles (black) present in the TiO_2 films as potential scatterers at $\lambda = 550$ nm. *Source:* (b) Tachibana et al. 2002 [17]. Reproduced with permission of American Chemical Society. (c) Zhao et al. 2016 [18] Reproduced with permission of Elsevier. (f) Hore et al. 2005 [19] Reproduced with permission of Royal Society of Chemistry.

4.5.2 Voids as Light Scatters

In DSSCs, spherical voids are used as scattering centers in TiO_2 and ZnO thin films. Hore et al. [19] reported that the spherical voids are incorporated into TiO_2 films to make spherical voids as scattering centers, as shown in Figure 4.3e. The polystyrene spheres were stabilized with the help of carboxyl to make spherical voids in TiO_2 films; here the carboxyl was used for anchoring on the TiO_2 surface [21, 22]. The polystyrene spheres were added to TiO_2 paste in the volume ratio of 1 : 5 and well mixed with the help of the ultrasonication process at room temperature. The polystyrene sphere mixed TiO_2 film was sintered at 450 °C to make voids due to the melting of the polystyrene sphere in the TiO_2 film (Figure 4.3e). The scattering of voids filled with electrolyte was calculated by Mie theory [23]. The scattering angle was very large that enhanced the absorbance of the TiO_2 film. The scattering cross section σ_S per volume is used to estimate the scattering effect of the per volume fraction. From Figure 4.3f, it is showing that the spherical voids at 50–100 nm act as very good backscatterers. The electrochemical impedance spectroscopy shows that the spherical voids help the ions for diffusion, and it increased the overall efficiency of the device by 25%. Similarly, several researchers used polystyrene sphere and carbon spheres to make spherical voids scattering layer in DSSCs to increase the overall efficiency [24–26]. The extremely porous structure of the topmost spherical voids layer increased the diffusion of I^-/I_3 electrolyte in the TiO_2 bilayer film [27]. T. Pham et al. [28, 29] also created different-sized cavities in TiO_2 particles with the help of ZnO template and improved 11% PCE compared to other sized cavities and conventional TNP photoanodes in DSSC. C. Kim et al. [30] reported that the effect of the scattering layer in ZnO based DSSC was improved by mixing of nanoparticles and nanoporous spheres in various ratios. The nanoporous spheres have the large surface area to increase the effect of scattering, but it has disadvantages due to large voids and point contacts between spheres. Even though, the carrier transport increased due to additional surface area and improving the efficiency of DSSC by enhanced the short circuit current (J_{SC}) and fill factor (FF).

4.5.3 Nano-Composites for Light Scattering

4.5.3.1 Nanowire–Nanoparticle Composite

Tan and Wu [31] also studied the solar cell performance using nanowire/nanoparticle composites at the photoanodes. Different wt% of TiO_2 had been grown by a hydrothermal method in which the 20 wt% of TiO_2 nanowires show an excellent efficiency of 8.6%. The 72 wt% of TiO_2 nanowires show an opaque property in the photoanodes. So the higher concentrations of nanowires decreased the current density in DSSCs. Baxter and Aydil [32] reported that the LHE was increased by depositing the ZnO nanoparticles in interstitial voids between ZnO nanowires.

4.5.3.2 Nanofiber–Nanoparticle Composite

The TiO_2 nanofiber/TNP composite had increased the efficiency by 44% than the use of nanoparticles alone. It also improved the LHE without any compromise in the dye adsorption capability. The intensity of light scattering had increased due to the optimal diameter of nanofiber at 200 nm or above [33]. Y. Dzenis [34] explained the preparation method of TiO_2 nanofiber nanoparticle composites. Here, the nanofibers were dispersed in anhydrous ethanol, then, it was mixed with a nanoparticle paste. Then the mixtures were sonicated followed by heating at 450 °C for one hour. The prepared nanofiber had the diameter of 200–300 nm and the length of tens to micron. The length of the nanofibers was increased due to the increase happening in the forward scattering. The 15% of nanofibers in composite increased the light scattering and reduced the transmission of light in the dye-sensitized composite. The 15% of nanofibers in 7.5 µm thick film was producing higher PCE (8.8%) in DSSC. Similarly,

TNP/NF–ZnO photoanode was prepared by M. Yang et al. [35] and investigated the PCE, which showed 6.54% PCE than NP/NF (5.17%) due to increase in scattering of light in DSSC.

4.5.3.3 SrTiO₃ Nanocubes–ZnO Nanoparticle Composite

A. Banik et al. [36] prepared a hybrid photoanode by mixing micron-sized SrTiO$_3$ (STO nanocube assembled micron-sized (NCMS)) into ZnO nanoparticles with different weight ratios. Herein, the 3% of STO NCMS mixed with ZnO nanoparticles hybrid photoanode provided double-fold increment in PCE compared to pristine ZnO NP-based DSSC due to multiple reflections that occurred in the mirror-like facets of STO NCMS and enhanced the light scattering from the individual entity.

4.5.3.4 Silica Nanosphere–ZnO Nanoparticle Composite

A. Banik et al. [37] prepared a hybrid photoanode by mixing silica nanosphere (SiO$_2$ nanosphere (NS)) with ZnO NPs with different weight ratios. The hybrid photoanode with 1 wt% of SiO$_2$ NS provided 22% enhancement compared to pristine ZnO NPs-based photoanode in DSSC due to increase in propagation length of incident light by multiple internal reflections that occurred in microsized SiO$_2$ NS.

4.5.3.5 SnO₂ Aggregate–SnO₂ Nanosheet Composite

D. Wang et al. [38] fabricated the hybrid photoanodes by mixing 50 wt% of SnO$_2$ aggregate and 50 wt% of SnO$_2$ nanosheet with each other, coated on the FTO substrate. This hybrid photoanode show 5.59% PCE, which was 60% higher than the bare SnO$_2$ NAs architecture in DSSC due to the enhancement of electron transport capability by SnO$_2$ nanosheets and provided large surface area as well as effective light scattering by SnO$_2$ nanoaggregates.

4.5.3.6 Ag (4,4′-Dicyanamidobiphenyl) Complex–TiO₂ NP Composite

N. Irannejad et al. [39] prepared nanocomposites that consist of TiO$_2$ suspension with different weight ratios of Ag (4,4′-dicyanamidobiphenyl) complex (Ag (DCBP)) and coated with FTO substrate by electrophoretic deposition (EPD). The 1.5% of Ag complex-based TiO$_2$ photoanode shows 5.17% of PCE, which was higher than the pure TiO$_2$ photoanode in DSSC.

4.6 Light Scattering Layers

Z.-S. Wang et al. [40] prepared DSSC using different morphology TiO$_2$ photoelectrode that significantly increases the efficiency from 7.6% to 9.8% by tuning the structure of the film from a monolayer to multilayer. The DSSC performance depends on the morphology of the film. So the nanoparticles are essential to increase the surface area and adsorption of dye, while large particles are required to increase the absorption of red light through light scattering techniques. Simultaneously, the increase of surface area and light scattering are impossible because they oppose each other. Therefore, there is a requirement of balance between them. This balancing could happen due to tuning the structure of the layer with the well-controlled manner by the screen printing technique. The best efficiency of 10.2% was achieved by using multilayer structure which was having anti-reflection film on the surface of the cell.

4.6.1 Surface Modified TiO₂ Particles in Scattering Layer

S. Choi et al. [41] prepared surface modified light scattering TiO$_2$ particles by Al$_2$O$_3$ coating on the surface of TiO$_2$ particles using the modified sol-gel method. The uniform distribution of Al$_2$O$_3$ increased the short-circuit photocurrent and PCE (8.71%) due to increased light scattering in DSSC.

4.6.2 Dual Functional Materials in DSSC

The TiO_2 aggregates are considered as an assemblage of nanocrystallites [42]. The nanobeads and nanoporous microspheres were also considered as TiO_2 aggregates. The TiO_2 nanobeads were prepared by sol–gel and solvothermal method [43, 44]. D. Chen et al. [43] prepared mesoporous TiO_2 beads with high surface areas ($108.0\,m^2/g$) and tunable pore size from 14.0 to 22.6 nm through a combination of sol–gel and solvothermal method to enhance the light harvesting and dye adsorbances in electrode films at DSSC. The PCE (7.20%) of TiO_2 nanobeads-based electrodes were higher than the PCE (5.66%) of standard P25 TiO_2 electrodes in DSSC. F. Sauvage et al. [44] also prepared mesoporous TiO_2 beads by using sol–gel and solvothermal method to increase the electron diffusion lengths and electron lifetimes compare to Degussa P25 titania-based electrodes due to fine interconnected and densely packed TiO_2 particles inside the beads. In addition to this, it also increased the dye adsorption with a careful selection of dye and light scattering properties that lead to enhancing the PCE (10.6% for Ru (II)-based dye C101 and 10.7% using C106) and light harvesting in DSSC. Similarly, several researchers used nanoporous and mesoporous TiO_2 spheres to increase light scattering and surface area for dye absorption due to the effect of pore diameter and particle size [45–47].

4.6.3 Double-Light Scattering Layer

Zhao et al. [48] prepared TiO_2 double light scattering layer (DLL) film containing THS as an above layer and TiO_2 nanosheets (NS) as below layer in TiO_2 photoelectrode film. It was found that the efficiency of TiO_2-DLL was 5.08% which was 23.3% higher than the TiO_2 HS film and 8.3% higher than the TiO_2 NS film due to higher surface area and better light scattering capability. In addition to this, the underlayer nanosheets make good electronic contact between TiO_2 and FTO conducting substrate.

4.6.4 Large Particles as Scattering Layers

The scattering layer, made up of large particles, was placed above the high surface area TiO_2 film, which enhanced the current density by 80% [15]. J.-K. Lee et al. [49] prepared two types of TiO_2 layer by the sol–gel method, one was 9 nm-TiO_2 layer for dense layer, and another one was 123-nm TiO_2 layer for light scattering layer. The observation specified that the multilayered TiO_2 film with the scattering layer has lesser transparency than the monolayered TiO_2 films. The efficiency of PEG electrolyte-based DSSC was achieved by 6.03% using 123 nm sized TiO_2 scattering layer. C.-S. Chou et al. [107] used a hybrid TiO_2 electrode (working electrode with scattering layer) in DSSC. The microcrystalline TiO_2 particles were prepared via a sol–gel method with the help of $TiCl_4$ ethanol solution used as a precursor. The hybrid TiO_2 electrode had 50% of TiO_2 particles (P-25), and 50% of TiO_2 particles with an average size of 268.7 nm for DSSC that show higher PCE (7.02%) than the conventional DSSC PCE (5.16%) due to the effect of scattering layer in the DSSC. L. Yang et al. [50] produced cauliflower-like TiO_2 rough and smooth spheres with different initial temperatures by hydrolysis of $Ti(OBu)_4$ with P105 ($EO_{37}PO_{56}EO_{37}$) or tri-block copolymer F68 ($EO_{78}PO_{30}EO_{78}$) as structural agents. It has the diameter of $\sim$275 nm and strong light scattering at the wavelength of 400–750 nm. The rough spheres have higher light scattering and higher BET surface area compared to the smooth spheres. The 25 wt% of large TiO_2 spheres added on the TiO_2 film ($\sim$20 nm of TiO_2 particles) to act as a scattering layer. The energy conversion efficiency of the rough spheres (7.36%) shows larger than the smooth spheres (6.25%).

4.6.4.1 TiO$_2$ Nanotubes

K. Nakayama et al. [51] prepared a bilayer structure that consists of light absorbing TNP layer, and light scattering TiO$_2$ nanotube (TNT) layer that consists of submicrons length and 20 nm diameter for DSSC. The light scattering TNT which was having a high aspect ratio was prepared by anodization of Ti sheet in the perchloric acid solution. Then the TNT was electrophoretically deposited on the sintered TNPs layer. The PCE of the bilayer structure (NP/α-TNT) delivers 7.53% which was higher than the PCE (6.91%) of conventional DSSC (NP/SP). In DSSC, the nanotubes film shows a higher current density than the nanoparticle-based DSSC as a result of light scattering in nanotubes. The PCE of both different morphologies shows more or less equal currents, which may be due to less fill factor in nanotube-based DSSC. The fill factor may be produced because of the insulating layer formed in between nanotubes and substrate by the anodization process. So the nanotubes were more favorable for the scattering layer because the formation of the insulating layer was reduced in between substrate and material interface.

4.6.4.2 TiO$_2$ Nanowires

Bakhshayesh et al. [52] prepared corn-like anatase TiO$_2$ nanowires by two consecutive hydrothermal operations. The surface morphology was produced by a surface modification process with the help of surface tension stress mechanism. In this double-layer DSSC, the corn-like TiO$_2$ nanowires (light scattering layer) were on the over-layer of the TNPs. The corn-like TiO$_2$ nanowires show higher conversion efficiency (7.11%) and short circuit current density (16.54 mA/cm^2) as compared to regular TiO$_2$ nanowires. So the novel TiO$_2$ morphology increases the photon capture in DSSC by using triple function mechanisms such as photo-generated charge carriers, dye sensitization, and light scattering. The maximum 12% cell efficiency was achieved by using corn-like TiO$_2$ nanowires due to slow charge recombination.

4.6.4.3 TiO$_2$ Nanospindles

Y. Qiu et al. [53] prepared different sizes of TiO$_2$ nanospindles by the hydrothermal method to construct the double-layer photoanode for DSSC. In the double layer, one layer is made up of large nanospindles and another layer is made up of small nanospindles. The large nanospindles increased the light scattering, while the small nanospindles increased the roughness factor to increase the dye adsorption. The double-layer structured photoanode was made up of two different sizes of TiO$_2$ nanospindles that show a higher energy conversion efficiency (8.3%) than the P25 TiO$_2$ photoanode.

4.6.4.4 TiO$_2$ Nanofibers

S. Chuangchote et al. [54] prepared TiO$_2$ nanofibers by an electrospinning method that directly act on to the thick nanoparticles electrodes. The nanofibers had an average diameter of 250 nm and a one-dimensional structure with high crystallinity after the calcination process. So the nanoparticles/nanofibers produced 85% of incident photon-to-current conversion efficiency (IPCE) at the wavelength of 540 nm with the conversion efficiency of 8.14% (0.25 cm^2 area) and 10.3% (0.052 cm^2 area), which was higher than the nanoparticles based electrodes at 15.5 μm thickness (80% IPCE). The similar kind of nanofiber-structured TiO$_2$ nanocrystals acted as a scattering layer and showed 6.0% PCE compared to conventional TiO$_2$ photoanode (402%) in DSSC [55].

4.6.4.5 TiO$_2$ Rice Grain Nanostructures

P. Zhu et al. [56] prepared a thin film of TiO$_2$ rice grain-like nanostructures and nanofiber nanostructures on the TNPs in DSSC by electrospinning method. The prepared rice grain-like nanostructures and nanofiber nanostructures act as an effective scattering layer, here wherein the rice grain-like nanostructures (15.7%) produced higher efficiency than the nanofibers structures (9.63%) due to the high surface area [57, 58], single crystallinity, and good packing density.

4.6.4.6 Nest-Shaped TiO$_2$ Structures

G. Zhu et al. [59] prepared nest-shaped TiO$_2$ nanostructures as an effective scattering layer on top of the TNPs by the electrospinning method in DSSC. The nest-shaped TiO$_2$ scattering layer increased the photocurrent via light scattering and decreased the electron transfer resistance due to improved light harvesting in DSSC. The nest-shaped TiO$_2$ increased the PCE (8.02%) compared to the conventional TNPs (P25) PCE (7.49%).

4.6.4.7 Nano-Embossed Hollow Spherical TiO$_2$

H.-J. Koo et al. [60] prepared nano-embossed hollow spherical (NeHS) TiO$_2$ as a bi-functional layer on top of the nanocrystalline TiO$_2$ particles in DSSC by using the solvothermal method without adding the templates or surfactants. The NeHS act as a scattering layer as well as generate the photo-excited electrons due to light scattering and efficient dye adsorption. The dye adsorption was five times higher than the normally used 400 nm diameter scattering particles. This NeHS TiO$_2$ enhanced the PCE (9.43%) compared to the conventional scattering layer-based electrodes and without scattering layer electrodes in DSSC.

4.6.4.8 Hexagonal TiO$_2$ Plates

W. Shao et al. [61] fabricated hierarchical TiO$_2$ nanoplates (Figure 4.4c) with tunable shell structures by a simple hydrolysis and deposition process with the help of Cd(OH)$_2$ nanoplate templates. W. Shao et al. prepared three kinds of scattering layers such as single shell, double shell, and three shells, which enhanced the PCE due to light harvesting capability from the excellent scattering (Figure 4.4a) and dye loading capabilities. The TiO$_2$ nanoplates with three shells show the 6.53% of PCE at air mass coefficient (AM) 1.5G condition compared to single cell and double cells, as shown in Figure 4.4b. The overall PCE was 24% times higher than the conventional solar cell which was using P25 TiO$_2$ and 14% higher than the commercial purchased 200 nm sized TiO$_2$ particles. W. Peng and L. Han [62] prepared hexagonal TiO$_2$ microplates (Figure 4.4f) by the hydrothermal method in a water-ethanol solvent. The TiO$_2$ microplates had superior light scattering in DSSC due to unique planar structures. When the TiO$_2$ microplates were used as scattering over layer in DSSC, these microplates enhanced the light harvesting and increased the photocurrent and PCE. The microplate scattering layer shows 7.91% of PCE, which was 16.3% higher than the commercially purchased scattering layer.

4.6.4.9 TiO$_2$ Photonic Crystals

S. Nishimura et al. [22] coupled TiO$_2$ photonic crystal layer to conventional TNPs film, which enhanced the LHE in DSSC photoelectrodes. The TiO$_2$ photonic crystal act as a dielectric mirror for wavelengths corresponding to the stop band and act as a medium for enhancing the light absorption at the long wavelength side of the stop band, shown in Figure 4.4d. The short-circuit photocurrent efficiency in the visible spectrum (400–750 nm) increased by 26% compared to ordinary DSSC, as shown in Figure 4.4e.

L.I. Halaoui et al. [63] investigated the mechanism of LHE in DSSC by coupling the TiO$_2$ inverse opals (Figure 4.5e) or disordered scattering layers to the conventional TNPs film. The investigated results revealed that bilayer architecture was responsible for the bulk of the gain in IPCE rather than the enhanced light harvesting within the inverse opal structures. Several mechanisms were involved in increasing the backscattering during light interact with these structures such as Bragg diffraction in the periodic lattice and multiple scattering events at disordered regions in the photonic crystal or disordered films. These structures increased the LHE

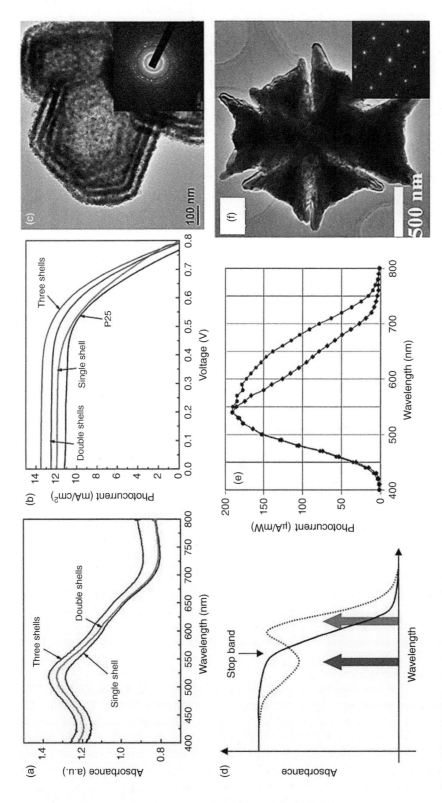

Figure 4.4 (a) UV–Vis absorbance spectra, (b) I–V curves, and (c) transmission electron microscopic (TEM) images of nanoplates with different shells used in DSSCs. (d) The dye absorbs strongly in the blue region but weakly in the red region (heavy line), the blue absorbance is reduced and the red absorbance is improved when the dye is confined to the high refractive index part of the photonic crystal (dotted line). (e) Wavelength-dependent of short-circuiting photocurrent in the bilayer electrode (upper curve) and the conventional nanocrystalline TiO₂ photoelectrode (lower curve). (f) TEM images of microplates [22]. *Source:* (c) Shao et al. 2011. [61] Reproduced with permission of Royal Society of Chemistry. (e) Nishimura et al. 2003 [22]. Reproduced with permission of American Chemical Society.

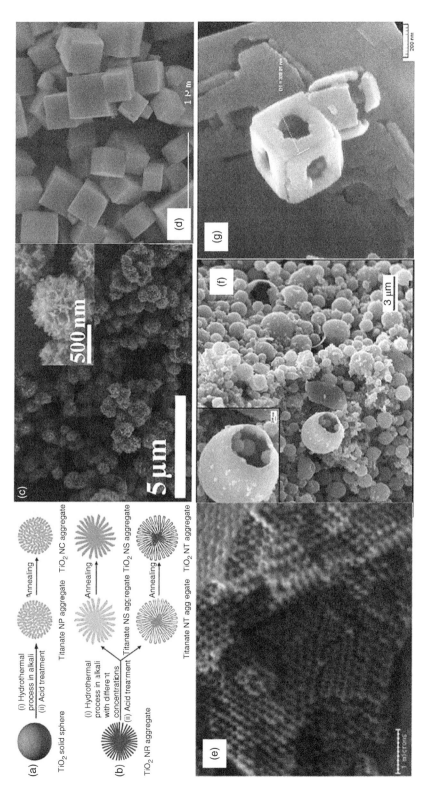

Figure 4.5 (a) The TiO$_2$ NC aggregates are came from TiO$_2$ solid sphere. (b) The TiO$_2$ NT and NS aggregates are synthesized using TiO$_2$ NR aggregates. (c) SEM images of HTFs after annealing. (d) SEM image of CeO$_2$ nanoparticles. (e) SEM image of TiO$_2$ opal structure. (f) SEM image of hollow microspheres. (g) TEM images of TiO$_2$ hollow cubes. *Source:* (a,b) Liu et al. 20' 3 [64]. Reproduced with permission of Royal Society of Chemistry. (c) Xu et al. 2019 [65]. Reproduced with permission of Elsevier. (d) Yu et al. 2012 [66]. Reproduced with permission of Royal Society of Chemistry. (e) Halaoui et al. 2005 [63] Reproduced with permission of American Chemical Society. (f) Sarvari and Mohammadi 2018 [67]. Reproduced with permission of Elsevier. (g) Dwivedi et al. 2013 [68]. Reproduced with permission of Elsevier.

in the red spectral region (600–750 nm) where the sensitizer was a poor absorber. A. Mihi et al. [69] studied the photocurrent of the spectral response of opal-based DSSC. The colloidal crystal with different configurations was introduced in DSSC for IPCE studies. The TiO_2 nanocrystalline electrode film was molded in the shape of an inverse opal that indicates the efficiency at the spectral region that was decreased when the photonic band gap opens up. However, the standard film of disordered titania nanocrystallites coupled with an inverse opal shows the mirror effect of the photonic crystal in the band gap frequencies that increased the LHE and IPCE of the DSSC. S. Guldin et al. [70] fabricated optically and electrically active 3D periodic TiO_2 Photonic Crystal (PC) on the high surface area mesoporous anatase TiO_2 for DSSC photoelectrode. From the earlier studies, this was opposite because the double-layer structure had the porosity at the mesoporous, microporous length scales and pore and electronic connectivity at all levels. This double-layer structure enables the effective dye sensitization, electrolyte infiltration, and charge collection at mesoporous and PC layers. The PC-induced resonances increased the light harvesting in the specific parts of the spectrum due to smooth layer interface and direct physical and electronic contact between the layers. This architecture may be suitable for a solid-state device where pore infiltration is the problem in poor absorbing organic photovoltaic devices.

4.6.4.10 Cubic CeO₂ Nanoparticles

H. Yu et al. [66] prepared cubic CeO_2 nanoparticles (shown in Figure 4.5d) with an average particle size of 400 nm by a hydrothermal method and the prepared nanoparticles were coated on to the mesoporous TNPs layer by the screen printing technique. The 1.5 μm thickness of CeO_2 top layer increased 17.8% of PCE compared to reference TiO_2 photoanodes due to the mirror-like light scattering, but the amount of dye loading was slightly reduced.

4.6.4.11 Spherical TiO₂ Aggregates

Z. Liu et al. prepared spherical TiO_2 aggregates from different building units such as nanorods (NRs), nanotubes, nanosheets, and nanocrystallites by the hydrothermal method, as shown in Figure 4.5a,b [64, 71]. From these different spherical TiO_2 aggregates, spherical TNTs aggregates exhibited higher PCE of 7.48%, which was higher than other spherical TiO_2 aggregates due to higher surface area, efficient light scattering, and better electron transport properties.

4.6.4.12 Hierarchical TiO₂ Submicroflowers

L. Xu et al. [65] fabricated photoanodes by mixing with hierarchical TiO_2 submicroflowers (HTFs) and anatase TNPs (Figure 4.5c). The HTFs provided a high surface area (98 m^2/g) and promising dimension (~660 nm) for increasing dye loading and light scattering to increase light harvesting capability in DSSCs. Due to high crystallinity and low defects or grain boundaries, the HTFs acted as efficient conductive pathways for the charge in photoanodes at DSSCs. The excess and very low incorporation of HTF after calcination provided poor connectivity in TiO_2 network and provided poor charge collection capability and cracks in the photoanodes. So the balance should be used to incorporate HTFs in TNPs. The 10 wt% HTFs embedded with photoanode provided 6.4% PCE which was 52.4% higher than the NP-based photoanodes.

4.6.4.13 SnO₂ Aggregates

P. Zhu et al. [72] prepared mesoporous SnO_2 aggregate particles with hierarchical structures and high surface areas by the facile molten salt method. The SnO_2 aggregates-based electrodes provide higher PCE (3.05% and 6.23% [$TiCl_4$ treated]) due to high dye-loading and efficient light scattering, which was confirmed by IPCE measurement and UV–Vis diffuse reflectance spectroscopy.

4.6.4.14 ZnO Nanoflowers

X. Chen et al. [73] prepared ZnO nanoflower-like structure by a hydrothermal method and fabricated photoanode, using this ZnO NF showed increase in specific surface area and light absorption via light scattering. This photoanode showed 5.96% PCE than normal ZnO NP (4.39%) in DSSC.

4.6.5 Core–Shell Nanoparticles for Light Scattering in DSSCs

S. Son et al. [74] studied the effect of size and the refractive index of SiO_2/TiO_2 core/shell nanoparticles (STCS-NPs) with different diameters include 110, 240, and 530 nm prepared by seed regrowth method for monodisperse SiO_2 nanoparticles and sol–gel method for TiO_2 coating over SiO_2. The prepared core/shell particles were compared with the SiO_2-NPs and TiO_2-NPs. The prepared core/shell particles provide efficient light scattering, which was having a low refractive index at the core and high refractive index at the shell. The core/shell particles with 240 nm provide higher PCE (7.9%) compared to other sized core/shell particles and SiO_2-NPs and TiO_2-NPs.

4.6.6 Double-Layer Photoanode

W. Zhang et al. [75] had prepared mesoporous TiO_2 microspheres using one plot hydrothermal method and coated light scattering layer on the P25 TNP layer to form double-layer thin-film photoanode in DSSC. This double-layer photoanode show 5.61% PCE compared to P25 TiO_2 film (2.67%) and TiO_2 microspheres film (2.14%), respectively, in DSSC. Similarly, several researchers used TiO_2 [76, 77] and zirconia (ZrO_2) [78] scattering layers deposited over on the TNP film to increase light scattering and PCE in DSSC. S. Hejazi and J.A. Mohandesi [79] fabricated the photoanode using a nanodiamond as a scattering layer at different thickness coated on the TiO_2 layer for DSSC. The thickness was controlled by spin coating because of the thickness of the layers affecting the efficiency of DSSC. The nanodiamond particles-based DSSC had 6.16% of PCE using 2000 rpm at spin coating for coating of nanodiamond, which was 15% higher than the nonscattering layer DSSC. Similarly, Zn-doped TiO_2 hollow fibers (HFs) were used as scattering layers prepared on the TNP by a co-axial electrospinning method and used as photoanode in DSSC. This photoanode-based DSSC show 3.122% PCE compared to single TiO_2 layer photoanode-based DSSC (1.293%) [80].

N. Sarvari and M.R. Mohammadi [67] prepared two kinds of photoanodes: one was a single-layer photoanode composed of TNPs and TiO_2 hollow cubes (HCs) (shown in Figure 4.5g), another one was double-layer photoanode that consists of TNP active layer and scattering layer made up of mixture TNPs and TiO_2 HC, respectively. In these photoanodes, the double-layer photoanode increased the light scattering and also balanced the electron transfer and dye sensitization. The double-layer based DSSC shows 9.39% PCE compared to pure TNPs (7.0%)-based DSSC. C. Dwivedi et al. [68] prepared THS (shown in Figure 4.5f) using continuous spray pyrolysis (CoSP) reactor, and it was used as a scattered layer deposited on the TNP by the screen printing technique. The TiO_2 HS layer-based DSSC showed 7.46% PCE which was higher than the single layer of TNP (7.1%)-based DSSC. Similarly, R.K. Chava et al. [81] prepared a photoanode using Hollow TiO_2 (H-TiO_2) NPs light scattering layer to increase the light scattering in DSSC. R. Gao et al. [82] prepared ZnO nanorods (ZnO NR) using the coprecipitation method. The prepared ZnO NR is deposited on top of the ZnO NPs to made double-layer photoanode in DSSC. This double-layer photoanode show 35% improvement in PCE compared to a single layer of ZnO NP DSSC. The scattering layer not only increased the light scattering, but it also decreased the charge recombination in DSSC, which was demonstrated by the results of increasing recombination resistance and decreasing dark current. M.J. Yun et al. [83] used 500 nm TNPs as a scattering layer to fabricate 3D micro-patterned photoanode for high-energy conversion efficiency (12.7%) and enhance the diffusivity and modification of photon distribution due to scattering layer in DSSC.

4.6.6.1 TiO$_2$ Aggregates

K. Al-Attafi et al. [42] prepared hierarchically structured TiO$_2$ aggregates as a scattering layer that consists of TNP by a one-step solvothermal method with the mixed solvent system (acetic acid and ethanol). This scattering-layer based DSSC show 9.1% of PCE than conventionally used scattering layer (7.4%).

4.6.6.2 Morphology-Controlled 1D–3D Bilayer TiO$_2$ Nanostructures

Morphology-controlled 1D–3D bilayer TiO$_2$ nanostructures had been synthesized by one step hydrothermal method [84]. From 1D–3D bilayer, 1D TiO$_2$ nanowires are used as a bottom layer while 3D TiO$_2$ dendritic microsphere are used as the top layer in photoelectrode of DSSC. The 1D nanowire layer provides high photoinjected collection efficiency, long electron diffusion length, and fast electron transport, while 3D TiO$_2$ dendritic microsphere provides the high surface to volume area, efficient light scattering effect, and high dye loading. The prepared bilayer possesses higher surface to volume ratio, quick electron transport, and efficient light scattering effect to increase the PCE in DSSC. The 1D–3D bilayer TiO$_2$ provides higher energy conversion efficiency of 7.2% after TiCl$_4$ treatment compared to 4.7% of commercial P25 TNP photoanode.

4.6.6.3 Quintuple-Shelled SnO$_2$ Hollow Microspheres

Z. Dong et al. [85] prepared quintuple-shelled SnO$_2$ Hollow Microspheres by the facile hard template method and used as an overlayer on the P25 film in the photoelectrodes. This bilayer (5S-SnO$_2$-HMSs-closed exterior double shells (CDS)) provides excellent PCE of 9.53%, which was 30% higher than the conventional P25 film in DSSC due to excellent light scattering.

4.6.6.4 Carbon-Based Materials for Light Scattering

B. Tang et al. [86] fabricated graphene ([reduced graphene oxide (RGO) nanosheets-reduced graphene oxide nanosheets] and [3DGNs – three-dimensional graphene networks])-assisted DSSC to increase the electron transportability and light scattering performance. The RGO-3DGNs-TiO$_2$ layer was coated on the RGO-TiO$_2$ layer, and it was used as photoanode in DSSC. In this photoanode, the RGO not only acted as the linker for electron transport between TNPs and graphene but also provided the better light scattering ability to improve 7.68% PCE in DSSC as compared to 3DGNs only based or RGO nanosheets only based DSSCs.

4.6.6.5 3D N-Doped TiO$_2$ Microspheres Used as Scattering Layers

Z. Cui et al. [87] prepared 3D N-doped TiO$_2$ microsphere through a simple step solvothermal method, and it was coated on the TNP with different ratios by the screen printing technique. This photoanode showed 8.08% PCE than a typical P25 NP (6.52%) due to increase in light scattering and photocurrent by the presence of nitrogen in TiO$_2$ microsphere.

4.6.6.6 ZnO Hollow Spheres and Urchin-like TiO$_2$ Microspheres

P. Zhao et al. [88] fabricated a photoanode using urchin-like TiO$_2$ microspheres (UTSs) as scattering layer and coated on the ZnO hollow spheres (ZHSs) to improve PCE of 8.67% due to strong light scattering in DSSC.

4.6.6.7 SnO$_2$ as Light-Scattering Layer

M. Batmunkh et al. [89] fabricated a bilayer photoanode using morphologically controlled SnO$_2$ layer coated on top of the nanocrystalline TiO$_2$ layer. By optimizing both layer thicknesses, the bilayer photoanode showed 7.8% PCE compared to nanocrystalline TiO$_2$ photoanode (5.6%).

4.6.7 Three-Layer Photoanode

B. Tang et al. [90] fabricated a photoanode with a three-layer structure and studied each layer role in DSSC. One layer acted as the transport layer (graphene–TiO$_2$), the second layer acted as the work layer (RGO and 3DGNs co-modified TiO$_2$), and the third layer acted as the scattering layer (RGO–TiO$_2$) in photoanodes. The scattering layer not only increased the light scattering but also increased the surface area for dye loading. This three-layer photoanode show 11.80% PCE compared to double-layer and single-layer photoanodes in DSSCs. S. Shital et al. [91] fabricated the photoanode with three layers, wherein the scattering layer (blend of NS TiO$_2$) was coated inbetween ~3.0-µm thick TiO$_2$ NS layer and 3.0 µm NS TiO$_2$ layers using the screen printing technique. The 40% blend in the scattering layer show 8.54% and 5.26% of PCE in the front side and counter side of the DSSC, respectively, which was higher than the standard double-layer film. M. Dissanayake et al. [92] fabricated tri-layer TiO$_2$ photoanode that consists of rice-grain shaped TiO$_2$ nanofibers as scattering layer coated in between two TNP layers. The TiCL$_4$ treated tri-layer photoanode showed 7.32% PCE compared to nontreated tri-layer (6.90%) and conventional photoanodes in DSSC due to enhanced surface area by filling the gaps between TNP film and TiO$_2$ nanostructures and scattering of light. A similar kind of tri-layer photoanode was prepared, here TiO$_2$ NF acted as the scattering layer and enhanced the light scattering and PCE in DSSC [93]. Similarly, several researchers prepared fabricated three-layer photoanode using TiO$_2$ with different structures to increase light scattering and PCE in DSSCs [94–97].

K. Susmitha et al. [98] fabricated three kinds of photoelectrodes where the scattering layer (200 nm TiO$_2$) was coated inbetween bottom active layer and top active layer in the three-layer photoanodes using the screen printing method (Figure 4.6a). This photoanode showed 9.61% PCE compared to standard scattering layer photoanode (8.72%) due to enhanced light scattering by increasing optical path length of incident light. Similarly, several researchers prepared multilayer photoanodes, wherein the scattering layer was used to increase the LHE and PCE in DSSC [99, 100].

4.6.8 Four-Layer Photoanode

A. Bakhshayesh et al. [101] fabricated a multilayer photoanode that consists of two thick layers (6 µm) of nanocrystalline TiO$_2$ particles and two thin layers (1 µm) of TiO$_2$ aggregates, which were alternatively deposited on each other, as shown in Figure 4.6b. The aggregate TiO$_2$ layers were deposited by the direct sol-gel method and prepared sponge-like and uniform scattering layer for DSSC applications. The aggregate layers prepared the combination of anatase and rutile phase while the nanocrystalline TiO$_2$ layers had anatase phase only. The multilayer photoanode shows 7.69% PCE through the results of LHE and less recombination due to the increase in light scattering at thin film scattering layers in DSSC.

4.6.9 Surface Plasmon Effect in DSSC

J. Qi et al. [103] prepared Ag@TNPs (core–shell) for increasing dye absorbance and decreased the thickness of photoelectrode to improve the electron collection and performance of the

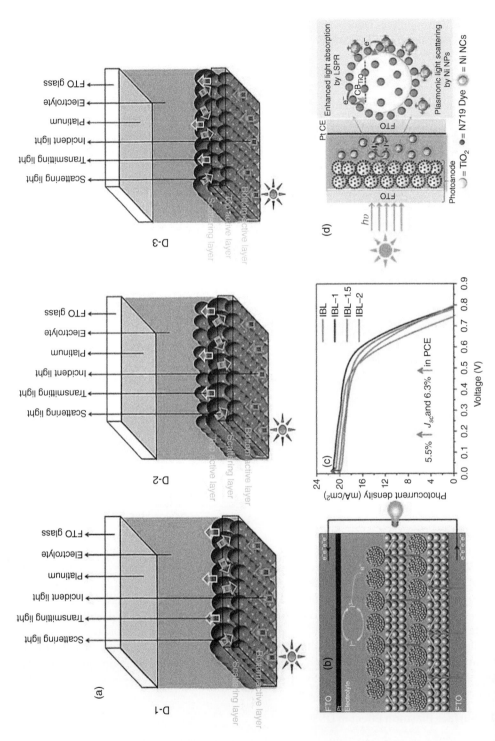

Figure 4.6 (a) Three kinds of photoelectrode prepared by K. Susmitha et al. (b) DSSC based on TiO_2 multilayered film (i.e. 4L electrode), containing two thin layers (i.e. 1 μm) of spherical aggregates as light scattering cites and two thick layers (i.e. 6 μm) of nanoparticles as light absorbers. (c) J–V curves of TiO_2 interconnected beads-like (IBL) structure. (d) Probable electron transfer mechanism via surface plasmon resonance. *Source:* (a) Susmitha et al. 2017 [98]. Reproduced with permission of Royal Society of Chemistry. (b) Bakhshayesh et al. 2015 [101]. Reproduced with permission of Springer Nature. (d) Krishnapriya et al. 2017 [102]. Reproduced with permission of Royal Society of Chemistry.

DSSC with the help of localized surface plasmons (LSPs) from Ag NPs. The dye molecule absorption increased the dye molecular dipole and the electric field from the surroundings of NPs. In addition to this, it also increased the light scattering due to LSPs to increase the optical path in DSSC. The Ag thin shell keeps the photoelectron from the recombination on the surface of shell NPs with dye and electrolyte and leads to increasing the stability of metal NPs. A small amount of Ag@TNPs (0.1 wt%) increased the efficiency from 7.8% to 9.0% because of the decreasing thickness of the photoanode by 25%. Similarly, several researchers prepared Ag-decorated ZnO nanoparticles [104] and Ag_2S-encapsulated Au nanorods (AuNRs@Ag$_2$S) [105] to increase the optical path length with the help of light scattering in surface plasmon effect-associated Ag NPs and longitudinal plasmon resonance occurs in AuNRs, respectively. J. Chen et al. [106] fabricated a series of Au NPs inlaid into TNT photoanodes by photoreduction process. The proper size distribution was increased in the PCE up to 4.63% compared to TNT PCE (3.89%) due to plasmonic enhancement by results of surface plasmon resonance (SPR) and scattering effects. R. Krishnapriya et al. [102] incorporated Ni NC into the electrolyte to increase light scattering in DSSC. In the different morphologies of TiO_2, the interconnected beads-like (IBL) structure showed good PCE (10.02%) due to efficient light scattering and light absorption because LSP was introduced on the surface of TiO_2 film due to Ni NC in DSSC, as shown in Figure 4.6c,d.

4.7 Conclusion

The different structure of light scattering materials was used as the increasing scattering of light. Generally, light scattering was improved in many ways such as mixing of large particles into TiO_2 nanocrystalline slurry, dual functional materials and light scattering layers that were used in the double layer, tri-layer, and multi-layer photoelectrodes in DSSCs. Light scattering was also improved by using core-shell particles, SPR effects, and composite materials to increase PCE in DSSC. Theoretical studies revealed that increases in thickness of the film would increase the light absorption, but due to electron diffusion length, the photoanode thickness is limited. Dual functional materials are used to scatter the light as well as providing high surface area for efficient dye absorption. The dual scattering layer was used to reduce the probability of outcoming or escaped light from the device. So the studies conclude that light scattering is a very important function in DSSC to improve the light harvesting and PCE in DSSC.

References

1 Wu, W.-Q., Liao, J.-F., and Kuang, D.-B. (2018). Layered-stacking of titania films for solar energy conversion: toward tailored optical, electronic and photovoltaic performance. *J. Energy Chem.* 27 (3): 690–702.

2 Ramachandran, A., Sreekala, C.O., Sreelatha, K.S., and Jinchu, I. (2018). Photovoltaic studies of dye sensitized solar cells fabricated from microwave exposed photo anodes. *IOP Conf. Ser. Mater. Sci. Eng.* 310 (1): 012151.

3 Peedikakkandy, L., Naduvath, J., Mallick, S., and Bhargava, P. (2018). Lead free, air stable perovskite derivative Cs2SnI6 as HTM in DSSCs employing TiO_2 nanotubes as photoanode. *Mater. Res. Bull.* 108: 113–119.

4 Sulaeman, U. and Zuhairi Abdullah, A. (2017). The way forward for the modification of dye-sensitized solar cell towards better power conversion efficiency. *Renewable Sustainable Energy Rev.* 74: 438–452.

5 Shakeel Ahmad, M., Pandey, A.K., and Abd Rahim, N. (2017). Advancements in the development of TiO$_2$ photoanodes and its fabrication methods for dye sensitized solar cell (DSSC) applications. A review. *Renewable Sustainable Energy Rev.* 77: 89–108.

6 Liu, G., Wang, M., Wang, H. et al. (2018). Hierarchically structured photoanode with enhanced charge collection and light harvesting abilities for fiber-shaped dye-sensitized solar cells. *Nano Energy* 49: 95–102.

7 Idígoras, J., Sobuś, J., Jancelewicz, M. et al. (2016). Effect of different photoanode nanostructures on the initial charge separation and electron injection process in dye sensitized solar cells: a photophysical study with indoline dyes. *Mater. Chem. Phys.* 170: 218–228.

8 Usami, A. (1997). Theoretical study of application of multiple scattering of light to a dye-sensitized nanocrystalline photoelectrichemical cell. *Chem. Phys. Lett.* 277 (1–3): 105–108.

9 Bhagwat, S., Dani, R., Goswami, P., and Kerawalla, M.A.K. (2017). Recent advances in optimization of photoanodes and counter electrodes of dye-sensitized solar cells. *Curr. Sci.* 113 (02): 228.

10 Zhang, Q., Myers, D., Lan, J. et al. (2012). Applications of light scattering in dye-sensitized solar cells. *Phys. Chem. Chem. Phys.* 14 (43): 14982.

11 Deepak, T.G., Anjusree, G.S., Thomas, S. et al. (2014). A review on materials for light scattering in dye-sensitized solar cells. *RSC Adv.* 4 (34): 17615–17638.

12 Ferber, J. and Luther, J. (1998). Computer simulations of light scattering and absorption in dye-sensitized solar cells. *Sol. Energy Mater. Sol. Cells* 54 (1–4): 265–275.

13 Rothenberger, G., Comte, P., and Grätzel, M. (1999). A contribution to the optical design of dye-sensitized nanocrystalline solar cells. *Sol. Energy Mater. Sol. Cells* 58 (3): 321–336.

14 Ferber, J. (1998). An electrical model of the dye-sensitized solar cell. *Sol. Energy Mater. Sol. Cells* 53 (1–2): 29–54.

15 Hore, S., Vetter, C., Kern, R. et al. (2006). Influence of scattering layers on efficiency of dye-sensitized solar cells. *Sol. Energy Mater. Sol. Cells* 90 (9): 1176–1188.

16 Vargas, W.E. and Niklasson, G.A. (2001). Optical properties of nano-structured dye-sensitized solar cells. *Sol. Energy Mater. Sol. Cells* 69 (2): 147–163.

17 Tachibana, Y., Hara, K., Sayama, K., and Arakawa, H. (2002). Quantitative analysis of light-harvesting efficiency and electron-transfer yield in ruthenium-dye-sensitized nanocrystalline TiO$_2$ solar cells. *Chem. Mater.* 14 (6): 2527–2535.

18 Zhao, J., Yang, Y., Cui, C. et al. (2016). TiO$_2$ hollow spheres as light scattering centers in TiO$_2$ photoanodes for dye-sensitized solar cells: the effect of sphere diameter. *J. Alloys Compd.* 663: 211–216.

19 Hore, S., Nitz, P., Vetter, C. et al. (2005). Scattering spherical voids in nanocrystalline TiO$_2$? Enhancement of efficiency in dye-sensitized solar cells. *Chem. Commun.* (15): 2011.

20 Lin, J., Zheng, L., Liu, X. et al. (2015). Assembly of a high-scattering photoelectrode using a hybrid nano-TiO$_2$ paste. *J. Mater. Chem. C* 3 (26): 6645–6651.

21 Huisman, C.L., Schoonman, J., and Goossens, A. (2005). The application of inverse titania opals in nanostructured solar cells. *Sol. Energy Mater. Sol. Cells* 85 (1): 115–124.

22 Nishimura, S., Abrams, N., Lewis, B.A. et al. (2003). Standing wave enhancement of red absorbance and photocurrent in dye-sensitized titanium dioxide photoelectrodes coupled to photonic crystals. *J. Am. Chem. Soc.* 125 (20): 6306–6310.

23 Ishimaru, A. (1991). Wave propagation and scattering in random media and rough surfaces. *Proc. IEEE* 79 (10): 1359–1366.

24 Sun, P., Zhang, X., Wang, L. et al. (2015). Bilayer TiO$_2$ photoanode consisting of a nanowire–nanoparticle bottom layer and a spherical voids scattering layer for dye-sensitized solar cells. *New J. Chem.* 39 (6): 4845–4851.

25 Hieu, H.N., Dao, V.D., Vuong, N.M. et al. (2014). Enhancement of dye-sensitized solar cell efficiency by spherical voids in nanocrystalline ZnO electrodes. *Korean J. Mater. Res.* 24 (9): 458–464.

26 Yang, G., Zhang, J., Wang, P. et al. (2011). Light scattering enhanced photoanodes for dye-sensitized solar cells prepared by carbon spheres/TiO$_2$ nanoparticle composites. *Curr. Appl. Phys.* 11 (3): 376–381.

27 Hejazi, S.M.H., Aghazadeh Mohandesi, J., and Javanbakht, M. (2017). The effect of functionally graded porous nano structure TiO$_2$ photoanode on efficiency of dye sensitized solar cells. *Sol. Energy* 144: 699–706.

28 Pham, T.T.T., Mathews, N., Lam, Y.-M., and Mhaisalkar, S. (2018). Influence of size and shape of sub-micrometer light scattering centers in ZnO-assisted TiO$_2$ photoanode for dye-sensitized solar cells. *Physica B* 532: 225–229.

29 Trang Pham, T.T., Bessho, T., Mathews, N. et al. (2012). Light scattering enhancement from sub-micrometer cavities in the photoanode for dye-sensitized solar cells. *J. Mater. Chem.* 22 (32): 16201.

30 Kim, C., Choi, H., Kim, J.I. et al. (2014). Improving scattering layer through mixture of nanoporous spheres and nanoparticles in ZnO-based dye-sensitized solar cells. *Nanoscale Res. Lett.* 9 (1): 1–6.

31 Tan, B. and Wu, Y. (2006). Dye-sensitized solar cells based on anatase TiO$_2$ nanoparticle/nanowire composites. *J. Phys. Chem. B* 110 (32): 15932–15938.

32 Baxter, J.B. and Aydil, E.S. (2006). Dye-sensitized solar cells based on semiconductor morphologies with ZnO nanowires. *Sol. Energy Mater. Sol. Cells* 90 (5): 607–622.

33 Joshi, P., Zhang, L., Davoux, D. et al. (2010). Composite of TiO$_2$ nanofibers and nanoparticles for dye-sensitized solar cells with significantly improved efficiency. *Energy Environ. Sci.* 3 (10): 1507.

34 Dzenis, Y. (2004). Material science: Spinning continuous fibers for nanotechnology. *Science* 304 (5679): 1917–1919.

35 Yang, M., Dong, B., Yang, X. et al. (2017). TiO$_2$ nanoparticle/nanofiber–ZnO photoanode for the enhancement of the efficiency of dye-sensitized solar cells. *RSC Adv.* 7 (66): 41738–41744.

36 Banik, A., Ansari, M.S., Alam, S., and Qureshi, M. (2018). Thermodynamic barrier and light scattering effects of nanocube assembled SrTiO$_3$ in enhancing the photovoltaic properties of zinc oxide based dye sensitized solar cells. *J. Phys. Chem. C* 122 (29): 16550–16560.

37 Banik, A., Ansari, M.S., Sahu, T.K., and Qureshi, M. (2016). Understanding the role of silica nanospheres with their light scattering and energy barrier properties in enhancing the photovoltaic performance of ZnO based solar cells. *Phys. Chem. Chem. Phys.* 18 (40): 27818–27828.

38 Wang, D., Liu, S., Shao, M. et al. (2018). Design of SnO$_2$ aggregate/nanosheet composite structures based on function-matching strategy for enhanced dye-sensitized solar cell performance. *Materials* 11 (9): 1774.

39 Irannejad, N., Rezaei, B., Ensafi, A.A., and Zandi-Atashbar, N. (2018). Photovoltaic performance analysis of dye-sensitized solar cell based on the Ag(4,4'-dicyanamidobiphenyl) complex as a light-scattering layer agent and linker molecule on TiO$_2$ photoanode. *IEEE J. Photovoltaics* 8 (5): 1230–1236.

40 Wang, Z.S., Kawauchi, H., Kashima, T., and Arakawa, H. (2004). Significant influence of TiO$_2$ photoelectrode morphology on the energy conversion efficiency of N719 dye-sensitized solar cell. *Coord. Chem. Rev.* 248 (13–14): 1381–1389.

41 Choi, S.C., Lee, H.S., Oh, S.J., and Sohn, S.H. (2012). Light scattering TiO_2 particles surface-modified by Al_2O_3 coating in a dye-sensitized solar cell. *Phys. Scr.* 85 (2): 025801.

42 Al-Attafi, K., Nattestad, A., Yamauchi, Y. et al. (2017). Aggregated mesoporous nanoparticles for high surface area light scattering layer TiO_2 photoanodes in dye-sensitized solar cells. *Sci. Rep.* 7 (1): 10341.

43 Chen, D., Huang, F., Cheng, Y.B., and Caruso, R.A. (2009). Mesoporous anatase TiO_2 beads with high surface areas and controllable pore sizes: a superior candidate for high-performance dye-sensitized solar cells. *Adv. Mater.* 21 (21): 2206–2210.

44 Sauvage, F., Chen, D., Comte, P. et al. (2010). Dye-sensitized solar cells employing a achieve power conversion efficiencies single film of mesoporous TiO_2 beads over 10%. *ACS Nano* 4 (8): 4420–4425.

45 Kim, Y.J., Lee, M.H., Kim, H.J. et al. (2009). Formation of highly efficient dye-sensitized solar cells by hierarchical pore generation with nanoporous TiO_2 spheres. *Adv. Mater.* 21 (36): 3668–3673.

46 Nedelcu, M., Guldin, S., Orilall, M.C. et al. (2010). Monolithic route to efficient dye-sensitized solar cells employing diblock copolymers for mesoporous TiO_2. *J. Mater. Chem.* 20 (7): 1261–1268.

47 Peining, Z., Yongzhi, W., Reddy, M.V. et al. (2012). TiO_2 nanoparticles synthesized by the molten salt method as a dual functional material for dye-sensitized solar cells. *RSC Adv.* 2: 5123–5126.

48 Zhao, L., Li, J., Shi, Y. et al. (2013). Double light-scattering layer film based on TiO_2 hollow spheres and TiO_2 nanosheets: improved efficiency in dye-sensitized solar cells. *J. Alloys Compd.* 575: 168–173.

49 Lee, J.-K., Jeong, B.-H., Jang, S. et al. (2009). Preparations of TiO_2 pastes and its application to light-scattering layer for dye-sensitized solar cells. *J. Ind. Eng. Chem.* 15 (5): 724–729.

50 Yang, L., Lin, Y., Jia, J. et al. (2008). Light harvesting enhancement for dye-sensitized solar cells by novel anode containing cauliflower-like TiO_2 spheres. *J. Power Sources* 182 (1): 370–376.

51 Nakayama, K., Kubo, T., and Nishikitani, Y. (2008). Electrophoretically deposited TiO_2 nanotube light-scattering layers of dye-sensitized solar cells. *Jpn. J. Appl. Phys.* 47 (8 PART 1): 6610–6614.

52 Bakhshayesh, A.M., Mohammadi, M.R., Dadar, H., and Fray, D.J. (2013). Improved efficiency of dye-sensitized solar cells aided by corn-like TiO_2 nanowires as the light scattering layer. *Electrochim. Acta* 90: 302–308.

53 Qiu, Y., Chen, W., and Yang, S. (2010). Double-layered photoanodes from variable-size anatase TiO_2 nanospindles: a candidate for high-efficiency dye-sensitized solar cells. *Angew. Chem. Int. Ed.* 49 (21): 3675–3679.

54 Chuangchote, S., Sagawa, T., and Yoshikawa, S. (2008). Efficient dye-sensitized solar cells using electrospun TiO_2 nanofibers as a light harvesting layer. *Appl. Phys. Lett.* 93 (3): 5–8.

55 Navarro-Pardo, F., Benetti, D., Benavides, J. et al. (2017). Nanofiber-structured TiO_2 nanocrystals as a scattering layer in dye-sensitized solar cells. *ECS J. Solid State Sci. Technol.* 6 (4): N32–N37.

56 Zhu, P., Nair, A.S., Yang, S. et al. (2011). Which is a superior material for scattering layer in dye-sensitized solar cells – electrospun rice grain- or nanofiber-shaped TiO_2? *J. Mater. Chem.* 21 (33): 12210–12212.

57 Nair, A.S., Shengyuan, Y., Peining, Z., and Ramakrishna, S. (2010). Rice grain-shaped TiO_2 mesostructures by electrospinning for dye-sensitized solar cells. *Chem. Commun.* 46 (39): 7421–7423.

58 Shengyuan, Y., Peining, Z., Nair, A.S., and Ramakrishna, S. (2011). Rice grain-shaped TiO$_2$ mesostructures – synthesis, characterization and applications in dye-sensitized solar cells and photocatalysis. *J. Mater. Chem.* 21 (18): 6541–6548.

59 Zhu, G., Pan, L., Yang, J. et al. (2012). Electrospun nest-shaped TiO$_2$ structures as a scattering layer for dye sensitized solar cells. *J. Mater. Chem.* 22 (46): 24326–24329.

60 Koo, H.J., Kim, Y.J., Lee, Y.H. et al. (2008). Nano-embossed hollow spherical TiO$_2$ as bifunctional material for high-efficiency dye-sensitized solar cells. *Adv. Mater.* 20 (1): 195–199.

61 Shao, W., Gu, F., Gai, L., and Li, C. (2011). Planar scattering from hierarchical anatase TiO$_2$ nanoplates with variable shells to improve light harvesting in dye-sensitized solar cells. *Chem. Commun.* 47 (17): 5046.

62 Peng, W. and Han, L. (2012). Hexagonal TiO$_2$ microplates with superior light scattering for dye-sensitized solar cells. *J. Mater. Chem.* 22 (38): 20773.

63 Halaoui, L.I., Abrams, N.M., and Mallouk, T.E. (2005). Increasing the conversion efficiency of dye-sensitized TiO$_2$ photoelectrochemical cells by coupling to photonic crystals. *J. Phys. Chem. B* 109 (13): 6334–6342.

64 Liu, Z., Su, X., Hou, G. et al. (2013). Spherical TiO$_2$ aggregates with different building units for dye-sensitized solar cells. *Nanoscale* 5 (17): 8177.

65 Xu, L., Xu, J., Hu, H. et al. (2019). Hierarchical submicroflowers assembled from ultrathin anatase TiO$_2$ nanosheets as light scattering centers in TiO$_2$ photoanodes for dye-sensitized solar cells. *J. Alloys Compd.* 776: 1002–1008.

66 Yu, H., Bai, Y., Zong, X. et al. (2012). Cubic CeO$_2$ nanoparticles as mirror-like scattering layers for efficient light harvesting in dye-sensitized solar cells. *Chem. Commun.* 48 (59): 7386.

67 Sarvari, N. and Mohammadi, M.R. (2018). Influence of photoanode architecture on light scattering mechanism and device performance of dye-sensitized solar cells using TiO$_2$ hollow cubes and nanoparticles. *J. Taiwan Inst. Chem. Eng.* 86: 81–91.

68 Dwivedi, C., Dutta, V., Chandiran, A.K. et al. (2013). Anatase TiO$_2$ hollow microspheres fabricated by continuous spray pyrolysis as a scattering layer in dye-sensitised solar cells. *Energy Procedia* 33: 223–227.

69 Mihi, A., Calvo, M.E., Anta, J.A., and Mı́, H. (2008). Spectral response of opal-based dye-sensitized solar cells. *J. Phys. Chem. C* 112 (1): 13–17.

70 Guldin, S., Hüttner, S., Kolle, M. et al. (2010). Dye-sensitized solar cell based on a three-dimensional photonic crystal. *Nano Lett.* 10 (7): 2303–2309.

71 Liu, Z.H., Su, X.J., Hou, G.L. et al. (2012). Enhanced performance for dye-sensitized solar cells based on spherical TiO$_2$ nanorod-aggregate light-scattering layer. *J. Power Sources* 218: 280–285.

72 Zhu, P., Reddy, M.V., Wu, Y. et al. (2012). Mesoporous SnO$_2$ agglomerates with hierarchical structures as an efficient dual-functional material for dye-sensitized solar cells. *Chem. Commun.* 48 (88): 10865–10867.

73 Chen, X., Tang, Y., and Liu, W. (2017). Efficient dye-sensitized solar cells based on nanoflower-like ZnO photoelectrode. *Molecules* 22 (8): 1284.

74 Son, S., Hwang, S.H., Kim, C. et al. (2013). Designed synthesis of SiO$_2$/TiO$_2$ core/shell structure As light scattering material for highly efficient dye-sensitized solar cells. *ACS Appl. Mater. Interfaces* 5 (11): 4815–4820.

75 Zhang, W., Gu, J., Yao, S., and Wang, H. (2018). The synthesis and application of TiO$_2$ microspheres as scattering layer in dye-sensitized solar cells. *J. Mater. Sci. Mater. Electron.* 29 (9): 7356–7363.

76 Yang, Y., Cui, L., Wang, B. et al. (2018). Enhanced photoelectrochemical performances in flexible mesoscopic solar cells: an effective light-scattering material. *ChemPhotoChem* 2 (11): 986–993.

77 Swathy, K.S., Abraham, P.A., Panicker, N.R. et al. (2016). Nanostructured anatase titania spheres as light scattering layer in dye-sensitized solar cells. *Procedia Technol.* 24: 767–773.

78 Nursam, N.M., Hidayat, J., Shobih et al. (2018). A comparative study between titania and zirconia as material for scattering layer in dye-sensitized solar cells. *J. Phys. Conf. Ser.* 1011 (1): 012003.

79 Hejazi, S.M.H. and Mohandesi, J.A. (2018). Using nanodiamond particles in photoanode of dye-sensitised solar cell. *Micro Nano Lett.* 13 (2): 154–156.

80 Arifin, Z., Suyitno, S., Hadi, S., and Sutanto, B. (2018). Improved performance of dye-sensitized solar cells with TiO$_2$ nanoparticles/Zn-doped TiO$_2$ hollow fiber photoanodes. *Energies* 11 (11): 2922.

81 Chava, R.K., Lee, W.-M., Oh, S.-Y. et al. (2017). Improvement in light harvesting and device performance of dye sensitized solar cells using electrophoretic deposited hollow TiO$_2$ NPs scattering layer. *Sol. Energy Mater. Sol. Cells* 161: 255–262.

82 Gao, R., Liang, Z., Tian, J. et al. (2013). A ZnO nanorod layer with a superior light-scattering effect for dye-sensitized solar cells. *RSC Adv.* 3 (40): 18537.

83 Yun, M.J., Sim, Y.H., Cha, S.I. et al. (2017). High energy conversion efficiency with 3-D micro-patterned photoanode for enhancement diffusivity and modification of photon distribution in dye-sensitized solar cells. *Sci. Rep.* 7 (1): 15027.

84 Sun, Z., Kim, J.H., Zhao, Y. et al. (2013). Morphology-controllable 1D-3D nanostructured TiO$_2$ bilayer photoanodes for dye-sensitized solar cells. *Chem. Commun.* 49 (10): 966–968.

85 Dong, Z., Ren, H., Hessel, C.M. et al. (2014). Quintuple-shelled SnO$_2$ hollow microspheres with superior light scattering for high-performance dye-sensitized solar cells. *Adv. Mater.* 26 (6): 905–909.

86 Tang, B., Ji, G., Wang, Z. et al. (2017). Three-dimensional graphene networks and reduced graphene oxide nanosheets co-modified dye-sensitized solar cells. *RSC Adv.* 7 (72): 45280–45286.

87 Cui, Z., Zhang, K., Xing, G. et al. (2017). Multi-functional 3D N-doped TiO$_2$ microspheres used as scattering layers for dye-sensitized solar cells. *Front. Chem. Sci. Eng.* 11 (3): 395–404.

88 Zhao, P., Wang, L., Yu, Z. et al. (2016). Bilayered photoanode consisting of zinc oxide hollow spheres and urchin-like titanium dioxide microspheres enables fast electron transport and efficient light-harvesting for improved-performance dye-sensitized solar cells. *RSC Adv.* 6 (21): 17280–17287.

89 Batmunkh, M., Dadkhah, M., Shearer, C.J. et al. (2016). Tin oxide light-scattering layer for titania photoanodes in dye-sensitized solar cells. *Energy Technol.* 4 (8): 959–966.

90 Tang, B., Yu, H., Peng, H. et al. (2018). Graphene based photoanode for DSSCs with high performances. *RSC Adv.* 8 (51): 29220–29227.

91 Shital, S., Swami, S.K., Barnes, P., and Dutta, V. (2018). Monte Carlo simulation for optimization of a simple and efficient bifacial DSSC with a scattering layer in the middle. *Sol. Energy* 161: 64–73.

92 Dissanayake, M.A.K.L., Sarangika, H.N.M., Senadeera, G.K.R. et al. (2017). Application of a nanostructured, tri-layer TiO$_2$ photoanode for efficiency enhancement in quasi-solid electrolyte-based dye-sensitized solar cells. *J. Appl. Electrochem.* 47 (11): 1239–1249.

93 Dissanayake, M.A.K.L., Divarathna, H.K.D.W.M.N., Dissanayake, C.B. et al. (2016). An innovative TiO$_2$ nanoparticle/nanofibre/nanoparticle, three layer composite photoanode for efficiency enhancement in dye-sensitized solar cells. *J. Photochem. Photobiol. A* 322–323: 110–118.

94 Wang, Y., Liu, X., Li, Z. et al. (2017). Constructing synergetic trilayered TiO_2 photoanodes based on a flexible nanotube array/Ti substrate for efficient solar cells. *ChemNanoMat* 3 (1): 58–64.

95 Li, M., Li, M., Liu, X. et al. (2017). Performance optimization of dye-sensitized solar cells by multilayer gradient scattering architecture of TiO_2 microspheres. *Nanotechnology* 28 (3): 035201.

96 Kang, C., Zhang, Z., Zhang, Y. et al. (2011). Enhanced efficiency in dye-sensitised solar cells using a TiO_2-based sandwiched film as photoanode. *Micro Nano Lett.* 6 (8): 579.

97 Wang, W., Yuan, H., Xie, J. et al. (2018). Enhanced efficiency of large-area dye-sensitized solar cells by light-scattering effect using multilayer TiO_2 photoanodes. *Mater. Res. Bull.* 100 (2010): 434–439.

98 Susmitha, K., Kumar, M.N., Gurulakshmi, M. et al. (2017). Novel photoanode architecture for optimal dye-sensitized solar cell performance and its small cell module study. *Sustainable Energy Fuels* 1 (3): 439–443.

99 Anjidani, M., Milani Moghaddam, H., and Ojani, R. (2017). Binder-free MWCNT/TiO_2 multilayer nanocomposite as an efficient thin interfacial layer for photoanode of dye sensitized solar cell. *Mater. Sci. Semicond. Process.* 71: 20–28.

100 De Marco, L., Manca, M., Giannuzzi, R. et al. (2013). Shape-tailored TiO_2 nanocrystals with synergic peculiarities as building blocks for highly efficient multi-stack dye solar cells. *Energy Environ. Sci.* 6 (6): 1791.

101 Bakhshayesh, A.M., Azadfar, S.S., and Bakhshayesh, N. (2015). Multi-layered architecture of electrodes containing uniform TiO_2 aggregates layers for improving the light scattering efficiency of dye-sensitized solar cells. *J. Mater. Sci. Mater. Electron.* 26 (12): 9808–9816.

102 Krishnapriya, R., Praneetha, S., and Vadivel Murugan, A. (2017). Microwave-solvothermal synthesis of various TiO_2 nano-morphologies with enhanced efficiency by incorporating Ni nanoparticles in an electrolyte for dye-sensitized solar cells. *Inorg. Chem. Front.* 4 (10): 1665–1678.

103 Qi, J., Dang, X., Hammond, P.T., and Belcher, A.M. (2011). Highly efficient plasmon-enhanced dye-sensitized solar cells through metal@oxide core–shell nanostructure. *ACS Nano* 5 (9): 7108–7116.

104 Manikandan, V.S., Palai, A.K., Mohanty, S., and Nayak, S.K. (2018). Surface plasmonic effect of Ag enfold ZnO pyramid nanostructured photoanode for enhanced dye sensitized solar cell application. *Ceram. Int.* 44 (17): 21314–21322.

105 Chang, S., Li, Q., Xiao, X. et al. (2012). Enhancement of low energy sunlight harvesting in dye-sensitized solar cells using plasmonic gold nanorods. *Energy Environ. Sci.* 5 (11): 9444–9448.

106 Chen, J., Guo, M., Su, H. et al. (2018). Improving the efficiency of dye-sensitized solar cell via tuning the Au plasmons inlaid TiO_2 nanotube array photoanode. *J. Appl. Electrochem.* 48 (10): 1139–1149.

107 Chou, C.-S., Guo, M.-G., Liu, K.-H., & Chen, Y.-S. (2012). Preparation of TiO_2 particles and their applications in the light scattering layer of a dye-sensitized solar cell. *Applied Energy.* 92: 224–233.

5

Function of Compact (Blocking) Layer in Photoanode

Su Pei Lim[1,2]

[1] *Xiamen University Malaysia, School of Energy and Chemical Engineering, Jalan Sunsuria, Bandar Sunsuria, Sepang Selangor Darul Ehsan, Malaysia*
[2] *Xiamen University, College of Chemistry and Chemical Engineering, Xiamen, China*

5.1 Introduction

Other than the charge recombination, back reaction in dye-sensitized solar cell (DSSC) is also another major problem for the low efficiency [1, 2]. The commonly used porous TiO_2 layer, which is essential to serve the purpose of collecting and transporting photoelectrons, could leave a portion of the transparent conducting oxide (TCO) surface uncovered. This will cause the percolation of redox electrolyte on the TCO surface that permits direct electrochemical reduction of I_3^- ($I_3^- + 2e \rightarrow 3I^-$), leading to the consumption of the photogenerated electron. Hence, the introduction of a blocking layer is important to prevent electron leakage to increase the overall efficiency of the device [3]. The use of a blocking layer has been shown to increase the solar cell's efficiency [4].

The general principle of the compact layer in DSSC can be described as follows. The compact or dense layer will be deposited as a blocking layer between TCO and photoanode. The deposition of the compact blocking layer will cover the surface of the TCO and could minimize the electron leakage. Furthermore, the compact layer also helps in improving the connectivity among the porous nanoparticles. The porous particles are well interconnected among each other and electron could transfer through in a shorter distance. The efficient electron pathway not only reduces the resistance and recombination rate but also prolong the electron lifetime in DSSC.

Apart from titanium dioxide (TiO_2) and tin oxide (SnO_2), literature data on metal oxides such as copper(II) oxide (CuO), cadmium oxide (CdO), aluminium oxide (Al_2O_3), nickel oxide (NiO), silicon dioxide (SiO_2), zinc Oxide (ZnO), and other rare earth oxides use as compact layers are also well reported. There are various method that have been reported for the fabrication of compact layer, which include dip coating [5], spray pyrolysis [6], sol–gel [7], electrodeposition [8], sputtering [9], electrophoretic [4], spin coating [10], pulsed laser deposition [5], and aerosol-assisted chemical vapor deposition (AACVD) [11].

5.2 Titanium Dioxide (TiO_2) and Titanium (Ti)-Based Material as a Compact Layer

In 2009, Lee et al. [12] reported on mill prepared Nb-doped TiO_2 (NTO) and undoped dense TiO_2 (d-TiO_2) as a compact layer for DSSC after deposition on fluorine doped tin oxide (FTO)

Interfacial Engineering in Functional Materials for Dye-Sensitized Solar Cells, First Edition.
Edited by Alagarsamy Pandikumar, Kandasamy Jothivenkatachalam and Karuppanapillai B. Bhojanaa.

using pulsed laser deposition method. It was then followed by the deposition of porous TiO_2 film. The field emission scanning electron microscopy (FESEM) image (Figure 5.1a–c) showed that the thickness of NTO and d-TiO_2 is 110 and 120 nm, respectively. The assembled DSSC show an efficiency of 7.69% compared with the bare one which is 6.54% (Figure 5.1g). The dark current–voltage characteristics revealed that the suppressed dark current of the NTO indicates that it is a good material as the compact layer in DSSC (Figure 5.1d) The enhancement is due to the shifting of the onset of the dark current to a more positive voltage and slowing the open-circuit voltage (V_{OC}) decay after introduction of NTO compact layer (Figure 5.1e). Furthermore, the interfacial resistance of porous TiO_2 was also reduced with the introduction of

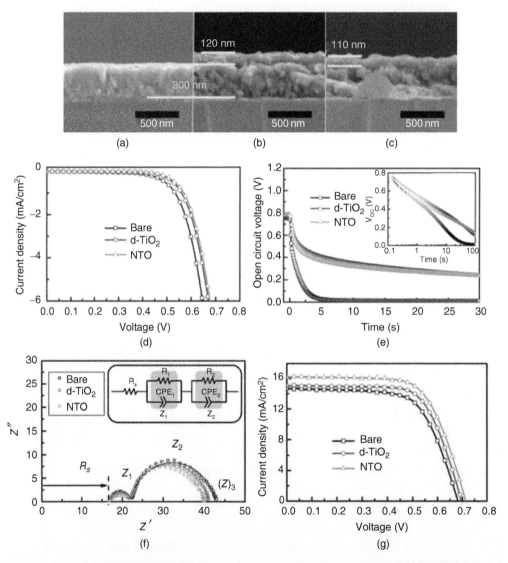

Figure 5.1 Cross-sectional FE-SEM images of (a) bare FTO, (b) d-TiO_2/FTO, and (c) NTO/FTO. (d) Dark current characteristics. (e) V_{OC} decay for the DSSCs. (f) Nyquist plots of the DSSCs employing bare-FTO, d-TiO_2/FTO and NTO/FTO. (g) J–V curve of bare-FTO, d-TiO_2/FTO and NTO/FTO. *Source:* Lee et al. 2009 [12]. Adapted with permission of The American Chemical Society.

the NTO. This indications showed that the NTO compact layer effectively suppressed the back electron transfer from the TCO to electrolyte.(Figure 5.1f). Moreover, the NTO also provide better contact between FTO and TiO_2 which allow efficient electron transfer and collection and finally enhance the device efficiency.

In another study, titanium organic sol was prepared at the room temperature for use as compact layer [13]. From this work, the organic sol was produced in low surface tension, low viscosity, and high mobility with the aim that the organic sol could effectively penetrate through the entire porous TiO_2 structure and reach the uncovered FTO surface. The scanning electron microscopy (SEM) and transmission electron microscopy (TEM) (Figure 5.2d) reveal that the large pores at the TiO_2 surface (Figure 5.2c) were reduced after deposition of the compact layer. This suggesting the improvement of the inter-particle connections and the increase of the contact area. As illustrated in Figure 5.2b, the presence of the compact layer will be able to prevent the electron leakage and low electron transport efficiency as compared to the one without compact layer (Figure 5.2a). The linear sweep voltammogram (Figure 5.2e) test provide information regarding the severity of the back electron transfer process in the devise. Their sample with the organic sol modification showed the increase of the onset potential and the reduction of the dark current. Consequently, the presence of the compact layer increased the cell efficiency by 28% (Figure 5.2f).

Liu and coworkers [14] utilized a chelating process of acetic acid on crystalline titania for use as a blocking layer. When comparing with the blocking layer prepared by titanium isopropoxide (TTIP), they found that acetic acid-treated titania showed better enhancement of photocurrent. The power conversion efficiency (PCE) of DSSC with treated TTIP increased from 5.23% to 7.86%, while the one treated with acetic acid increased to 10.49%. Regardless of TTIP or acetic acid-treated titania for blocking layer, smaller impedances are obtained from the Nyquist plot. The smaller impedance represents better charge transfer at the FTO/TiO_2 interface and TiO_2/electrolyte interface. The DSSC treated with acetic acid displays longer electron lifetime compared to TTIP treated and pure TiO_2. This result also proved that acetic acid-treated blocking layer showed better efficiency. The specific surface area increase after the introduction of the blocking layer and also can increase the dye loading, which is essential to enhance the photocurrent and therefore the overall cell efficiency. However, it is vital to control the amount of acetic acid loading on the titania. High amount of acetic acid could cause the cracking on the mesoporous titania surface due the significant shrinking induced by condensation reaction of acetic acid with TiO_2. The cracking will increase the probability of charge recombination and charge transport resistance and hence deteriorating the cell performance.

In addition to that, titanium nitride (TiN) was employed as blocking layer in DSSC by Nishio's group [15]. The TiN was deposited on the Ti metal mesh sheet by laser surface modification approach before being coated with mesoporous titania and used as photoanode. The use of the Ti metal mesh in this study also replaces the transparent conducting glass as shows in Figure 5.3a. The electrons are first collected by the Ti/TiN interface before being passed on to the counter electrode. The SEM image (Figure 5.3b) showed that the skin layer was observed as the color changes after laser irradiation. The thickness of the compact TiN increased from 300 to 1500 nm when the laser power increased from 40 to 92 J/cm^2. The presence of TiN blocking layer with the conduction band of -3.1 eV will help in blocking the charge recombination between Ti and I_3^- (Figure 5.3c,d). As a consequence of the effective charge suppression by the TiN blocking layer, the efficiency increases from 2.2% to 5.98% (Figure 5.3e). The increase of electron lifetime (Figure 5.3f) also suggests the increase in the DSSC efficiency. Other than this, electrodeposited TiO_2 for the use as compact layer have been reported by Jang et al. [8]. Among the samples, the TiO_2 compact layer with the 450 nm thickness showed the optimum efficiency after being assembled with the TiO_2 paste. By applying the compact layer to the DSSC, the cell

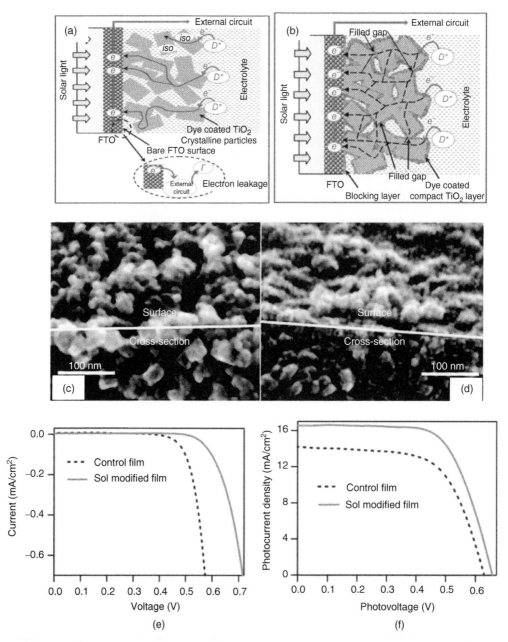

Figure 5.2 Schematic diagrams of electron pathways of the (a) porous structure and (b) sol modified structure (c, d) SEM and TEM micrographs of control film and sol modified film: the cross-sectional micrographs of surface structure for control film and sol modified film, respectively; (e) dark currents as a function of applied bias for DSSCs employing the control film and sol modified film; (f) photocurrents as a function of photovoltage for DSSCs employing the control film and sol modified film. *Source:* Yu et al. 2009 [13]. Adapted with permission of The American Chemical Society.

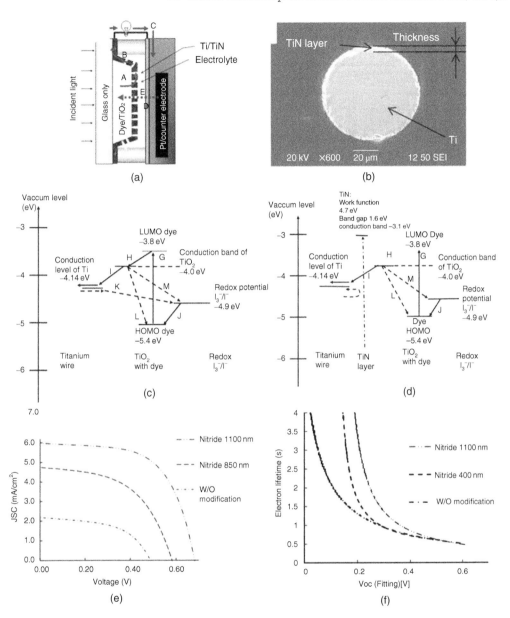

Figure 5.3 (a) Structure of TCO-less DSSC. (b) SEM image of cross sectional view of Ti wire exposed to 1064 nm laser under Ar:N$_2$ = 30 : 70 atm. Working principle for the DSSC (c) without TiN (d) with TiN. (e) Photovoltaic performance of DSSC with and without TiN. (f) Electron lifetime of DSSC with and without TiN. *Source:* Nishio et al. 2016 [15]. Adapted with permission of Springer Nature.

efficiency is 59.34% improved compared to the one without compared layer. A zinc ferrite-based photoanode has shown around 98% enhancement in the efficiency after being added in the titania thin film as a compact layer [16].

In a recent study, Hadi et al. [17] demonstrated a one-step laser technique to fabricate mesoporous and compact TiO$_2$ layer on indium tin oxide (ITO) by using millisecond-pulsed fiber laser with a wavelength of 1070 nm in ambient temperature. They compared the performance of the laser-prepared film with the conventional furnace one, and they found out that the laser processed film gave a better performance. The SEM images (Figure 5.4a–d) of cross-sectional and surface views of compact and mesoporous TiO_2 show that the thickness of the untreated and spin-coated compact layer is around 60 nm (Figure 5.4a) and reduced to 50 nm after one-step furnace treatment (Figure 5.4b). After two-step furnace treatment, the thickness of the compact layer is further reduced to 40 nm (Figure 5.4c,d) which have a similar thickness with the laser-treated mesoporous and compact TiO_2 layer. This suggested that the laser process offered a significant reduction of the fabrication time from approximately five hours to two minutes and yet obtained similar thickness. The efficiency obtained for one-step and two-step furnace treatment is 4.18% and 4.62%, respectively, while the laser-treated film gave an improvement in efficiency and achieved 6.01% (Figure 5.4e). The improved efficiency could be due to the longer electron lifetime (Figure 5.4f) achieved by the laser-treated sample. They confirmed the electron lifetime using the intermediate peak from the bode plot. The increase in the electron lifetime is due to the better interconnection of the TiO_2 and the addition of the compact TiO_2 which effectively prevent the back electron transfer and recombination.

5.3 Zinc Oxide (ZnO) as a Compact Layer

Other than TiO_2, utilizing ZnO thin film as a blocking layer seem to be highly feasible due to its more negative conduction band edge compared to TiO_2, it will prevent the electron recombination reaction and improve the V_{OC} [18–21]. The use of the ZnO-blocking layer on the transparent conducting film also can act as an alternative to the tedious and hazards preparation using $TiCl_4$ pretreatment which is commonly used in the preparation of the TiO_2-based blocking layer. For example, Kouhestanian et al. [19] fabricated the ZnO thin film using electrochemical deposition method. They prepared ZnO-blocking layer through the chroamperometric technique using $ZnCl_2$ precursor in KCl and poly(vinyl alcohol) (PVA). The deposition of ZnO thin film (60 seconds deposition) produced an average particle size of about 30 nm (Figure 5.5a). Then, the mesoporous TiO_2 with the average particle size of 25 nm (Figure 5.5b) was deposited on the blocking layer using doctor-blade method. They obtained the TiO_2 thickness of around 3.4 μm (Figure 5.5c). For comparison, they have prepared bare and $TiCl_4$ pretreated TiO_2 DSSC. The J–V characteristics (Figure 5.6d) show that their electrodepostion ZnO-blocking layer have a better photovoltaic result compared to the bare and $TiCl_4$ pretreated TiO_2 DSSC. However, the performance of the electrodepositon ZnO is dependent on the deposition time. The optimum time for the deposition is 60 seconds. Higher deposition time will cause too large an amount of ZnO thin film covering on the FTO and have a suppressing effect on the charge recombination due to the increase in FTO resistance. The open-circuit voltage-decay (OCVD) measurements show that the ZnO thin film could act better as a blocking layer to prevent the charge recombination compared to the bare and $TiCl_4$ pretreated TiO_2 DSSC (Figure 5.6e). The ZnO/TiO_2-based DSSC is having the slowest V_{OC} among all the samples.

Apart from that, ZnO as the blocking layer for TiO_2-based DSSC was studied by Ameri group [18]. The ZnO blocking layer was prepared using zinc acetate dehydrate through sol–gel method before being spin coated onto the FTO glass with the aim of preventing the electron back flow toward the electrolyte as shown in Figure 5.6a. In the studies, different precursor concentrations were investigated and found that 0.05 M concentration is the optimum concentration. The SEM and atomic force microscope (AFM) images (Figure 5.6b,c) show the ZnO blocking layer with 0.05 M concentration that is composed of a dense packing surface and a homogenous and crack-free surface. They also confirm the transparency of the thin film after applying ZnO-blocking layer through transmittance spectra (Figure 5.6d). The average transmittance for

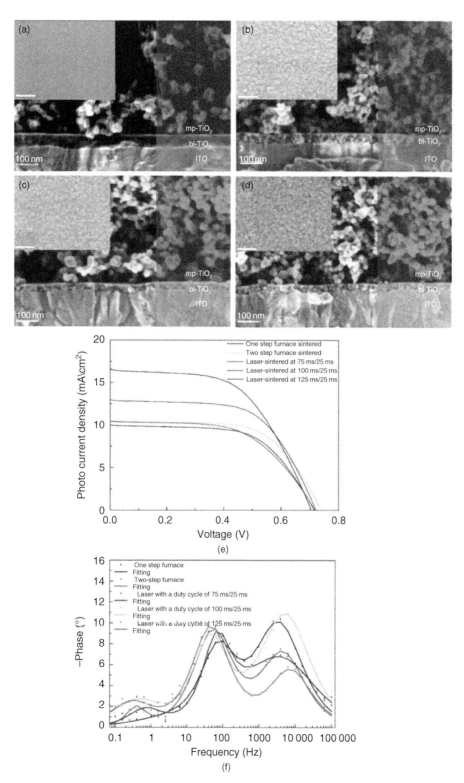

Figure 5.4 FESEM images of surface and cross-sectional views of (a) un-treated, (b) one-step furnace-treated, (c) two-step furnace treated, (d) laser-treated mesoporous and compact TiO$_2$ layers at 85 W/cm^2 with a duty cycle of 125 ms/25 ms. (e) Current density–voltage and (f) Bode plots of the DSSCs with photoanodes fabricated by different methods. *Source:* Hadi et al. 2018 [17]. Adapted with permission of American Chemical Society.

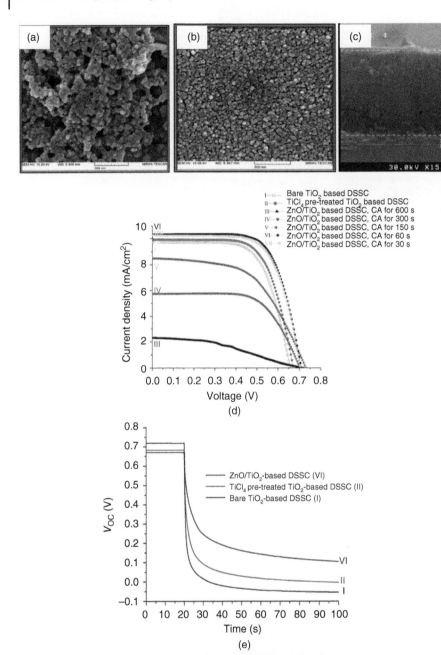

Figure 5.5 SEM images of (a) ZnO electrodeposited on FTO glass; (b) bare TiO$_2$ film prepared by doctor blade method; (c) TiO$_2$ film. (d) The *J–V* characteristic curves of the DSSCs fabricated various ZnO/TiO$_2$ electrodes. (e) The open-circuit voltage decays for the bare TiO$_2$(I) TiCl$_4$ pre-treated TiO$_2$(II) and ZnO/TiO$_2$ (VI) based DSSC. *Source:* Kouhestanian et al. 2016 [19]. Adapted with permission of Elsevier.

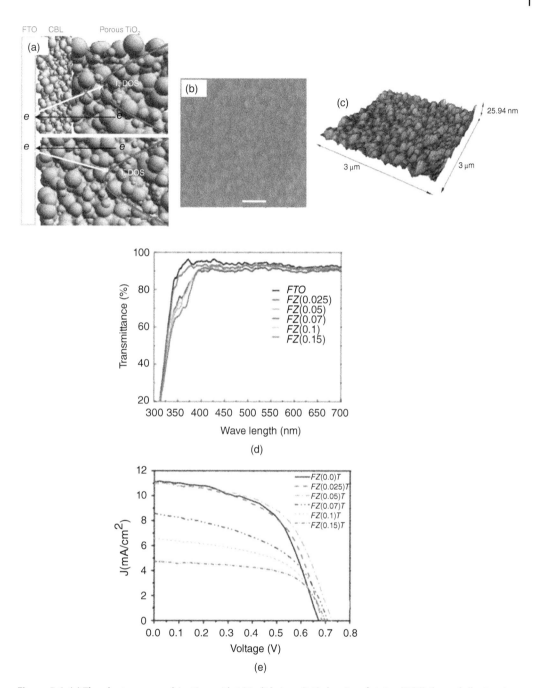

Figure 5.6 (a) The electron recombination with tri-iodide ions (I_3^-) density of states (DOS) through (bottom) bare FTO/TiO$_2$ interface and (top) FTO/CBL TiO$_2$ which is blocked by incorporating a transparent compact (b) SEM and (c) AFM images ZnO blocking layer with 0.05 M concentration (d) Transmittance spectra of current–voltage curves under (e) illumination and (f) dark. The *J–V* characteristic curves of the DSSCs fabricated various thickness of MoO$_3$. (e) Thickness dependent variation in efficiency of DSSCs. (f) Nyquist and (g) the OCVD data and (b) related carrier lifetimes as a function of cell voltage for DSSCs. *Source:* Ameri et al. 2016 [18]. Adapted with permission of The Institute of Physics.

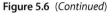

Figure 5.6 (*Continued*)

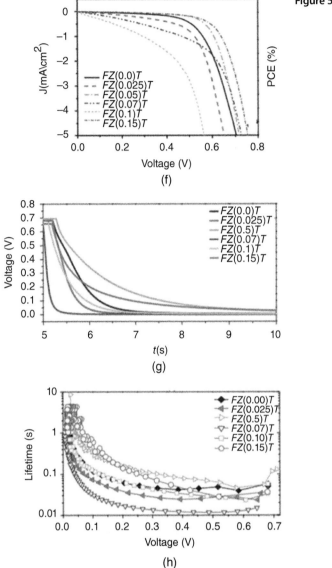

(f)

(g)

(h)

ZnO-blocking layer decreased by 5% compared with FTO and will not significantly affect the light-harvesting properties of the thin film. Current–voltage (J–V) was carried out under light and dark conditions to evaluate the cell efficiency and the functionality of the ZnO-blocking layer, respectively. Form the J–V curve under light illumination (Figure 5.6e), it was observed that the value of J_{SC} slightly decreased after incorporation of ZnO blocking layer, and this could be due to ZnO-blocking layer which acts as an energy barrier and consequently slows down the electron transport toward the FTO. Although, there is a decrease for the J_{SC}, however, the V_{OC}, fill factor (FF), and power conversion efficiency (PEC) increase. The best performance was obtained by the ZnO-blocking layer (0.05 M concentration) with J_{SC}, V_{OC}, FF, and PEC of 10.95 mA/cm^2, 730 mV, 60%, and 4.55% compared with the bare TiO$_2$ of 11.10, 660 mV, 56%, and 4.05%, respectively. A further increase of the ZnO concentration will deteriorate the devise performance because the electrode encounter stronger energy barrier. The reduction in the dark current (Figure 5.6f) for the optimum cell suggested the charges recombination was successfully suppressed. The OCVD data and the electron lifetime (Figure 5.6g,h) show that the photovoltage response and electron lifetime are the highest for the ZnO blocking layer with 0.05 M concentration. Furthermore, Ding et al. [22] fabricated electrospun TiO_2 nanofiber DSSCs, with the application of the 15 nm ZnO as the compact layer, and the short current density and efficiency capable of increasing to 17.3 mA/cm^2 and 8.01%, respectively. The intensity-modulated photovoltage (IMVS) measurement shows that the electron lifetime for the cell without the ZnO compact layer increased from 119.5 to 345.5 ms after an addition of the compact layer. This increase suggested that the compact layer plays a role in preventing the charge recombination between electrons from the FTO substrate and I_3^- ions in the electrolyte.

Lately, Priyanka et al. [21] fabricated a ZnO-blocking layer by dip coating using ZnO sol. They have prepared different morphologies of ZnO-based photoanode using hydrothermal method such as micro-rod and nano-tips-decorated micro-rod samples. From their work, the introduction of the ZnO blocking the nano-tips-decorated mirco rod samples produced better efficiency because of the morphological advantage of the sample. The decoration of the uneven zigzag pattern of nano-tips over the mirco rod serves as a good platform for a large amount of dye absorbing and effective scattering of incident light. Besides, the introduction of the ZnO blocking layer also further increases the devise efficiency from 0.86% to 1.29%. The blocking layer not only improves the interconnectivity of the photoanode for better electron transport but also prevents the charge recombination at the electrode/electrolyte interface.

5.4 Less Common Metal Oxide as a Compact Layer

Indeed, all the literature data are commonly reported on the metal oxides such as TiO_2 and ZnO, however, there are various less common metal oxides such as hafnium (IV) oxide (HfO_2), tin oxide (SnO_2), and molybdenum trioxide (MoO_3) which have also been explored as blocking layers. The concept of these blocking layers is the same as with the commonly used metal oxides. For example, the blocking layers should be compact and highly dense; not defective and should be hole free so that it could effectively prevent the back transfer of electrons from the conducting transparent glass to the electrolyte. In view of the advantages of the electronic properties, Braden et al. [23] investigated the effect of HfO_2 as a compact layer by using two different syntheses route. The compact layer was prepared through the atomic layer deposition (ALD) method and produced better efficiency compared with the conventional sol–gel method. This is because the ALD method provides a higher compactness of film and can efficiently suppress the back transfer of the electrons. The suppressed dark current density and lower capacitance in the low bias region in the ALD-produced films also account for the improvements. The DSSCs assembled with the ALD compact layers show J_{SC} = 14.3 mA/cm^2, V_{OC} = 680 mV, and η% = 6.0%.

Another approach is the use of SnO_2 as a compact layer for DSSC by the Duong group [24]. Various thicknesses of SnO_2 have been prepared using the nanocluster deposition technique. They found that the film with a 120 nm thickness provided the best results among their samples without changing the transmittance of the film structure after the deposition. The Nyquist plots show that the compact layer contributes to decreasing the sheet resistance and results in an increase in the short-circuit density. From IV curve, the V_{OC} decayed slowly with the addition of compact layer compared to the bare FTO sample with compact layer compared to the bare FTO sample. This suggested that the compact layer increases the electron lifetime by preventing the back electron movement. They have achieved around 19% improvement with the addition of compact layer. CuO nanofibres also have become a promising blocking layer in DSSC. Sahay et al. reported on the electrospinning CuO nanofibers as the blocking layer in ZnO-based DSSC [25]. It showed 25% increment in the current density with the application of CuO.

In addition, inspired by the works of efficient hole injection characteristics of MoO_3 in TCO and organic hole transporting materials [26, 27], Aditya et al. [28] fabricated MoO_3 thin film as the blocking layer via a physical vapor deposition method. The deposition is carried by the reactive DC sputtering using oxygen as bombarding gas and molybdenum as the metal target. It was found that the packing fraction of the MoO_3 particles (Figure 5.7b) is higher than the TiO_2 nanoparticles (Figure 5.7a) with no porosity, which is essential for a material to perform as a blocking layer to effectively prevent the back transfer electrolyte from the electrolyte to the FTO. They also have prepared different thickness of the MoO_3 thin film by applying different deposition time. To get a thicker film, deposition time have to be increased. The film deposited for 5, 10, 15, and 20 minutes will give a thickness of 70, 120, 170, and 240 nm, respectively (Figure 5.7c). From the $J-V$ curves, the five-minute deposition of MoO_3-based TiO_2 results in 5.2% efficiency compared to reference TiO_2 (4.5%) (Figure 5.7d). It could be found that, the J_{sc} values improved after introducing MoO_3 blocking layer and proved that there is efficient charge transport at the FTO/TiO_2 interface. However further increase deposition time will increase the thickness of blocking layer and cause the decrease in the J_{sc} values due to the reduction of the photo-electron transport through the blocking layer and hence deteriorate the devise efficiency. The optimum thickness of the blocking layer used in their work is around 80 nm (Figure 5.7e). Therefore, controlling the thickness of the blocking is also one of the important factors in order to improve the cell efficiency. In their studies, the deposition time of 5 minutes is the optimum one where it gave the highest efficiency. The bigger semicircle that is obtained in the Nyquist and low frequency shift in the bode plots (Figure 5.7f,g) show that the MoO_3 blocking layer could assure that the charge transfer process is more efficient compared to the reference DSSC. In short, the incorporating of MoO_3 at FTO/TiO_2 interface can act as a barrier for recombination between photo-generated carriers and hole present in the electrolyte that diffuses through the mesoporous TiO_2 layer (Figure 5.7i) and prevent the charge recombination process (processes 2 and 3) (Figure 5.7h). Table 5.1 shows the summary of selected types of blocking layer for DSSC application.

5.5 Conclusion

Over the past 10 years, the effort dedicated to the different types of blocking layer has led to a rich knowledge and database for the fabrication, characterizations, and their applications. As reported, the role of the blocking layer was clearly observed enhancing the DSSC efficiency by reducing the charge recombination in the devise. In this regards, TiO_2-based blocking layers still yield higher PEC values, and this is also the reason why TiO_2-based blocking layer gained much research interest compared to other metal-based blocking layer. In order to achieve higher PCE values, efforts have been made in adding in other alternative materials, such as graphene and Nb. Without a doubt, it is also necessary to take into account the parameters, such as the concentration of the precursor, the thickness, and the morphology structure of the blocking layer, which could ultimately affect the final result if the DSSCs' efficiency. High concentration of the precursor will affect the transmittance of the thin film and the increase of the energy barrier which will have reverse effect on the device performance. On the other hand, the thick blocking layer will also lead to the increase of the devise resistance and suppressing effect of the charge recombination. Therefore, controlling some of the critical parameters are crucial to develop an efficient DSSC.

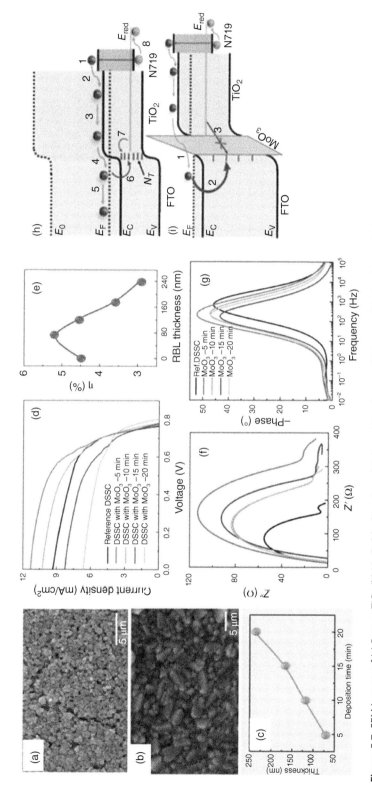

Figure 5.7 SEM images of (a) Porous TiC₂ (b) MoO₃ blocking layer (c) deposition time vs. thickness of MoO₃ obtained (d) The *J–V* characteristic curves of the DSSCs fabricated various thickness of MoO₃ (e) thickness dependent variation in efficiency of DSSC. Energy level alignment of DSSC (h) without MoO₃ (i) with MoO₃. *Source:* Ashok et al. 2017 [28]. https://pubs.rsc.org/en/content/articlelanding/2017/ra/c7ra08988k#!divAbstract. Licensed under CCBY 3.0 https://creativecommons.org/licenses/by/3.0/.

Table 5.1 Comparison of different blocking layers for DSSC.

Material for blocking/compact layer	Precursor used	Preparation method	Efficiency after addition of compact layer (%)	Percentage of improvement with compact layer (%)	Ref. (Year)
Nb-doped TiO_2	Nb_2O_5 and TiO_2 powder	Milling with zirconia ball	7.69	17.58	[12] (2009)
TiO_2	Titanium organic sol (tetrabutyl titanate, diethanolamine, and absolute ethanol)	Stirring	7.32	28	[13] (2009)
TiO_2	Titanium organic sol (tetrabutyl titanate, diethanolamine, and absolute ethanol)	Stirring	6.80	107.95	[29] (2010)
TiO_2	$TiCl_4$	Soaking	2.71	85.61	[30] (2010)
Hafnium oxide (HfO_2)	Hafnium tetra chloride gas and water vapor	Atomic layer deposition	6	66.67	[23] (2011)
TiO_2	$TiCl_3$	Anodization	8.2	12.32	[31] (2011)
TiO_2	—	RF sputtering	5.11	27.43	[32] (2011)
TiO_2	TiO_2 colloid	Spin coating	5.94	25.3	[33] (2011)
TiO_2	TiO_2 colloid	Layer-by-layer self assembly technique	7.38	24.45	[34] (2012)
TiO_2	Peroxotitanium solution	Dipping	7.33	29.94	[35] (2012)
TiO_2	TTIP	Spin coating	5.00	60.25	[36] (2013)
ZnO	$Zn(C_2H_5)_2$	Atomic layer deposition	7.12	16.53	[22] (2013)
TiO_2	TiO_2	RF reactive magnetron sputtering	6.71	26.84	[37] (2013)
TiO_2	TiO_2 sol	Dip coating	7.1	17.94	[38] (2013)
TiO_2	$TiCl_3$	Electrodeposition	5.38	55.04	[39] (2013)
TiO_2	$TiCl_4$	Atomic layer deposition	8.4	20	[40] (2013)
TiO_2	TiO_2 sol	Spin coating	6	36.05	[41] (2013)
TiO_2	TiO_2	Sputtering	3.9	227.73	[42] (2014)
TiO_2	TTIP	AACVD	4.63	25.82	[11] (2014)
TiO_2: Nb	NTO sol	Dip coating	5.1	90.30	[43] (2015)

(Continued)

Table 5.1 (Continued)

Material for blocking/compact layer	Precursor used	Preparation method	Efficiency after addition of compact layer (%)	Percentage of improvement with compact layer (%)	Ref. (Year)
TiO_2	—	RF magnetron sputtering	5.69	154.02	[44] (2015)
Titania chelated by acetic acid	TTIP, acetic acid	Screen printing	3.70	107.87	[14] (2015)
TiN	Ti wire	Laser irradiation process	4.82	120.45	[15] (2016)
ZnO	$Zn(CH_3CO_2)_2$	Spin coating	4.55	12.35	[18] (2016)
ZnO	$ZnCl_2$	Electrodeposition	4.88	18.73	[19] (2016)
TiO_2	$TiCl_4$	Immerse	7.33	29.05	[45] (2016)
ZnO	ZnO	RF sputtering	3.19	38.10	[20] (2016)
MoO_3	Mo	DC sputtering	5.20	15.56	[28] (2017)
ZnO	TiO_2/graphene	Doctor blade	1.81	38.17	[46] (2017)
TiO_2	$[(NH_4)_2 TiF_6$	Liquid phase deposition	0.75	74.42	[47] (2018)
ZnO	$Zn(NO_3)_2$	Hydrothermal	1.29	79.17	[21] (2018)
Nb-TiO_2	TTIP and $NbCl_5$	Spraying using the ultrasonic atomizer	7.50	4.74	[48] (2017)
TiO_2	$C_{16}H_{32}O_6Ti$	Laser fabrication	6.01	47.31	[17] (2018)

References

1 Bach, U., Lupo, D., Comte, P. et al. (1998). Solid-state dye-sensitized mesoporous TiO_2 solar cells with high photon-to-electron conversion efficiencies. *Nature* 395: 583–585.

2 Cameron, P.J., Peter, L.M., and Hore, S. (2004). How important is the back reaction of electrons via the substrate in dye-sensitized nanocrystalline solar cells? *J. Phys. Chem. B* 109: 930–936.

3 Haque, S.A., Palomares, E., Cho, B.M. et al. (2005). Charge separation versus recombination in dye-sensitized nanocrystalline solar cells: the minimization of kinetic redundancy. *J. Am. Chem. Soc.* 127: 3456–3462.

4 Li, X., Qiu, Y., Wang, S. et al. (2013). Electrophoretically deposited TiO_2 compact layers using aqueous suspension for dye-sensitized solar cells. *Phys. Chem. Chem. Phys.* 15: 14729–14735.

5 Hart, J., Menzies, D., Cheng, Y.-B. et al. (2006). Microwave processing of TiO_2 blocking layers for dye-sensitized solar cells. *J. Sol-Gel Sci. Technol.* 40: 45–54.

6 Cameron, P.J. and Peter, L.M. (2003). Characterization of titanium dioxide blocking layers in dye-sensitized nanocrystalline solar cells. *J. Phys. Chem. B* 107: 14394–14400.

7 Yu, H., Zhang, S., Zhao, H. et al. (2009). An efficient and low-cost TiO_2 compact layer for performance improvement of dye-sensitized solar cells. *Electrochim. Acta* 54: 1319–1324.

8 Jang, K.-I., Hong, E., and Kim, J. (2012). Effect of an electrodeposited TiO$_2$ blocking layer on efficiency improvement of dye-sensitized solar cell. *Korean J. Chem. Eng.* 29: 356–361.

9 Meng, L. and Li, C. (2011). Blocking layer effect on dye-sensitized solar cells assembled with TiO$_2$ nanorods prepared by dc reactive magnetron sputtering. *Nanosci. Nanotechnol. Lett.* 3: 181–185.

10 Lellig, P., Niedermeier, M.A., Rawolle, M. et al. (2012). Comparative study of conventional and hybrid blocking layers for solid-state dye-sensitized solar cells. *PCCP* 14: 1607–1613.

11 Lim, S.P., Huang, N.M., Lim, H.N., and Mazhar, M. (2014). Aerosol assisted chemical vapour deposited (AACVD) of TiO$_2$ thin film as compact layer for dye-sensitised solar cell. *Ceram. Int.* 40: 8045–8052.

12 Lee, S., Noh, J.H., Han, H.S. et al. (2009). Nb-doped TiO$_2$: a new compact layer material for TiO$_2$ dye-sensitized solar cells. *J. Phys. Chem. C* 113: 6878–6882.

13 Yu, H., Zhang, S., Zhao, H. et al. (2009). High-performance TiO$_2$ photoanode with an efficient electron transport network for dye-sensitized solar cells. *J. Phys. Chem. C* 113: 16277–16282.

14 Liu, B.-T., Chou, Y.-H., and Liu, J.-Y. (2016). Extremely enhanced photovoltaic properties of dye-sensitized solar cells by sintering mesoporous TiO$_2$ photoanodes with crystalline titania chelated by acetic acid. *J. Power Sources* 310: 79–84.

15 Nishio, Y., Yamaguchi, T., Yamguchi, T. et al. (2016). Transparent conductive oxide-less dye-sensitized solar cells (TCO-less DSSC) with titanium nitride compact layer on back contact Ti metal mesh. *J. Appl. Electrochem.* 46: 551–557.

16 Habibi, M.H., Habibi, A.H., Zendehdel, M., and Habibi, M. (2013). Dye-sensitized solar cell characteristics of nanocomposite zinc ferrite working electrode: effect of composite precursors and titania as a blocking layer on photovoltaic performance. *Spectrochim. Acta, Part A* 110: 226–232.

17 Hadi, A., Chen, Q., Curioni, M. et al. (2018). One-step fiber laser fabrication of mesoporous and compact TiO$_2$ layers for enhanced performance of dye-sensitized solar cells. *ACS Sustainable Chem. Eng.* 6: 12299–12308.

18 Ameri, M., Samavat, F., Mohajerani, E., and Fathollahi, M.-R. (2016). Facile realization of efficient blocking from ZnO/TiO$_2$ mismatch interface in dye-sensitized solar cells and precise microscopic modeling adapted by circuit analysis. *J. Phys. D: Appl. Phys.* 49: 225601.

19 Kouhestanian, E., Mozaffari, S., Ranjbar, M. et al. (2016). Electrodeposited ZnO thin film as an efficient alternative blocking layer for TiCl$_4$ pre-treatment in TiO$_2$-based dye sensitized solar cells. *Superlattices Microstruct.* 96: 82–94.

20 Chou, J.-C., Lin, Y.-J., Liao, Y.-H. et al. (2016). Photovoltaic performance analysis of dye-sensitized solar cell with ZnO compact layer and TiO$_2$/graphene oxide composite photoanode. *IEEE J. Electron. Devices Soc.* 4: 402–409.

21 Das, P., Mondal, B., and Mukherjee, K. (2018). Improved efficiency of ZnO hierarchical particle based dye sensitized solar cell by incorporating thin passivation layer in photo-anode. *Appl. Phys. A* 124: 80.

22 Ding, J., Li, Y., Hu, H. et al. (2013). The influence of anatase-rutile mixed phase and ZnO blocking layer on dye-sensitized solar cells based on TiO$_2$ nanofiberphotoanodes. *Nanoscale Res. Lett.* 8: 9.

23 Bills, B., Shanmugam, M., and Baroughi, M.F. (2011). Effects of atomic layer deposited HfO$_2$ compact layer on the performance of dye-sensitized solar cells. *Thin Solid Films* 519: 7803–7808.

24 Duong, T.-T., Choi, H.-J., He, Q.-J. et al. (2013). Enhancing the efficiency of dye sensitized solar cells with an SnO$_2$ blocking layer grown by nanocluster deposition. *J. Alloys Compd.* 561: 206–210.

25 Sahay, R., Sundaramurthy, J., Kumar, P.S. et al. (2012). Synthesis and characterization of CuO nanofibers, and investigation for its suitability as blocking layer in ZnO NPs based dye sensitized solar cell and as photocatalyst in organic dye degradation. *J. Solid State Chem.* 186: 261–267.

26 Matsushima, T., Kinoshita, Y., and Murata, H. (2007). Formation of ohmic hole injection by inserting an ultrathin layer of molybdenum trioxide between indium tin oxide and organic hole-transporting layers. *Appl. Phys. Lett.* 91: 253504.

27 Lee, H., Cho, S.W., Han, K. et al. (2008). The origin of the hole injection improvements at indium tin oxide/molybdenum trioxide/N, N′-bis (1-naphthyl)-N, N′-diphenyl-1, 1′-biphenyl-4, 4′-diamine interfaces. *Appl. Phys. Lett.* 93: 279.

28 Ashok, A., Vijayaraghavan, S., Nair, S.V., and Shanmugam, M. (2017). Molybdenum trioxide thin film recombination barrier layers for dye sensitized solar cells. *RSC Adv.* 7: 48853–48860.

29 Sharma, G., Suresh, P., and Mikroyannidis, J.A. (2010). Quasi solid state dye-sensitized solar cells with modified TiO_2 photoelectrodes and triphenylamine-based dye. *Electrochim. Acta* 55: 2368–2372.

30 Sandquist, C. and McHale, J.L. (2011). Improved efficiency of betanin-based dye-sensitized solar cells. *J. Photochem. Photobiol., A* 221: 90–97.

31 Wu, M.-S., Tsai, C.-H., Jow, J.-J., and Wei, T.-C. (2011). Enhanced performance of dye-sensitized solar cell via surface modification of mesoporous TiO_2 photoanode with electrodeposited thin TiO_2 layer. *Electrochim. Acta* 56: 8906–8911.

32 Chou, J.-C., Chiu, Y.-Y., Yu, Y.-M. et al. (2011). Research of titanium dioxide compact layer applied to dye-sensitized solar cell with different substrates. *J. Electrochem. Soc.* 159: A145–A151.

33 Kovash, C.S. Jr., Hoefelmeyer, J.D., and Logue, B.A. (2012). TiO_2 compact layers prepared by low temperature colloidal synthesis and deposition for high performance dye-sensitized solar cells. *Electrochim. Acta* 67: 18–23.

34 Yuan, S., Li, Y., Zhang, Q., and Wang, H. (2012). Anatase TiO_2 sol as a low reactive precursor to form the photoanodes with compact films of dye-sensitized solar cells. *Electrochim. Acta* 79: 182–188.

35 Yang, S., Hou, Y., Zhang, B. et al. (2013). Highly efficient overlayer derived from peroxotitanium for dye-sensitized solar cells. *J. Mater. Chem. A* 1: 1374–1379.

36 Abdullah, M., Saurdi, I., and Rusop, M. (2013). *2013 IEEE Regional Symposium on Micro and Nanoelectronics (RSM)*, 163–166. IEEE.

37 Wang, Y., Chen, C., Cui, X. et al. (2013). The functions of compact TiO_2 blocking layers in dye-sensitized solar cells investigated by direct photoelectrochemical methods. *Nanosci. Nanotechnol. Lett.* 5: 293–296.

38 Guai, G.H., Song, Q.L., Lu, Z.S. et al. (2013). Tailor and functionalize TiO_2 compact layer by acid treatment for high performance dye-sensitized solar cell and its enhancement mechanism. *Renewable Energy* 51: 29–35.

39 Jang, K.-I., Hong, E., and Kim, J.H. (2013). Improved electrochemical performance of dye-sensitized solar cell via surface modifications of the working electrode by electrodeposition. *Korean J. Chem. Eng.* 30: 620–625.

40 Kim, D.H., Woodroof, M., Lee, K., and Parsons, G.N. (2013). Atomic layer deposition of high performance ultrathin TiO_2 blocking layers for dye-sensitized solar cells. *ChemSusChem* 6: 1014–1020.

41 Gu, Z.-Y., Gao, X.-D., Li, X.-M. et al. (2014). Nanoporous TiO_2 aerogel blocking layer with enhanced efficiency for dye-sensitized solar cells. *J. Alloys Compd.* 590: 33–40.

42 Alberti, A., Pellegrino, G., Condorelli, G. et al. (2014). Efficiency enhancement in ZnO: Al-based dye-sensitized solar cells structured with sputtered TiO$_2$ blocking layers. *J. Phys. Chem. C* 118: 6576–6585.

43 Parthiban, S., Anuratha, K., Arunprabaharan, S. et al. (2015). Enhanced dye-sensitized solar cell performance using TiO$_2$: Nb blocking layer deposited by soft chemical method. *Ceram. Int.* 41: 205–209.

44 Huang, C., Chang, K., and Hsu, C. (2015). TiO$_2$ compact layers prepared for high performance dye-sensitized solar cells. *Electrochim. Acta* 170: 256–262.

45 Sivakumar, R., Ramkumar, J., Shaji, S., and Paulraj, M. (2016). Efficient TiO$_2$ blocking layer for TiO$_2$ nanorod arrays-based dye-sensitized solar cells. *Thin Solid Films* 615: 171–176.

46 Huang, C.-Y., Chen, P.-H., Wu, Y.-J. et al. (2017). Enhanced performance of ZnO-based dye-sensitized solar cells using TiO$_2$/graphene nanocomposite compact layer. *Jpn. J. Appl. Phys.* 56: 045201.

47 Huang, J.-J., Wu, C.-K., and Hsu, C.-F. (2017). Characterization of LPD-TiO$_2$ compact layer in ZnO nano-rods photoelectrode for dye-sensitized solar cell. *Appl. Phys. A* 123: 741.

48 Koo, B.-R., Oh, D.-H., and Ahn, H.-J. (2018). Influence of Nb-doped TiO$_2$ blocking layers as a cascading band structure for enhanced photovoltaic properties. *Appl. Surf. Sci.* 433: 27–34.

6

Function of TiCl$_4$ Posttreatment in Photoanode

T.S. Senthil and C.R. Kalaiselvi

Department of Physics, Erode Sengunthar Engineering College, Perundurai, India

6.1 Introduction

Energy is the key factor for any living creature to exist in this universe. The advent of industrialization and increase in population have led to a surge in the crisis for energy. The reduction of our dependence on fossil fuels (oil, coal, and natural gas), as well as the evolution towards a cleaner future, requires the large deployment of sustainable renewable energy sources. Among them solar energy is the most abundant and also available throughout the year. Moreover, the solar energy has the greatest potential to fulfill the thirst for energy and the need for innovation of clean and eco-friendly technologies. In this perspective, developing solar cells is one of the best approaches to convert solar energy into electrical energy based on photovoltaic effect. Solar cells based on crystalline silicon and thin film technologies are often referred to as first- and second-generation solar cells. The demerits in that are the limited availability and the cost of silicon. An emerging third-generation photovoltaics have been developed as an alternate to it. These include Dye-sensitized solar cells (DSSCs), organic photovoltaic, quantum dots and recently perovskite solar cells. DSSCs based on nanocrystalline TiO$_2$ as a photo-anode have attracted a lot of scientific and technological interest since their breakthrough in 1991 [1]. The two main functional aspects of charge generation and transport are no longer combined in one material but separated in different materials, i.e. a sensitizing dye, a wide-band-gap semiconductor (TiO$_2$), and a liquid redox electrolyte [2].

The TiO$_2$ passivating layer plays a significant role in realizing highly efficient DSSCs as it improves the adhesion of the TiO$_2$ to the transparent conducting oxide (TCO) and provides a larger contact area thereby enhances the effective electron transfer by preventing electron recombination [3]. A well-known method to improve the performance is the posttreatment using several TiO$_2$ precursors and methods: TiCl$_3$ electrodeposition and titanium isopropoxide and titanium tetrachloride (TiCl$_4$) posttreatment [2]. Among them TiCl$_4$ surface treatment causes improvement in the electron transport and dye anchoring, resulting in enhanced efficiency of the solar cells [4]. Titanium tetrachloride (TiCl$_4$) treatment processed by chemical bath deposition is usually adopted as pre- and posttreatment for nanocrystalline titanium dioxide thin film deposition in DSSC technology. Pretreatment positively influences the bonding strength between the fluorinated tin oxide (FTO) substrate and the porous TiO$_2$ layer, blocking the charge recombination at the interface [5].

The effect on the cell performance is revealed by an increase in incident photon to current conversion efficiency (IPCE) and short-circuit current density (J_{SC}) of typically 10–30% relative to untreated films. The IPCE of a solar cell is influenced by three factors: light harvesting

Interfacial Engineering in Functional Materials for Dye-Sensitized Solar Cells, First Edition.
Edited by Alagarsamy Pandikumar, Kandasamy Jothivenkatachalam and Karuppanapillai B. Bhojanaa.
© 2020 John Wiley & Sons, Inc. Published 2020 by John Wiley & Sons, Inc.

efficiency (LHE) of the colored TiO_2 film, electron injection efficiency of the excited dye into the TiO_2, and the collection efficiency of the injected electrons to the TCO [6].

A widely known method to improve the performance of DSSC is the surface treatment of TiO_2 film with an aqueous solution of titanium tetrachloride ($TiCl_4$). Gratzel group reported on the influence of different FTO surface treatments ($TiCl_4$ or compact TiO_2 under layer) on dark and the photocurrent behavior in DSSCs. The correlations between the surface morphologies of $TiCl_4$-treated FTO glass and the cell performances were investigated and an optimal condition for maximizing the DSSC performance was also obtained [7]. Here the various hypotheses pertaining to the treatment of $TiCl_4$ on the TiO_2 layer for enhancing the performance of DSSCs reported by various authors are summarized.

6.2   Role of TiCl₄ Posttreatment in Photo-Anode

Improvements to the TiO_2 electrode in the DSSC have been made in terms of light absorption, light scattering, charge transport, suppression of charge recombination, and improvement of the interfacial energetics. For the state-of-the-art DSSCs, the employed architecture of the mesoporous TiO_2 electrode is as follows:

(a) A TiO_2 blocking layer (thickness ~ 50 nm), a coating on the FTO plate prevent the contact between the redox mediator in the electrolyte and the FTO, is prepared by chemical bath deposition, spray pyrolysis, or sputtering.

(b) A light absorption layer consisting of a $\sim 10\,\mu m$ thick film of mesoporous TiO_2 with ~ 20 nm particle size that provides a large surface area for sensitizer adsorption and good electron transport to the substrate.

(c) A light scattering layer on the top of the mesoporous film, consisting of a $\sim 3\,\mu m$ porous layer containing ~ 400 nm sized TiO_2 particles.

(d) An ultrathin over coating of TiO_2 on the whole structure, deposited by means of chemical bath deposition (using aqueous $TiCl_4$), and followed by heat treatment.

The $TiCl_4$ treatment increases the injection of electrons into the TiO_2 and thus increases the current delivered by the solar cell. Two hypotheses can explain this improvement of the injection after treatment:

(i) Small particles are nucleated on the surface of the electrode thereby the surface area and the amount of dye adsorbed increases, or

(ii) The electron percolation in the TiO_2 mesoporous film is improved. It is analyzed from the data that the $TiCl_4$ treatment decreases the surface area of the film [8].

Therefore, the first hypothesis can be rejected. In addition, the $TiCl_4$ treatment decreases the average pore size and the porosity. All this data suggests that the titanium complexes that are present in the $TiCl_4$ solution condense at the interparticle neck. This hypothesis is consistent with the pore-size reduction, the surface-area loss, and the densification that are observed in the films after the $TiCl_4$ treatment.

6.3   Effect of Posttreatment of TiCl₄ on Various Perspectives

6.3.1   TiO₂ Morphology, Porosity, and Surface Area

$TiCl_4$ posttreatment and treatment temperature shows a high impact on the morphology of the film. It was clearly discussed by Kim et al. [9], the effect of $TiCl_4$ posttreatment on the embedded-type TiO_2 nanotube-based DSSCs. Their result showed that the posttreatment makes TiO_2 nanoparticle layers on TiO_2 nano tubes (NTs) surface. Further the TiO_2 NT arrays

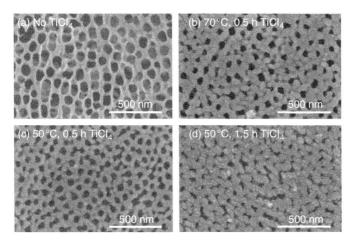

Figure 6.1 FE-SEM images (top-view) of (a) TiO$_2$ NT arrays without TiCl$_4$, (b) TiO$_2$ NT arrays treated by TiCl$_4$ at 70 °C for 0.5 hour, (c) at 50 °C for 0.5 hour, and (d) at 50 °C for 1.5 hours.

treated by TiCl$_4$ at 70 °C for 0.5 hour makes a uniform and dense TiO$_2$ nanoparticle (NP) layer on the surface of the TiO$_2$ NTs and is shown in Figure 6.1. In order to achieve the charge transfer between the electrolyte and the dyes on TiO$_2$ NT, they reduced TiCl$_4$ treatment temperature to decrease the reaction rate. TiCl$_4$ reaction rate at the temperature below 50 °C was slow enough. Therefore, TiCl$_4$ posttreatment was performed at 50 °C for 1.5 hour, and they observed, a rough and island-like TiO$_2$ NP layer formed on the surface of the TiO$_2$ NT. On further increasing the TiCl$_4$ treatment time at 50 °C, the pore size of TiO$_2$ NT arrays decreases. The result shows that the treatment time at 50 °C closes the pores of the TiO$_2$ NT arrays and prevents the infiltration of dyes and electrolytes into the TiO$_2$ NT arrays. As a result of this, the surface area of TiO$_2$ NTs increased after the TiCl$_4$ posttreatment. The increased surface area enhances the adsorption of dyes and thereby increases the performance of the cell.

The TiO$_2$ NT arrays treated by TiCl$_4$ at 50 °C for 1.5 hours showed the better cell efficiency than the arrays treated at 70 °C for 0.5 hour, although the pore size of the TiCl$_4$-treated NT arrays was smaller. The dense TiO$_2$ NP layer formed by the TiCl$_4$ treatment at 70 °C for 0.5 hour prevented the charge transfer from the electrolyte to the TiO$_2$ NT. Accordingly, the power conversion efficiency (PCE) of the TiO$_2$ NT arrays treated by TiCl$_4$ at 50 °C for 1.5 hours was higher than the TiCl$_4$ treated at 70 °C for 0.5 hour due to the increase of fill factor (FF), although the current density (J_{SC}) is lower which was shown in Table 6.1. A rough and island-like TiO$_2$ NP layer allowed the transfer from the electrolyte to the TiO$_2$ NT with TiCl$_4$ treatment at a temperature lower than 70 °C, which was a conventional temperature in the TiCl$_4$ treatment of TiO$_2$ NP photoelectrode to be more effective.

Table 6.1 Comparison of the DSSC performance according to the TiCl$_4$ treatment condition.

Sample	J_{SC} (mA/cm^2)	V_{OC} (V)	FF	η (%)
Without TiCl$_4$ treatment	11.9	0.71	0.73	5.59
TiCl$_4$ treatment at 70 °C, 0.5 h	14.22	0.70	0.59	5.97
TiCl$_4$ treatment at 50 °C, 0.5 h	12.39	0.70	0.70	6.16
TiCl$_4$ treatment at 50 °C, 1.5 h	13.36	0.71	0.68	6.52

Table 6.2 Change of TiO$_2$ characteristics after TiCl$_4$ treatment.

Sample	BET (m^2/g)	TiO$_2$ mass	Electrode surface area (m^2)	Porosity (%)	Particle diameter (nm)	
					From TiO$_2$ mass	From TEM
Without TiCl$_4$ treatment	72	1.56	0.449	70	Not applicable	13.1
After TiCl$_4$ treatment	53	2.00	0.424	63	Not applicable	15.7

The influence of TiCl$_4$ posttreatment on nanocrystalline TiO$_2$ thin film electrodes in DSSCs was also reported by Sommeling et al. [6]. They reported that compared with the nontreated films, TiCl$_4$-treated films show enhanced performance due to higher photocurrents in treated films. On a microscopic scale, TiO$_2$ particle growth was observed in the order of 1 nm. In spite of this, some obvious effects of the TiCl$_4$ treatment noted by them are a decrease in the Brunauer–Emmett–Teller (BET) surface area and an increase in particle diameter and mass of the treated-TiO$_2$ films. Despite their observation on the substantial decrease in BET surface area, the loss in actual electrode surface area after TiCl$_4$ treatment is only 6% (from 0.449 to 0.424 m^2) because of the increase in mass of TiO$_2$ on the electrode. Further the increase in particle diameter of 8.6%, as shown in Table 6.2, was calculated by the assumption that the particles are spherical. By using transmission electron microscopic (TEM) measurements an increase in average particle diameter of 20% was observed. Finally, they concluded that TiCl$_4$ treatment on the TiO$_2$ morphology is due to the enhanced current density and dye absorption. The enhanced current originated from the shift in conduction band edge of TiO$_2$ upon TiCl$_4$ treatment provided the path for an improved charge injection into the TiO$_2$.

Sedghi and Miankushki examined the effect of TiCl$_4$ pre- and posttreatment on TiO$_2$ electrodes for the performance of DSSCs. They reported the effect of sponge-like morphology on TiO$_2$ film and particle necking for electron diffusion. Further, the surface of the TiO$_2$ film was smaller than that in other electrodes because these particles are smaller. From SEM image, they pointed out that the necks between particles and particle size were increased after TiCl$_4$ treatment. As a result, TiO$_2$ particle necking decreased the electron transport path and charge recombination between TiO$_2$ and the oxidant in the electrolyte. In addition to this, it enhances the surface area of TiO$_2$ films and PCE of the device.

TiCl$_4$ treatment also enhanced the ruthenium dye adsorption, and it was reported by Ito et al. [10]. Due to the TiCl$_4$ treatment, the specific surface area of nanoporous TiO$_2$ film decreases to 79.7 m^2/g from 86.0 m^2/g and hence results in an increased absorbance at 540 nm. This disagreement between the variation of the dye absorption and the specific surface area was explained by the increase of TiO$_2$ weight of 0.173 ± 0.003 mg/cm^2/μm. The roughness factor was calculated by multiplying the specific surface area and TiO$_2$ weight. In spite of the fact that the specific surface area with TiCl$_4$ treatment decreases, the weight of TiO$_2$ increases with the increase of roughness factor. Further, they signified that the ratios of (nano-TiO$_2$)/(TiCl$_4$-treated nano-TiO$_2$) in roughness factor and absorbance at 540 nm were 1.19 and 1.16, respectively. This coincidence in relationship between roughness factor and absorbance suggested that the enhancement of TiO$_2$ surface area by TiCl$_4$ treatment thereby increases the photocurrent.

Lee and Ahn [11] discussed the effects of morphology of TiO$_2$ nanocrystalline films with different concentrations of TiCl$_4$ (5–500 mM). Table 6.3 summarized the specific surface area, porosity, weight gain, and total surface area derived from BET analyses of the TiCl$_4$-treated TiO$_2$ films.

Table 6.3 Morphological Changes of TiO$_2$ films after treatment at different TiCl$_4$ concentrations.

Parameters	Standard	5 mM	15 mM	50 mM	100 mM
BET area (m^2/g)	85.4	87.7	95.5	73.9	64.9
Average pore width (nm)	22.4	21.8	19.6	15.3	12.2
TiO$_2$ mass gain (wt%)	0.0	4.4	13.2	33.1	58.2
Total surface area (%)	0.0	7	27	15	20

From Table 6.3, it was observed that the mass of TiO$_2$ was higher when the TiCl$_4$ concentrations were increased from 5 to 100 mM, but no essential change in the film thickness was noted. Further in this concentration range, the average pore width got decreased and the specific surface area of TiCl$_4$-treated TiO$_2$ electrodes increased gradually. This leads to surface roughening in the TiO$_2$. They also noted that TiCl$_4$ concentration was increased further than the specified range, the specific surface area got decreased because of significant pore narrowing. The total surface area of the films treated with TiCl$_4$ at concentrations greater than 50 mM was still higher than that of the untreated standard film, despite the decreased pore width and BET specific surface area of the film. They concluded that the DSSCs treated with 15–50 mM TiCl$_4$ exhibit largest PCE. On further increasing the TiCl$_4$ concentration to 500 mM, the film pores got narrowed thereby resulting in the slow transport of electrons and the overall cell performance got declined.

The effect of TiCl$_4$ treatment on nanoporous TiO$_2$ films for the influence of film surface area was also reported by Yue et al. [12]. They reported that after the treatment with TiCl$_4$, the pore diameter of the TiO$_2$ film got decreased. In the course of treatment they noticed that nanoparticles from the hydrolysis of TiCl$_4$ solution filled those pores and the surface area of those films decreased from 65.7 to 51.3 m^2/g. Thus resulted in easy electron transfer and hence improved the charge collection efficiency, η_e (the percentage of electrons from the conduction band to TCO glass).

Lee et al. [13] optimized the morphology of electrospun TiO$_2$-nanorod (NR)-based photo-electrodes with TiCl$_4$ posttreatment. As a result of this, the charge transport and sensitizer adsorption were also optimized to achieve the high efficiency (>11%) and also a comparative study was made with TiO$_2$-NP-based photo-electrodes. The specific surface area and pore volume of TiO$_2$-NR and TiO$_2$-NP photoelectrodes were estimated by BET and Barrett–Joyner–Halenda (BJH) analysis. From these analyses, they measured the specific surface areas of TiO$_2$-NR photoelectrodes (123 m^2/g) that were approximately two times greater than that observed for TiO$_2$-NP photoelectrodes (56.0 m^2/g), thereby encountered the TiO$_2$-NR photoelectrodes for large sensitizer surface coverage than the traditional TiO$_2$-NP photoelectrodes. The pore volume of TiO$_2$-NRs and TiO$_2$-NPs were calculated as 1.28 and 0.61 cm^3/g, respectively. Finally, they concluded that the electrospun TiO$_2$-NR-based photoelectrodes had twice the pore volume of TiO$_2$-NP-based photoelectrodes and gave ~2.5 times higher sensitizer surface coverage at the same weight of TiO$_2$ and thereby exhibits higher energy conversion efficiencies than NP-DSSCs.

6.3.2 Dye Adsorption and Photocurrent Generation

TiCl$_4$ posttreatment greatly enhances the absorption of the wavelength range from 400 to 800 nm and dye adsorption with improved photocurrent. Lee et al. [14] achieved the enhancement in the performance of DSSCs with nanoporous TiO$_2$ nanotubes (TNTs) with different TiCl$_4$ posttreatment time from 30 to 120 minutes. More adsorption of the dye molecule

Table 6.4 The integral photocurrent density (J_{SC}), open-circuit voltage (J_{OC}), fill factor (FF), and efficiency (η) of dye-sensitized solar cells fabricated using TiCl₄ treatment on TiO₂ nanotubes film.

Sample	J_{SC} (mA/cm²)	V_{OC} (V)	FF (%)	η (%)
Bare TNTs film	12.53	0.61	60.78	4.69
TiCl₄ 90 mM (30 min)	18.94	0.60	62.57	7.16
TiCl₄ 90 mM (60 min)	20.09	0.60	62.50	7.59
TiCl₄ 90 mM (90 min)	20.72	0.61	63.90	8.08
TiCl₄ 90 mM (120 min)	20.26	0.60	66.92	7.71

was connected to the increase in light harvesting and J_{SC} as noted from Table 6.4. From the absorbance spectra, they noted that at TiCl₄ treatment for 120 minutes, the absorbance value was lower than TiCl₄ treatment for 90 minutes. Moreover, the reduction of the TNT film porosity and increased nucleation in the nanoparticle can cause the reduction of dye absorption of TNT film.

From Table 6.4, the authors analyzed that the DSSCs with bare TNTs film exhibits a J_{SC} of 12.53 mA/cm² and η of 4.69%. However, TiCl₄ treatment on TNTs film enhanced the energy conversion efficiency due to the increment in dye adsorption and charge transport. When the TNTs film was electrochemically dipcoated, the values of η increases with TiCl₄ treatment time and reached a maximum efficiency of 8.08% at 90 minutes and thereafter η value decreases. Finally, they proved that the amount of dye loading on the photo-anode enhances the efficiency of the cells.

The effect of TiCl₄ posttreatment with variations in the concentration of TiO₂ electrode was also explained by Eom et al. [3]. They found that the absorbance value increases with increasing TiCl₄ at the wavelength 400–500 nm and was highest for the sample treated with TiCl₄ (80 mM) and was low for TiCl₄ (120 mM). Therefore, the TiO₂ layer with more dye molecules adsorbed leads to more incident light harvesting, as well as larger photocurrent. For the DSSCs prepared without a posttreatment, it was noted that they had the short-circuit current density (J_{SC}) of 6.43 mA/cm², an open-circuit potential (V_{OC}) of 0.69 V, and a cell conversion efficiency of 3.03%. TiCl₄ posttreatment at 80 mM was observed as having a higher conversion efficiency range, with a photocurrent density (J_{SC}) of 10.43 mA/cm², an open-circuit potential (V_{OC}) of 0.67 V, and a cell conversion efficiency of 5.04%, and above this range at 120 mM, the value decreases. Hence, they finalized from this work that the posttreatment of TiCl₄ on TiO₂ was an effective method to improve the efficiency in DSSCs.

Sedghi and Miankushki reported on the effect of dye loading on pre- and posttreatment with TiCl₄ onTiO₂ electrodes in DSSCs and inferred that the capacity of dye loading exerts a profound influence on the photocurrent. From the UV–Vis optical transmittance spectra, it was noted that the maximum transmittances of untreated, posttreated, and pre- and posttreated TiO₂ films in the visible range were about 97%, 87%, and 85%, respectively. It was shown that with TiCl₄ treatment, the transmittance of the films decreases because the films become thicker and coagulated, whereas the surface area of the TiO₂ film and the number of adsorbed photons in the visible range increases. The TiCl₄-treated electrodes adsorb about 11% more photons than the untreated electrodes, which was attributed to the availability of more specific binding sites on the TiO₂ surface upon TiCl₄ treatment. On the basis of the findings made, it was stated that the short-circuit current was 7.82, 9.31, and 9.78 mA/cm² and efficiency was 3.75%, 4.7%, and 5.1% for untreated, TiCl₄ posttreated and pretreated films.

Table 6.5 Change in the parameters that influence the performance of solar cell after TiCl$_4$ treatment.

Sample	J_{SC} (mA/cm^2)	V_{OC} (V)	FF	η (%)
Untreated TiO$_2$ film	9.5 ± 0.2	0.68 ± 0.01	0.67 ± 0.004	4.3 ± 0.1
TiO$_2$ after TiCl$_4$ treatment	11.2 ± 0.2	0.69 ± 0.01	0.66 ± 0.004	5.1 ± 0.1
% of change	+18	+1.5	−1.5	+19

The influence of TiCl$_4$ posttreatment on nanocrystalline TiO$_2$ film electrodes in DSSCs are compared to nontreated films by Sommeling. They experimentally verified that in TiCl$_4$ treated and untreated TiO$_2$ films, the dye has been adsorbed and desorbed from the electrodes by treating with diluted NH$_3$, resulting in a dye solution of which a UV–Vis spectrum was recorded. It was noted that the difference between TiCl$_4$-treated and nontreated TiO$_2$ is distinctive, showing a 7.6% higher dye loading for TiCl$_4$-treated electrodes. Further TiO$_2$ film revealed that there is hardly any visible effect of the TiCl$_4$ treatment in the absorption maximum (550 nm) of the dye. This was due to the fact that the light absorption is already saturated around this maximum range, even without TiCl$_4$ treatment. In Table 6.5, they marked out that the increase in current due to the TiCl$_4$ treatment is 18%, whereas for the films, a higher dye loading of 7.6% was observed. It was obvious that the increase in current is much more than can be ascribed due to the increase in dye loading.

The dye adsorption mechanism in DSSCs was fabricated using the rutile nanorods that were differently treated with TiCl$_4$ solution and the rise in energy conversion efficiency was examined by Yang et al. [15]. They estimated the amount of adsorbed dye by the intensity of the absorption peak in UV–Vis spectra around 308 nm. Further, the actual surface area occupied by the dye was obtained based on the assumption that a single N719 dye molecule is around 1 nm^2 as N3. Roughness factor was accessed from the surface area and the values are 243, 346, and 491 for pure, 30 minutes treated, and 120 minutes treated nanorods, respectively. Hence, the authors confirmed that as the roughness factor is doubled after TiCl$_4$ treatment, and it is evident that the increase in the amount of the adsorbed dye significantly contributes to the increase in the photocurrent by the TiCl$_4$ treatment, it was noted from Table 6.6.

The importance of the temperature during posttreatment was studied by Nath et al. They reported that high thermal treatment was required for more effective and higher dye-loading of the compact TiO$_2$ overlayers with TiCl$_4$ posttreatment. TiCl$_4$-treated TiO$_2$ photo-electrodes result in an increase in light-absorption over wavelengths range from 325 to 600 nm in UV–Vis spectra. This indicated that the net surface area of TiO$_2$ increases with the thickness of the TiO$_2$ over layer, which was dependent on the TiCl$_4$ posttreatment conditions. Therefore, light absorption was maximum on the electrode PE/TiCl$_4$(HT)/N719/TiCl$_4$(LT)/N719 because of an increase in both dye-loading of 7.54 μM/cm^2 and light scattering in the TiO$_2$ photo-electrode. The larger absorption was demonstrated by the PE/TiCl$_4$(HT)/N719 electrode compared to that of the PE/N719/TiCl$_4$(LT)/N719 electrode, and suggests that the high-temperature annealing (500 °C) induces higher dye-loading, which was associated with the improved crystallinity.

Table 6.6 Change in photovoltaic parameters of TiO$_2$ nanorods with and without TiCl$_4$ treatment.

Sample	J_{SC} (mA/cm^2)	V_{OC} (V)	FF	η (%)
TiO$_2$ nanorod without TiCl$_4$	2.55	0.85	0.60	1.31
TiO$_2$ nanorod with TiCl$_4$ for 2 h	8.13	0.72	0.63	3.7

Table 6.7 Electrochemical impedance spectroscopy (EIS) parameters of DSSCs with various electrodes.

Photo-electrodes	Charge transport resistance, R_t (Ω)	Chemical capacitance, C (μF $\times 10^4$)	Charge recombination resistance, R_r (Ω)	Electron diffusion coefficient, D_n (cm²/s $\times 10^4$)
PE/N719	15.83	6.82	10.28	1.33
PE/N719/TiCl₄ (LT)/N719	16.51	8.55	11.46	1.02
PE/TiCl₄(HT)/ N719	3.51	5.76	17.01	7.12
PE/TiCl₄(HT)/ N719/TiCl₄ (LT)/N719	5.36	8.19	22.98	3.28

6.3.3 Electron Transport and Diffusion Coefficient

The effect of compact TiO_2 over layers, deposited on TiO_2 photo-electrodes through the hydrolysis of TiCl₄ posttreatment for electron transport and its performance was reported by Nath et al. [16]. They estimated the electron diffusion kinetic parameters in the TiO_2 photo-electrodes after TiCl₄ posttreatment by using electrochemical impedance spectroscopic (EIS). Table 6.7 summarizes all the corresponding EIS parameters. The PE/TiCl₄(HT)/N719 exhibits a lower C_μ than PE/N719 photo-electrodes without TiCl₄ treatment. Thus, the conventional TiCl₄ posttreatment reduces the surface states below the conduction band of TiO_2 nanoparticles and decreases the charge transport resistance (R_t) through the well-established TiO_2 nanoparticle networks. Furthermore, the reduced defect states result in an increased electron diffusion coefficient (D_n) in the TiO_2 photo-electrodes. Moreover, the increase in charge recombination resistance (R_r) at the TiO_2 electrolyte interface for the electrode PE/TiCl₄(HT)/N719 results in the suppression of charge recombination with electrolytes, while the cell with electrode PE/N719/TiCl₄(LT)/N719 shows higher C_μ than that without TiCl₄. This suggests that the increase in surface states due to crystal defects in the TiO_2 layer, caused by incomplete crystallization, increases the value of Rt through the TiO_2 nanoparticle networks. This study also concluded that the temperature of the thermal treatment and the thickness of the TiO_2 over layers on the TiO_2 nanocrystalline film affects the distribution of crystallinity and the density of surface states of the over layer itself and hence, influences the electron transport process and the overall recombination kinetics during cell operation.

The electron transport and recombination of the carriers in the porous coating layer on TiO_2 nanorods for TiCl₄ treatment at different time duration was reported by Yang et al. [15]. They examined the detailed change in the electron diffusion coefficients and lifetimes of the photo-generated electrons in the microscale by using step-light induced transient measurement of photo current and voltage (SLIM-PV) method. The diffusion coefficient is very sensitive to the surface state of rutile nanorods. As TiCl₄ treatment time increases, the diffusion coefficient decreases. When the nanoparticles directly coated on FTO substrate were used to fabricate DSSCs, the short-circuit current, FF, and energy conversion efficiency of the device became less. This indicates that the nanorods successfully extract electrons from the surface-coated nanoparticle layer. When the electrons passed through the nanoparticle electrodes, the carriers repeatedly experience trapping and de-trapping processes with increased diffusion time. The comparison made with the rutile nanoparticles and rutile nanorod. The

Table 6.8 Measured ($J_{SC\,AM1.5}$) and calculated ($J_{SC\,calc.}$) photocurrents for two cells ($d = 13.5\,\mu m$) prepared with and without the TiCl₄ treatment under "front" and "back" Illumination (substrate-electrode side (SE) and electrolyte-electrode side (EE) sides).

Cell illumination direction	$J_{(SC\,AM1.5)}$ [measured] (mA/cm²)	L_{IPCE} (µm)	L_{TRANS} (µm)	$\overline{\eta_{LH}}$ [AM1.5]	η_{inj}	$\overline{\eta_{col}}$ [L_{IPCE}]	$\overline{\eta_{col}}$ [L_{TRANS}]	$J_{SC\,calc.}$ [L_{IPCE}] (mA/m²)	$J_{SC\,calc.}$ [L_{IPCE}] (mA/m²)
No TiCl₄ SE	5.2	8.3	20	0.18	0.63	0.67	0.91	5.0	7.1
No TiCl₄ EE	3.8	8.3	20	0.16	0.63	0.49	0.85	3.4	5.9
With TiCl₄ SE	8.9	28	55	0.20	0.69	0.95	0.99	8.9	9.3
With TiCl₄ EE	7.8	28	55	0.18	0.69	0.91	0.98	7.7	8.3

TiCl₄-treated nanorod indicates that the electron diffusion is very effective in the rutile nanorods. In addition, the introduction of the rutile nanoparticle in the electron transport path produces the trapping sites and prevents the fast diffusion of the electron. The final observation stated that the dramatic improvement in the diffusion coefficient was achieved with single crystalline nanorod that has less trapping sites such as necks.

Barnes et al. [17] derived the value of diffusion length for TiO₂ electrode with and without TiCl₄ treatment in DSSCs from IPCE measurements. From the measurements for the cells without and with treatment, the fitted diffusion lengths of 8 and 22 µm corresponds to a difference in J_{SC} AM1.5 photocurrents of 8.6 and 11.4 mA/cm², respectively. This indicates that the addition of a TiO₂ layer to the film has indeed increased the photocurrent, and this is correlated with the observed increase in diffusion length (L). In Table 6.8, it was shown to have one more example, where J_{SC} in the solar simulator at "1 sun" increases from 5.2 to 8.9 mA/cm² with this treatment. In this casestudy, for a given charge concentration, the addition of the TiCl₄ treatment reduces the value of diffusion coefficient (D_n) by a factor of approximately 3. It was contradicted in a number of studies which conclude that the treatment increases the effective diffusion coefficient due to increase in the neck diameter between neighboring particles. The diffusion length of electrons was shown to increase in TiO₂ films treated with TiCl₄, and this change was attributed primarily to an increase in the effective electron lifetime despite a reduction in the diffusion coefficient. One of the significant conclusions drawn from the work done by them was that the electron diffusion length had a substantial influence on the device photocurrent at short circuit even when illuminating the TiO₂ side of the cell.

The treatment of TiCl₄ with nanocrystalline TiO₂ film enhances the electron transport in back contact dye-sensitized solar cell (BCDSC) which was reported by Fuke et al. [18] They reported that the impedance spectra of BCDSC resistances related to each impedance (Z_1, Z_2, and Z_3) are 0.63, 2.80, and 1.05 Ω cm² without TiCl₄ treatment and 0.63, 3.16, and 1.16 Ω cm² with TiCl₄ treatment, respectively. Generally, at a sintering temperature below 500 °C, the electron transport in the TiO₂ film reduced with decreasing sintering temperature due to remaining grain boundaries between TiO₂ particles. This new impedance element was therefore assigned to electron transport properties between TiO₂ nanoparticles. The TiCl₄ treatment reduces the flat regions due to the presence of an extra layer of TiO₂ in the sintered TiO₂ particles. As a result of this, TiCl₄ treatment enhances the electron transport properties and electron collection efficiency in the nanocrystalline TiO₂ film, especially between TiO₂ nanoparticles.

Lee et al. [13] reported the charge transport characteristics of DSSC based on electrospun TiO₂ nanorod photo-electrode with TiCl₄ posttreatment. The efficiency of charge collection was determined by the competition between recombination and transport of electrons within the photo-anode. For most of liquid electrolyte based DSSCs, charge transport was dominated by the diffusion of electrons in the absence of a significant electric field because of the photo-electrode dimensions and the high dielectric constant of the TiO_2 photo-electrode. The charge collection efficiency, η_{cc}, was 21.5% greater for nanorod (NR)-DSSCs than that for NP-DSSCs at a light intensity (incident photon flux) of 8.1×10^{16} cm^{-2}/s. The diffusion coefficient (D_n) of posttreated NR-DSSCs (P-NR-DSSC) was 51% higher than that of untreated NR-DSSCs at a light intensity of 8.1×10^{16} cm^{-2}/s, in accordance with the observed increase (19%) in charge collection efficiency. The TiO_2 posttreatment presumably enhances the inter particle connectivity at NR grain boundaries and made the electron transport less obstructed by attenuating the scattering of free carriers. Finally, they inferred that an efficiency of 9.52% was achieved for nanorod-based DSSCs.

6.3.4 Recombination Losses at Short Circuit

$TiCl_4$ treatment on TiO_2 nanocrystalline films greatly reduces the recombination rate of electrons in the working electrode. The transport lifetime and the recombination lifetime under short-circuit conditions were clearly explained by Sommeling et al. From their report, it was clear that the transport under short-circuit conditions was much faster than the recombination and thus no losses to recombination occurs at short circuit. More quantitatively, the recombination lifetime was 40 ms, where the transport half times were about 900 and 700 μs. Under those conditions about 2% of the charges were lost to recombination, so $\Phi COLL = K_t/(K_t + K_r)$ was close to unity. Thus, even if large improvements in transport had been measured, they could not have been responsible for the increase in short-circuit photocurrent. It was concluded by them that $\Phi COLL$ was not increased by the $TiCl_4$ treatment. This only leaves ΦINJ as a candidate for improvement of IPCE/current. It was noted that since the $TiCl_4$ treatment only increases the TiO_2 mass by a factor of 1.4 an overall increase in traps of a factor 2.3 implied that the $TiCl_4$-treated TiO_2 had a trap state density that is 3.5 times higher than the original TiO_2 particles. Moreover, such a large increase in trap density had a chance to slow the transport and also to increase the amount of charge stored in the film under short-circuit conditions that were not observed. A shift in band edge was the most probable explanation for that change. It may be noted that a downward shift in the conduction band should give a lower V_{OC}. This does not occur because the $TiCl_4$ treatment also acts too strongly to reduce the recombination rate. The reduction in the recombination rate allows the 1 sun flux of injected electrons to build up a higher density of charge in the TiO_2. They concluded that the reduction in the recombination rate was responsible for the downward shift in the conduction band that leads to an enhancement in the photocurrent.

Regan et al. [19] investigated the quantification of recombination losses at the short circuit for the TiO_2 films treated with $TiCl_4$ posttreatment. They noted that the decreased recombination rate at V_{OC} is implied for the short circuit. This reduced recombination losses at short circuit in the $TiCl_4$-treated cell relative to the non-$TiCl_4$ cell causes an increase in the photocurrent. However, this increased photocurrent was not larger than the existing recombination losses in the non-$TiCl_4$ cell. The decay time of the photovoltage was 40 ms, corresponding to the transient recombination lifetime at short circuit. Also showed that the photocurrent decays at short circuit for the same cell, which has a lifetime of <1 ms. Moreover, the implication from the experiment was that recombination losses at short circuit in the non-$TiCl_4$ cell are ≤2.5%. They also estimated the recombination losses from the steady-state charge density at short circuit. It shows that the excess charge in the TiO_2, at short circuit was ~10 μC/cm^2, which corresponds to a charge density of 5×10^{17}/cm^3. The recombination rate followed the same function of charge density at open circuit and at short circuit; and this gave 43 ms for the recombination lifetime at short circuit under 1 sun illumination.

Table 6.9 Properties of TiO$_2$ passivating layer with various dipping time and different concentration with TiCl$_4$.

Sample		J_{SC} (mA/cm^2)	V_{OC} (V)	FF (%)	η (%)
40 mM	Bare	6.068	0.671	69.0	2.81
	30 min	6.772	0.677	69.6	3.19
	60 min	7.235	0.687	70.2	3.49
	90 min	7.264	0.719	69.6	3.62
	120 min	7.394	0.739	59.8	3.27
20 mM		7.032	0.700	68.1	3.35
40 mM		7.264	0.719	69.6	3.62
80 mM		7.956	0.736	65.2	3.82
120 mM		6.278	0.710	80.4	3.58

6.3.5 Concentration and Dipping Time of TiCl$_4$

The performance of DSSCs was highly affected by TiCl$_4$ treatment time and concentration of the solution. Eom et al. reported that the solar cells without a passivating layer had a short-circuit current density (J_{SC}) of 6.068 A/cm^2, an open-circuit potential (V_{OC}) of 0.671 V, and a cell conversion efficiency of 2.81%. The DSSCs fabricated with the 90-min dipped TiO$_2$ passivating layer substrate shows the efficiency of (3.62%) as a result of increased photocurrent density. A higher conversion efficiency of the DSSC with the 80-mM TiCl$_4$ treated passivating layer shows a photocurrent density (J_{SC}) of 7.956 A/cm^2, an open-circuit potential (V_{OC}) of 0.710 V and a cell conversion efficiency of 3.82%. Table 6.9 summarizes the various properties of the TiO$_2$ passivating layers with various dipping times and also at different concentrations of TiCl$_4$ in DSSCs. The J_{SC} and V_{OC} increase significantly owing to the effects of the TiCl$_4$ treatment, because a suitable amount of TiCl$_4$ in the film provided a large surface area for dye adsorption. In fact, TiCl$_4$ concentrations higher than 120 mM favored an increase in the nucleation of nanoparticles through reduced film porosity. Consequently, the inefficient charge-transfer paths increase the recombination rate of electrons, resulting in a low photocurrent density and conversion efficiency.

Lee et al. [11] reported the effects of TiCl$_4$ treatment (5–500 mM) on TiO$_2$ nanocrystalline films for the film morphology, charge carrier dynamics, and the performance of DSSCs. The TiCl$_4$ treatments results in about 10–40% improvement of short-circuit current density (J_{SC}), depending on the TiCl$_4$ concentration. On the other hand, the FF and V_{OC} of the TiCl$_4$-treated DSSCs gradually decreases with increasing TiCl$_4$ concentrations with the exception of the 5 mM TiCl$_4$-treated sample, where the V_{OC} was slightly larger (by about 20 mV) than that of the untreated sample. The DSSCs treated with 15–50 mM TiCl$_4$ exhibits the largest PCE and results primarily from the increases in J_{SC}. In the case of films treated with TiCl$_4$ concentrations in the range of 200–500 mM, however, the cell efficiencies were lower than that of the DSSC containing the untreated film owing to the lower FF and V_{OC}, which more than compensated for the higher J_{SC}. They concluded that the TiCl$_4$ concentration effects the surface treatment on the TiO$_2$ film morphology and electrical properties were valuable for developing more effective nanostructured electrodes for various applications.

The effect of TiCl$_4$ posttreatment with various concentrations on DSSCs which were constructed with composite films made of TiO$_2$ nanoparticles and TiO$_2$ nanotubes (TNTs) was reported by Yang et al. [20]. They reported that the FF was higher due to the treatment. In

Table 6.10 Different parameters of DSSCs fabricated using TiO_2 particles/TNTs 10 wt% (bare), and those fabricated using $TiCl_4$ posttreatment.

Sample	J_{SC} (mA/cm^2)	V_{OC} (V)	FF (%)	η (%)
Bare TNT (10 wt%)	14.86	0.68	58.79	5.95
30 mM	15.45	0.67	61.37	6.42
60 mM	16.02	0.70	63.75	7.16
90 mM	17.37	0.70	63.84	7.83
120 mM	11.83	0.68	68.53	5.55
Bare TiO$_2$	6.43	0.69	67	3.03
30 min	9.80	9.80	70	4.80
60 min	7.88	7.88	71	3.82
90 min	7.23	7.23	70	3.49
20 mM	8.90	8.90	65	4.19
40 mM	9.80	9.80	70	4.80
80 mM	10.43	10.43	71	5.04
120 mM	6.27	6.27	80	3.58

Source: Adapted from Tae Sung Eom et al. 2014 [3].

addition, solar cells with a high FF had a stable output voltage and current compared to the cell with the same V_{OC} and J_{SC}, which produced more power. The photovoltaic properties of all posttreated films were summarized in Table 6.10. The J_{SC} value increases with the amount of $TiCl_4$ until the $TiCl_4$ concentration was 90 mM, beyond this limitation, J_{SC} decreases. The improved value of J_{SC} was due to the increase of dye adsorption on TiO_2 films, which results in an improvement of J_{SC}, and a charge-transfer resistance at interfaces was decreases. In the case of $TiCl_4$ at 120 mM, the J_{SC} value decreases due to low absorption of dye from the TiO_2 film to the photo-electrode, but FF increases from 58% to 68% after $TiCl_4$ posttreatment. The results showed that the DSSCs using a $TiCl_4$ (90 mM) posttreatment gave the maximum conversion efficiency of 7.83% due to effective electron transport and enhanced adsorption of dye on TiO_2 surface.

The influence of $TiCl_4$ posttreatment on various concentrations and dipping time such as 20– 150 mM and 30–120 minutes was reported by Eom et al. The photovoltaic properties of all dipping time and concentration of the cell were summarized in Table 6.10. The DSSC prepared without a posttreatment had a short-circuit current density (J_{SC}) of 6.43 mA/cm^2, an open-circuit potential (V_{OC}) of 0.69 V, and a cell conversion efficiency of 3.03%. The DSSCs fabricated on the 30 minutes posttreatment substrate showed the efficiency (4.80%) value due to increased photocurrent density. The DSSC using $TiCl_4$ posttreatment at 80 mM was observed as the higher conversion efficiency, with a photocurrent density (J_{SC}) of 10.43 mA/cm^2, an open-circuit potential (V_{OC}) of 0.67 V, and a cell conversion efficiency of 5.04%. As a result, higher V_{OC} and the short-circuit current (J_{SC}) was obtained for 80 mM and 30 minutes posttreatment $TiCl_4$ compared to the bare cell due to effective electron transport.

6.4 Conclusion

Though $TiCl_4$ treatment leads to some improvement, the degree of improvement is quite variable, and depends strongly on the starting TiO_2 material to which it is applied. Up to now, a real optimization of the TiCl$_4$ treatment has been impossible, in part due to the lack of knowledge concerning the mechanism of action. The authors made various perspects of TiCl$_4$ posttreatment and its significant effects on surface morphology, short-circuit current, recombination, dye adsorption, concentration, and possibly some effects on transport too. These effects have to be optimized separately for enhancing the efficiency of dye-sensitized cells.

References

1 O'Regan, B. and Grätzel, M. (1991). A low-cost, high-efficiency solar cell based on dye-sensitized colloidal TiO$_2$ films. *Nature* 353: 737.

2 Sedghi, A. and Miankushki, H.N. (2013). Influence of TiCl$_4$ treatment on structure and performance of dye-sensitized solar cells. *Jpn. J. Appl. Phys.* 52: 075002.

3 Eom, T.S., Kim, K.H., Bark, C.W., and hoi, H.W. (2014). Influence of TiCl$_4$ post-treatment condition onTiO$_2$ electrode for enhancement photovoltaic efficiency of dye-sensitized solar cells. *J. Nanosci. Nanotechnol.* 14: 7705–7709.

4 Dhiraj Saxena, M., Mridul Kumar, M., Sharma, G.D., and Roy, M.S.(2012). Effect of surface treatment of nanostructured-TiO$_2$ on the efficiency of dye-sensitized solar cell based on iron thalocyanine. *International Conference on Electrical and Electronics Engineering*, Bangkok, June 16-17.

5 Eom, T.S., Kim, K.H., and Choi, H.W. (2014). Enhancing performance of dye-sensitized solar cells by TiCl$_4$ treatment at different concentrations. *Jpn. J. Appl. Phys.* 53: 06JG10.

6 Sommeling, P.M., O'Regan, B.C., Haswell, R.R. et al. (2006). Influence of a TiCl$_4$ post-treatment on nanocrystalline TiO$_2$ films in dye-sensitized solar cells. *J. Phys. Chem. B* 110: 19191–19197.

7 Kim, J.K., Shin, K., Lee, K.S., and Park, J.H. (2010). Influence of a TiCl$_4$ treatment condition on dye-sensitized solar cells. *J. Electrochem. Sci. Technol.* 1 (2): 81–84.

8 Barbe, C.J., Arendse, F., Comte, P. et al. (1997). Nanocrystalline titanium oxide electrodes for photovoltaic applications. *J. Am. Ceram. Soc.* 80: 3157–3171.

9 Kim, K.-P., Kim, J.-H., Hwang, D.-K. et al. (2015). Effect of TiCl$_4$ post-treatment on the embedded-type TiO$_2$ nanotubes dye-sensitized solar cells. *J. Nanosci. Nanotechnol.* 15: 7845–7847.

10 Ito, S., Liska, P., Comte, P. et al. (2005). Control of dark current in photoelectrochemical (TiO$_2$/I$^-$ -I3$^-$) and dye-sensitized solar cells. *Chem. Commun.* 34: 4351–4353.

11 Lee, S.-W. and Ahn, K.-S. (2012). Effects of TiCl$_4$ treatment of nanoporous TiO$_2$ films on morphology, light harvesting, and charge-carrier dynamics in dye-sensitized solar cells. *J. Phys. Chem. C* 116: 21285–21290.

12 Long-Yue, Z., Song-Yuan, D., Kong-Jia, W. et al. (2004) Mechanism of enhanced performance of dye-sensitized solar cell based TiO$_2$ films treated by titanium tetrachloride. *Chin. Phys. Lett.* 21: 1835–1837.

13 Lee, B.H., Song, M.Y., Jang, S.-Y. et al. (2009). Charge transport characteristics of high efficiency dye-sensitized solar cells based on electrospun TiO$_2$ nanorod photoelectrodes. *J. Phys. Chem. C* 113: 21453–21457.

14 Lee, J.S., Kim, K.H., Kim, C.S., and Choi, H.W. (2015). Achieving enhanced dye-sensitized solar cell performance by TiCl$_4$/Al$_2$O$_3$ doped TiO2 nanotube array photoelectrodes. *J. Nanomater.*: 1–6.

15 Yang, M., Ding, B., Lee, S., and Lee, J.-K. (2011). Carrier transport in dye-sensitized solar cells using single crystalline TiO$_2$ nanorods grown by a microwave-assisted hydrothermal reaction. *J. Phys. Chem. C* 115: 14534–14541.

16 Deb Nath, N.C., Subramanian, A., Hu, R.Y. et al. (2015). Effects of $TiCl_4$ post-treatment on the efficiency of dye-sensitized solar cells. *J. Nanosci. Nanotechnol.* 15: 8870–8875.

17 Barnes, P.R.F., Anderson, A.Y., Koops, S.E. et al. (2009). Electron injection efficiency and diffusion length in dye-sensitized solar cells derived from incident photon conversion efficiency measurements. *J. Phys. Chem. C* 113: 1126–1136.

18 Fuke, N., Katoh, R., Islam, A. et al. (2009). Influence of $TiCl_4$ treatment on back contact dye-sensitized solar cells sensitized with black dye. *Energy Environ. Sci.* 2: 1205–1209.

19 O'Regan, B.C., Durrant, J.R., Sommeling, P.M., and Bakker, N.J. (2007). Influence of the $TiCl_4$ treatment on nanocrystalline TiO_2 films in dye-sensitized solar cells. 2. Charge density, band edge shifts, and quantification of recombination losses at short circuit. *J. Phys. Chem. C* 111: 14001–14010.

20 Yang, J.H., Bark, C.W., Kim, K.H., and Choi, H.W. (2014). Characteristics of the dye-sensitized solar cells using TiO_2 nanotubes treated with $TiCl_4$. *Materials* 7: 3522–3532.

7

Doped Semiconductor as Photoanode

K. S. Rajni and T. Raguram

Department of Sciences, Amrita School of Engineering, Amrita Vishwa Vidyapeetham, Coimbatore, India

7.1 Introduction

Dye-sensitized solar cells (DSSCs) are currently attracting widespread academic and commercial interests for the conversion of sunlight into electricity [1–8]. DSSCs are third-generation solar cells, which overcome the Shockley–Queisser limit of power efficiency for single band gap solar cells. These devices can be operated in extensive range of lighting conditions that make them suitable for various arrays of shaded and diffused light locations, without suffering from angular dependence of sunlight or light. The performances of these cells are not affected by temperature, direct sunlight, climate, etc. The main advantages of these solar cells are that they are highly efficient, low cost, easy fabrication, and eco-friendly. As a result they are versatile and can be incorporated into a wide variety of products. The key components of DSSC are transparent conducting oxide (TCO) fluorine-doped tin oxide/indium-doped tin oxide (FTO/ITO), semiconducting oxide materials such as TiO_2, ZnO, SiO_2 which acts as photoanodes, dyes such as inorganic/organic dyes as sensitizers, LiI/I as an electrolyte and platinum counter electrode. For an efficient DSSC,

- TCO should have high conductivity and high transparency.
- Photoanode material should have high surface area, high porosity, high conductivity, and high stability.
- Sensitizers should have wide range of absorption, high absorption coefficient, high-anchoring property, high stability, and optimum redox potential.
- Electrolyte system must be a stable redox material, optimum redox potential, and high interaction with dye, high stability and good solvent.
- Counter electrode material of high surface area, high catalytic activity, and high stability.

The working principle of DSSC is, when the device is illuminated, the dye is excited, promoting electrons from the highest occupied molecular orbital (HOMO) to the lowest unoccupied molecular orbital (LUMO). This excited state injects electrons into the conduction band (CB) of photoanode, which diffuses through the mesoporous scaffold to the TCO. The holes are transferred from the HOMO of the dye to the hole transporting material (HTM) and from there to the back electrode [9].

Interfacial Engineering in Functional Materials for Dye-Sensitized Solar Cells, First Edition.
Edited by Alagarsamy Pandikumar, Kandasamy Jothivenkatachalam and Karuppanapillai B. Bhojanaa.
© 2020 John Wiley & Sons, Inc. Published 2020 by John Wiley & Sons, Inc.

7.2 Photoanode

The dye-sensitized nanocrystalline semiconductor electrode is the heart of the device and plays an important role in determining the DSSC performance. To develop high-performance DSSCs, many scientists have made great efforts in this field [10–13]. Conventionally, TiO_2 is an electron transporting material consisting of mesoporous metal oxide coated onto the TCO, and is used as photoanode. Several other metal oxides such as SnO_2, ZnO, and $SrTiO_3$ have also been explored, but none of them perform as good as TiO_2 [9]. Until now, nanocrystalline TiO_2 film is the most successful photoanode of DSSCs because they show unique photoelectronic and photochemical properties, including high chemical stability, ideal position of the CB edge, and environment friendly [14]. The large band gap, suitable band edge levels for charge injection and extraction, long lifetime of excited electrons, exceptional resistance to photo-corrosion, non-toxicity and low cost have made TiO_2 a popular material for solar energy applications [15–17].

TiO_2 occurs in three crystalline forms such as anatase (tetragonal), rutile (tetragonal), and brookite (orthorhombic). For DSSCs, anatase is the most commonly used phase due to its superior charge transport. The tetragonal anatase crystal structure is made up of a chain of distorted TiO_6 octahedrons, which results in a unit cell containing four Ti atoms and eight O atoms [18–20]. DSSCs employing rutile TiO_2 generally suffer from a lower CB compared to anatase, leading to lower open-circuit voltage (V_{OC}). In addition, reduced dye adsorption and charge transport, lower the short circuit current (J_{SC}). Because of these complications, rutile is not frequently used in DSSCs. The band gap of anatase TiO_2 is approximately 3.2 eV and the resistivity is 10^{15} Ω cm [21]. The lower edge of the CB is made up of vacant Ti^{4+} 3d bands and the upper edge of the valence band (VB) is made up of filled O^{2-} 2p bands [22]. Bulk oxygen vacancies, titanium interstitials, and reduced crystal surfaces generate shallow electron traps that can enhance the conductivity of TiO_2 [9]. An effective way of modifying the electronic properties of TiO_2 is doping, the deliberate insertion of impurities into the TiO_2 lattice. Doping can be achieved by either replacing the Ti^{4+} cation that affects the CB structure or the O^{2-} anion that affects the VB structure. Cationic dopants are typically metals, whereas anionic dopants are nonmetals. The atomic radius of the dopant should not differ much from the ion it replaces to prevent lattice distortion, introducing new defects that may hamper device performance. Because dye molecules anchor to Ti atoms, the replacement of Ti with another cation also affects dye adsorption due to different binding strengths between the dye and the dopant, or induced oxygen vacancies by the dopants [23]. Dopants often inhibit the growth rate of the TiO_2 nanoparticles (NPs), resulting in smaller particles. The increased surface area accommodates more dye, leading to higher light absorption and current densities. The main advantage of high light absorption is that thinner films can be used in photovoltaic devices, resulting in a reduction of recombination, which benefits both J_{SC} and V_{OC} [9].

The dopant can be introduced into TiO_2 lattice in a number of ways: The most common method is simply mixing a dopant precursor with the TiO_2 precursor solution. This method can be employed in the sol–gel, hydrothermal, solvothermal, spray pyrolysis, atomic layer deposition, electrochemical deposition, sonochemical, microwave, and electrospinning methods [19].

The dopants can be grouped into separate categories that share common electronic configurations. These are earth alkali metals, metalloids, nonmetals, transition metals, posttransition metals, and lanthanides.

In some cases, codoping with two or more dopants is applied to further increase device performance. Each dopant can separately enhance device properties [23]. One dopant can reinforce the effect of the other dopant [24], or one dopant may counteract some of the detrimental effects caused by the other dopants [25].

7.3 Characterization

Computational modeling can be used to predict and design material properties. Density functional theory (DFT) has become a popular method due to its low computational cost, making it possible to quickly screen dopants for their suitability to increase TiO_2 properties [9]. X-Ray diffraction (XRD) is a precise method to determine the crystal structure of TiO_2. From XRD, we calculated the average crystallite size using Scherrer equation, to determine lattice constants, microstrain, and dislocation density of the prepared samples. The amount of dye adsorbed on the TiO_2 surface is typically a good measure to determine how many photons will be absorbed. It depends on the porosity of the mesoporous of TiO_2, which can be measured by gas adsorption. Gas adsorption isotherms are typically analyzed by the Brunauer–Emmett–Teller (BET) theory [26]. From a spectroscopic technique such as UV–Vis spectroscopy, X-ray photoelectron spectroscopy, FTIR spectroscopy, and Raman spectroscopy studies, the absorbance, band gap, compositional, and functional group of the material is analyzed. Electromagnetic measurements such as Hall effect and electron paramagnetic resonance analysis are used to study the interaction between magnetic fields, and free charges in the sample makes it possible to determine the type of conduction, carrier concentration, and detect trap states. The photo-electrochemical measurements such as photovoltaic properties (I–V curves), electrochemical impedance spectroscopy (EIS), and flat band potential. From I–V curves, open-circuit voltage (V_{OC}), short-circuit current density (J_{SC}), fill factor (FF), and Efficiency (η) can be calculated. V_{OC} is determined by the difference between the redox potential of the electrolyte and the quasi-Fermi level of the semiconductor oxide photoelectrode. From impedance analysis, the frequency domain ratio of the voltage with respect to the current and is a complex value. The resulting function can be visualized with a Nyquist plot, where the real part of the function is plotted on x-axis and the imaginary part on the y-axis. The system can be described by an equivalent circuit consisting of parallel and series connected elements. From this charge transfer and transport processes and capacitance can be extracted [27, 28].

7.4 Doped TiO₂ Photoanodes

7.4.1 Alkali Earth Metals-doped TiO₂

Alkali metals such as lithium, magnesium, and calcium are considered as cationic dopants for TiO_2 as the outer electron shell of these metals can be easily donated [9].

7.4.1.1 Lithium-doped TiO₂

Lithium increases the conductivity of the TiO_2 electrodes as the Li insertion in the TiO_2 lattice causes the electrons to enter the CB of TiO_2 to keep the overall charge neutrality. Alagesan Subramanian et al. reported when Li is doped with TiO_2, half of the interstitial octahedral sites are randomly distributed by Li ions of different polymorphs of TiO_2: anatase phase accommodates different ratios of Li and rutile and brookite phase accommodate small amount of Li [29]. Also Li^+ insertion into anatase phase results in transition from tetragonal TiO_2 (space group $I4/amd$) to orthorhombic $Li_{0.5}TiO_2$ (space group *Imma*) [30]. The fundamental process of the lithium reaction mechanism in nanocrystalline anatase particles is not well understood, such that the smaller particle showed high lithium storage, whereas the bigger particles showed low lithium storage capacity [28–32]. The comparison between (i) P25 & LiP25, (ii) P90 & LiP90, and (iii) MW170 & LiMW170 shown by Alagesan Subramanian et al. [29]. The influence of crystallite size with the Li incorporation is well established in this study. The decreased crystallinity and the increased d spacing show the Li^+ incorporation into the TiO_2 lattice. The unit cell of the anatase phase increases by ~0.3% upon lithium insertion was reported [31]. Though no peak shift is observed for P90 and MW170 on Li doping, a slight intensity reduction was found for the LiP90 sample, and intensity increases for the LiMW170 sample. Thus, the storage capability of anatase nanoparticles is strongly dependent on the particle size, namely, the smaller the particle size, and the higher the capacity [28–30]. This clearly indicates that Li ions can be inserted into both the surface and bulk of the nanostructured anatase particles, based on different mechanisms [28]. From EIS, the ohmic resistance in the high-frequency region corresponds to the electrolyte/FTO resistance, resistance in the middle frequency region belongs to TiO_2/electrolyte interface, and Nernstian diffusion within the electrolyte is in the low-frequency region [33, 34]. The diameter of the semicircle in the 1–100 Hz region in the spectra is a representative of charge-transfer resistance at the TiO_2/dye/electrolyte interface [35]. In Figure 7.1, the first semicircle represents the electrolyte/FTO resistance, and the different initial resistance values were observed in the impedance plots. The diameter of the second

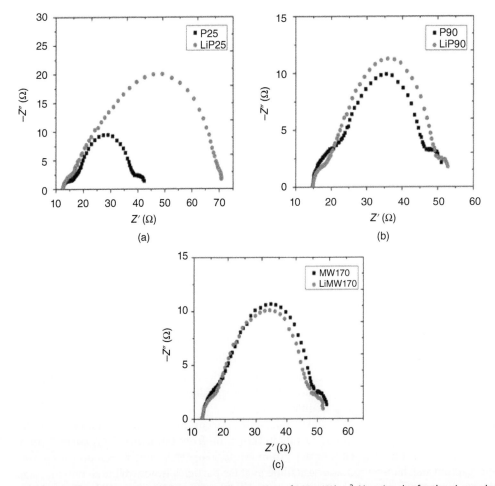

Figure 7.1 EIS analysis of the DSSCs under an illumination of 100 mW/cm^2. Nyquist plot for the electrodes made from (a) P25 & LiP25, (b) P90 & LiP90, and (c) MW170 & LiMW170, respectively. *Source:* Subramanian et al. 2012 [29]. Reproduced with the permission of Elsevier.

semicircle increases for the Li-doped P25 and Li-doped P90 samples (LiP25 & LiP90) and such an increase in mid-frequency domain is the result of loss of reduction in electron recombination rate at the TiO_2/dye/electrolyte interface that consequently, influences the conversion efficiencies of the solar cells. However, the loss of electrons at the TiO_2 dye/electrolyte interface for the LiP90 sample is smaller compared to that of the LiP25 due to some surface Li^+ on the TiO_2 nanoparticles and hence, some temporary electron trap occurs on the surface. Comparing the MW170 and LiMW170 samples, the diameter of the second semicircle of the LiMW170 is smaller than that of the MW170 sample. This proves that large number of electrons are available in the case of LiMW170 sample due to surface Li^+ ion for the facile transport of the electrons and reduce the corresponding resistance at the interface. The $I-V$ characteristics study shows the efficiency, V_{OC}, J_{SC}, and FF of P25 and is found to be 4.18%, 0.80 V, 7.62 mA/cm², and 0.69 with thickness10 μm, and doped samples are shown in Table 7.1.

7.4.1.2 Magnesium-doped TiO₂

$MgTiO_3$ and MgTi have higher band gaps than pure TiO_2 [36, 37].The doping of Mg shift in the optical absorption of Mg–TiO_2 is shown by Shinji Iwamoto et al. [38]. The observed shifts suggest the formation of Mg–Ti mixed oxides; however, the local structure of Mg ions is not elucidated sufficiently. Photoelectrochemical properties of DSSCs with Mg(x)–TiO electrodes using N719 shows V_{OC} values have increased slightly using Mg-containing TiO_2 electrodes. V_{OC} values for the DSSC using TiO_2 electrodes and I^-/I_3^- electrolytes are known to be around 0.9 V, and Mg–Ti shows that values higher than 0.9 V indicate the Fermi levels of these electrodes under working conditions are negatively shifted as compared to those of pure TiO_2, which is consistent with the shift of the flat-band potential estimated from the band gap energy. A very small amount of the barrier layers, about 0.5 wt%, is effective to function as a blocking layer, which prohibits the recombination reaction, hence increasing V_{OC}, whereas increased amounts of MgO coating result in a drastic decrease in J_{SC} [39, 40].

7.4.1.3 Calcium-doped TiO₂

Doping of calcium (Ca) in TiO_2 results in the shift to lower theta values because of the larger effective radius of Ca^{2+} (1.0 Å) compared to that of Ti (0.61 Å) reported by Qiuping Liu et al. [41]. TiO_2 and Ca-doped TiO_2 have a similar morphologies and uniform particle size distribution. The UV–Vis absorption spectra indicate that the doping of Ca boron induces the shift and an increase in visible light absorption due to a small change of the band gap when Ca boron was in interstitial positions [42]. The current–voltage curves of the DSSCs based on the TiO_2 and four different Ca-doped TiO films under light intensity of 100 mW/cm² at AM1.5 was analyzed. The photovoltaic parameters J_{SC}, V_{OC}, FF, and η for undoped TiO_2 is 16.7 mA/cm², 675 mV, 0.65, and 7.33% and after 2% of Ca-doping, the performance of DSSC based on TiO_2 was enhanced to 19.2 mA/cm², 0.649 mV, 0.67, and 8.35% (Table 7.1). Also, the dye adsorption has no influence on the enhancement of photocurrent The increased J_{SC} in Ca-doped TiO_2 film based on DSSCs can be ascribed to two main factors: (i) the divalent Ca^{2+} ions was doped into the TiO_2 lattice and occupies the qua-divalent Ti, which causes an increased net in the electron concentration, and thus increases the electrical conductivity of Ca-doped TiO_2. (ii) The injection efficiency was faster than the relaxation of the excited state of the sensitizer dye.

7.4.2 Metalloids-doped TiO₂

The elements that belong to the metalloid group have properties that lie between metals and nonmetals, offering the possibility to combine the positive doping effects of metals and nonmetals. The metalloids normally used for doping are boron, silicon, germanium, and antimony [9].

Table 7.1 Photovoltaic parameters of doped TiO_2.

No.	Groups	Dopants	V_{OC} (V)	J_{SC} (mA/cm^2)	FF	η (%)	References
1	(Earth) Alkali metals	P25	0.80	7.62	0.69	4.18	[26]
		Lithium	0.84	4.53	0.67	2.54	[26]
		Magnesium	0.75	7.79	0.75	4.4	[32]
		Calcium	0.64	19.2	0.67	8.35	[41]
2	Metalloids	Boron	0.69	14.1	0.62	6.1	[43]
		Silicon	0.60	0.55	0.54	0.18	[46]
		Germanium	0.69	4.9	0.71	2.4	[48]
		Antimony	0.63	18.72	0.68	8.13	[14]
3	Nonmetals	Carbon	0.50	0.25	0.47	4.42	[50]
		Nitrogen	0.66	6.49	0.39	1.68	[55]
		Fluorine	0.79	13.69	0.76	8.24	[58]
		Sulfur	0.571	7.50	0.482	2.08	[61]
		Iodine	0.71	14.1	0.67	7.0	[62]
4	Transition metals	Scandium	0.752	19.10	0.675	9.6	[67]
		Vanadium	0.687	17.6	0.65	7.80	[68]
		Chromium	0.705	11.34	0.69	6.35	[69]
		Manganese	0.656	4.24	0.66	1.85	[75]
		Iron	0.66	12.62	0.687	5.76	[76]
		Cobalt	0.600	3.12	0.57	1.06	[75]
		Nickel	0.653	10.01	0.618	4.04	[77]
		Copper	0.64	1.80	0.38	0.44	[81]
		Zinc	0.56	15.81	0.64	3.122	[85]
		Yttrium	0.81	23.9	0.472	9.18	[86]
		Zirconium	0.64	10.66	0.61	4.16	[91]
		Niobium	0.84	20.58	0.49	8.53	[68]
		Molybdenum	0.76	17.51	0.53	7.16	[98]
		Silver	0.724	13.04	0.62	5.85	[99]
		Tantalum	0.665	19.7	0.65	8.18	[68]
5	Posttransition metals	Aluminum	0.64	18.0	0.63	7.26	[101]
		Gallium	0.75	13.4	0.79	8.1	[102]
		Indium	0.73	16.38	0.66	7.96	[103]
		Tin	0.710	15.92	0.72	8.14	[107]
6	Lanthanides	Lanthanum	0.68	15.0	0.689	7.0	[114]
		Cerium	0.74	12.5	0.76	7.12	[117]
		Neodymium	0.85	9.63	0.76	6.19	[120]
		Samarium	0.863	14.0	0.50	6.08	[123]
		Europium	-	0.260	0.33	1.01	[125]
7	Co-doped TiO_2	Erbium and ytterbium	0.773	18.9	0.615	8.98	[126]
		Chromium and gadolinium	0.530	6.82	0.55	1.99	[127]
		Nitrogen and lanthanum	0.727	10.518	0.69	5.33	[128]
8	Tri-doped TiO_2	Ho^{3+}, Yb^{3+}, F^-	0.73	21.60	0.57	8.93	[129]

7.4.2.1 Boron-doped TiO$_2$

The XRD result implies that the boron-doped and undoped samples consist of anatase-phase of TiO$_2$ and no boron phase, which could be attributed to the uniform distribution of boron among the anatase crystallites, or the quantity of the doping boron being too low is observed by Huajun Tian et al. [43]. The (101) anatase peak intensity of B–TiO$_2$ is larger than that of undoped TiO$_2$, indicating the improved crystallinity of the doped TiO$_2$. Boron doping results in a blue shift of the absorption spectra due to a small increase of the band gap when boron is in interstitial positions and the following process could be happening upon boron doping:

$$B + 3Ti^{4+} \rightarrow B^{3+} + 3Ti^{3+} \tag{7.1}$$

Boron doping reduces the valence of the titanium ion and concomitantly suggests an increase in the oxygen vacancy [44]. Gopal et al. also proposed that boron doping favors formation of an oxygen vacancy with two excess electrons, which would further reduce two Ti^{4+} ions to form Ti^{3+} [45]. The Nyquist and Bode plots (ZView software) of DSSCs show that boron doping is measured at a forward bias of 0.72 V in the dark. It is observed that the semicircles at the frequency range become smaller with increasing boron content, owing to the enhanced electron conductivity between the TiO$_2$ particles, and the fitted Nyquist plot data indicate the interfacial charge recombination resistance R of the undoped and the boron-doped DSSCs. In the Bode plots, it could be seen that the addition of boron into the TiO$_2$ electrode shifts the mid-frequency peak to higher frequencies a little, and increased their amplitude. The calculated electron lifetimes of the undoped and the boron-doped DSSCs are 88.0, 82.7, and 80.0 ms. The decrease of electron lifetime could be explained by the drop of V_{OC} in boron-doped DSSCs, which is consistent with the photovoltaic data (Table 7.1). It is indicated that the conductivity between the TiO$_2$ particles in the TiO$_2$ electrode could be improved after B doping, and the boron doping (<1 at.%) into the TiO$_2$ photoanode increases the electron recombination at the TiO$_2$/dye/electrolyte interface slightly.

7.4.2.2 Silicon-doped TiO$_2$

From XRD analysis, the intensity of anatase peak of electrodeposited TiO$_2$/SiO$_2$ film is observed to be much higher than that of bare TiO$_2$ film, and this is reported by The-Vinh Nguyen et al. [46]. Among the 50% of silica contained in the bath, only 10% of silica is deposited due to the fact that positively charged titania particles might be deposited easily as compared to negatively charged silica particles in the bath (pH of 5). The superior photovoltaic performance is significant in the DSSC fabricated with TiO$_2$/SiO$_2$-based DSSCs compared to the bare TiO$_2$ film counterparts calcined at 450 °C. Consistency between the short-circuit current and the photocurrent density is observed on all of the cells, suggesting that the electronic property of thin film is correlated well with the photovoltaic performance of resulting DSSC. When the TiO$_2$ based films are calcined at 450 °C, the nanoparticles in TiO$_2$-based network are subjected to partial sintering that renders them chemically bonded. This process not only increases the charge transport rate due to the enhanced number of inter connections between the particles but also improves the dispersion of photo-excited electrons in the TiO$_2$ network [47]. The suppression of charge recombination brings about the increase in the density of photo-excited electrons in the TiO$_2$-based network and therefore negatively shifts its quasi-Fermi level [47]. The photovoltaic parameters of the silicon doped TiO$_2$ is shown in Table 7.1.

7.4.2.3 Germanium-doped TiO$_2$

The dopant atoms niobium (Nb), germanium (Ge), and zirconium (Zr) with a relatively low level of doping (5 mol%) are suggested to sit substitutionally on Ti sites in the anatase crystal structure, yielding the homogeneous distribution of the doped atoms in the anatase nanocrystalline structure and is noted by Hiroshi Imahori et al. [48]. No rutile phase is observed in the XRD. It should be noted here that all the samples exhibit similar XRD patterns. This implies that the TiO_2 anatase nanocrystalline structure is retained after doping a small amount (5 mol%) of Nb, Ge, or Zr in the TiO_2 structure. The effect of the film thickness on the photovoltaic properties was examined to improve the cell performance of the Ge-added TiO_2 cell, which exhibits the highest cell performance of 2.4% efficiency among the four cells under the same conditions (Table 7.1).

7.4.2.4 Antimony-doped TiO_2

XRD analysis shows that the presence of Sb^{3+} ions introduced into the TiO_2 crystal lattice did not noticeably influence the particle size of TiO_2 and this is reported by Min Wang et al. [14]. Sb has a 3+ valence electrons, and it has one less electron than that of Ti. Therefore, the introduction of Sb decreases the electron concentration. The E_{fb} (flat band potential) shift positively with the increase of doping amount, whereas the E_{fb} at high doping amount was almost the same with pure TiO_2. The positive shift of the E_{fb} increases the energy gap between the LUMO of the dye and CB of TiO_2, which results in an increased injection driving force of electrons and then improve the electron injection efficiency from the LUMO of the dye to the TiO_2 CB. The electron life time can be calculated from $\tau_n = (2\pi f_{min})^{-1}$ [49], where f_{min} is the frequency of the minimum point of the semicircle in the intensity-modulated photovoltage spectroscopy (IMVS) plot. The electron life time is decreased compared to undoped TiO_2. The short-circuit photocurrent density (J_{SC}), the fill factor, and the photoelectric conversion efficiency increase with the Sb-doping to reach a maximum at a Sb/Ti ratio of 1 at.% and then decrease. The increased photocurrents of the Sb-doped DSSCs probably result from efficient electron transport. The similar dye-loading amount for Sb-doped TiO_2 and pure TiO_2 films indicate that the enhancement of the photocurrent for Sb-doped TiO_2 is not due to the increase of the dye absorption. Another reason for the increase of photocurrents after Sb-doping may be the positive shift of the flat band. The positive shift of the E_{fb} increases the energy gap between the LUMO of dye and CB of TiO_2, which results in an increased injection driving force electrons and improves the electron injection efficiency from the LUMO of the dye to the TiO_2 (V_{OC} of CB). The open-circuit photovoltage decreases with increase in the percentage of Sb:Ti. The percentage of efficiency of Sb–TiO_2 is as shown in Table 7.1.

7.4.3 Nonmetals-doped TiO_2

Characteristic properties of nonmetals are high ionization energies and high electro negativity. Because of these properties nonmetals usually gain electrons when reacting with other compounds, forming covalent bonds. Among the nonmetals, the anionic dopants have a strong influence on the VB. Nonmetal dopants are carbon, nitrogen, fluorine, sulfur, and iodine [9].

7.4.3.1 Carbon-doped TiO_2

A wide absorption peak in the carbon doped nanostructures (ns)-TiO_2 films at 605 nm and extends to near-infrared region compared with ns-TiO_2 and this is observed by Daobao Chu et al. [50]. It reveals that the new powerful absorption at 400–800 nm is related to the high carbon content. It is clear that an optical absorption threshold lies in about 980 nm in the infrared region. Nie and Sohlberg show [51] that carbon-doping gives rise to two band gaps by theoretical calculation, which was confirmed by experiments of Khan et al. [52]. The finding of three band gaps can be attributed to the higher carbon content in the carbon-doped TiO_2 films. It is evident that the amazing photo response observed from the carbon-doped TiO_2 films was attributed to the band gap lowering and the formation of the intragap band [53, 54]. XRD pattern of carbon-doped TiO_2 nanoparticles are of homogeneous anatase structure with low crystallinity. The open-circuit voltage (V_{OC}), short-circuit photocurrent (I_{SC}), and the fill factor (FF) of the fabricated cell are found to be 0.50 V, 0.256 mA/cm², and 0.47, respectively. The η obtained from the cell system is 4.42% (Table 7.1). The results indicated that incorporation of carbon into nanoporous space of TiO_2 is very successful.

7.4.3.2 Nitrogen-doped TiO₂

XRD patterns of nitrogen (N)-doped TiO_2 calcined at different temperatures (500, 600, 700, 800, and 900 °C) and show no significant change in the structure of TiO_2 and the anatase phase dominates at a specific calcination temperature and this is reported by Wanichaya Mekprasrt et al. [55]. At the elevating calcination temperature, XRD peak of anatase phase distinctively decreases, while strong crystalline of rutile phase occurs significantly [56]. However, no significant change of TiO_2 structure was observed by the influence of nitrogen doping. $I–V$ characteristics shows the maximum conversion efficiency of 1.68%, current density of 6.49 mA/cm², and open-circuit voltage of 0.66 V for working electrodes modified by N-doped TiO_2 calcined at 800 °C (Table 7.1). The tendency of the enhancement in its efficiency is found by varying the calcination temperature. This improvement can be originated from the formation of N–O species and oxygen vacancies in the system. The amount of oxygen vacancies associated with nitrogen doping can be increased by increasing temperature and the efficiency can be further improved [57].

7.4.3.3 Fluorine-doped TiO₂

The effect of F-doping (0.12, 0.35, 0.53, 1.18, 5.89, and 11.88 wt%) on TiO_2 show both the anatase and rutile phase reported by Chin Yong Neo and Jianyong Ouyang [58]. The crystallinity of the anatase phase increases with increasing amount of LiF when the LiF loading is up to 2.36 wt%. The change in the lattice parameter of anatase phase is negligible after the LiF doping. This is because of the similarity in size of both the fluoride and oxide ions. However, at too high LiF doping concentration (11.88 wt%), the intensity of the anatase (004) peaks drops drastically, probably due to the disordered structure induced by excess LiF. The Nyquist plots and the Bode plots of DSSCs with TiO_2 photoanodes doped with various weight percentages of LiF is noted. The interfacial resistance and the electron diffusion coefficient are higher for the DSSCs with 0.53 wt% LiF in TiO_2. Thus, the F-doping of TiO_2 reduces the charge recombination and facilitates the electron diffusion as a result of the F-doping-induced improvement in the anatase crystallinity and decrease in the electron trap sites. The inhibition of the charge recombination can lead to the increase in the photocurrent [59, 60]. At 11.88 wt% LiF, the interfacial resistance, electron lifetime, and electron diffusion coefficient remarkably decrease. This is consistent with the disordered TiO_2 structure due to the excess LiF. The photovoltaic parameters of fluorine doped TiO_2 is shown in Table 7.1.

7.4.3.4 Sulfur-doped TiO₂

Xin Zheng et al. reported that the characteristics of sulfur-doped TiO_2 films are prepared by the spin coating technique [61]. The XRD pattern for doped TiO_2 lattice planes are attributed to the signals of the anatase phase. There is no sulfur diffraction peaks in the XRD spectra, which shows the doping elements dispersed in TiO_2 with a high degree form and which also indicates that doping nonmetallic elements can significantly inhibit the growth and increase the specific surface area of nano-TiO_2 grains, which is very helpful for the absorption of dyes on S–TiO_2 electrodes. The photovoltaic parameters of S–TiO_2 is shown in Table 7.1.

7.4.3.5 Iodine-doped TiO$_2$

The iodine-doped TiO$_2$ nanopowders were prepared using autoclave with thermal treatment at 180 °C for 12 hours, and this is reported by Qian Hou et al. [62]. Structural characterization of the prepared particles shows anatase nanocrystalline structure which is retained after doping. The peak shift is observed, which is due to the increasing the contents of iodine doped in TiO$_2$ crystal matrix. By knowing that the ionic radii of I^{5+} (0.095 nm) is larger than that of Ti^{4+} (0.060 nm), it is reasonably deduced that iodine should be incorporated into the crystal structure [63, 64]. A diffraction peak shift has also been observed in other doping systems such as Nd-doped TiO$_2$ nanoparticles or Cl-doped ZnO nanowire arrays [65, 66].The optical properties of prepared samples shows the absorbance stopping edge at 380 nm. With increasing iodine contents, the absorbance around 400–550 nm is red shifted. The band gap of iodine-doped TiO$_2$ shows a narrowing range from 2.9 to 2.2 eV with increasing iodine ratios. The V_{OC}, J_{SC}, FF, and η of the iodine-doped TiO$_2$ are shown in Table 7.1.

7.4.4 Transition Metals-doped TiO$_2$

The incorporation of transition metals into TiO$_2$ gives rise to the formation of a wide range of new energy levels close to the CB arising from their partially filled d-orbitals. This makes transition metals suitable materials to tune the CB structure of TiO$_2$ [9]. The transition metals doped with TiO$_2$ are scandium, vanadium, chromium, manganese, iron, cobalt, nickel, copper, zinc, yttrium, zirconium, niobium, molybdenum, silver, tantalum, and tungsten.

7.4.4.1 Scandium-doped TiO$_2$

A widespread and careful work was done to establish that the scandium (Sc)-doped TiO$_2$ anatase is nanostructured powders, and this is reported by Alessandro Latini et al. [67]. From a structural analysis, Sc-doped TiO$_2$ shows anatase phase, and the peaks are shifted toward a lower angle which is due to the presence of scandium. The morphology of the prepared sample shows the beads of spherical shape at microscopic level and rice-grain shape at nanoscopic level (Figure 7.2). The photovoltaic parameters of 0.2% of Sc-doped TiO$_2$ shows 9.6% of efficiency (Table 7.1).

7.4.4.2 Vanadium, Niobium, and Tantalum-doped TiO$_2$

Jia Liu et al. reported vanadium (V), niobium (Nb), and tantalum (Ta)-doped TiO$_2$ are prepared through the hydrothermal method [68]. The XRD patterns of prepared particles show that the anatase phase was found and doping did not affect the crystal form (Figure 7.3A). The average crystallite size is found to be around 15 nm. Figure 7.3B shows the surface morphology of prepared films after sintering at 450 °C for 30 minutes. The SEM image shows TiO$_2$ nanocrystalline particles with narrow size distribution due to the hydrothermal method. The aggregation of TiO$_2$ particles in the film is flat and uniform. The size of the nanoparticles is in the range of 13–16 nm, which is in accordance with the XRD results. After doping, it induces the positive shift of E_{fb} of TiO$_2$, which can greatly improve the injecting efficiency from the LUMO of dye to the CB of TiO$_2$. The energy levels of prepared samples are shown in Figure 7.4. Moreover, charge transport was faster after doping into TiO$_2$. The photovoltaic parameters of the prepared samples are listed in Table 7.1.The efficiency of the V, Nb, and Ta-doped TiO$_2$ photoanodes are found to be 7.80%, 8.33%, and 8.18%.

7.4.4.3 Chromium-doped TiO$_2$

The XRD patterns of the undoped and chromium (Cr)-doped TiO$_2$ powders which show reflections of anatase TiO material is reported by Yana Xie et al. [69]. It is indicated that the

Figure 7.2 FEG-SEM images of synthesized 670 nm diameter bead particle with 0.25 at.% of Sc in anatase. The bead is constituted by a large number of nanoparticles randomly oriented but organized as a sphere to minimize the surface energy. The insets (a) and (b) on the upper corners of the figure show beads doped with Sc at 1% and 10%, respectively. EHT, electron high tension; WD, working distance. *Source:* Latini et al. 2013 [67]. Reproduced with the permission of American Chemical Society.

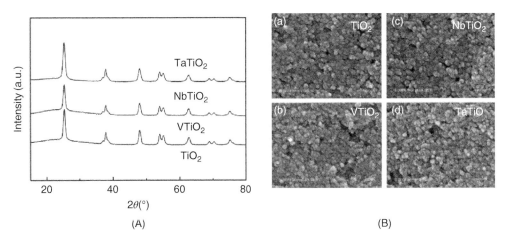

Figure 7.3 (A) XRD patterns (B) SEM images of TiO₂ and 1 mol% V$_B$ doped TiO₂. *Source:* Liu et al. 2013 [68]. Reproduced with the permission of Elsevier.

anatase–rutile phase transformation occurs from Cr doping. All the films have good crystallinity with particle size of 20–30 nm with typical porosity and good uniformity for DSSC. But the surface topography of the films is almost unchanged before and after modification of Cr doping. It is exhibited that the visible light adsorption of the Cr-doped TiO_2 films is slightly enhanced compared to the undoped TiO_2 film. It might be that owing to a downward movement of the CB edge of Cr-doped TiO_2, the band gaps of Cr-doped TiO_2 become slightly smaller than of undoped TiO_2. The EIS shows the resistance (R) of the electron transport

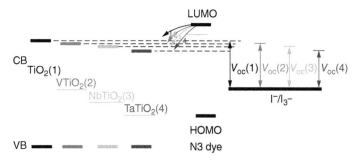

Figure 7.4 Energy level of TiO_2, V_B-doped TiO_2, dye and I^-/I_3^-. *Source:* Liu et al. 2013 [68]. Reproduced with the permission of Elsevier.

within each mesoporous layer that is indicated by the large semicircle remarkably decreases with increasing Cr content and ultimately increases with further Cr doping. The smallest R is achieved when Cr content is 50 ppm, which promotes more efficient charge transport in the mesoporous layer. Owing to the sites of recombination and severe defects introduced when there is too much Cr doping, the recombination speed within mesoporous layer dominates the charge transfer processes [70–73]. Therefore, the value of R_{ct} should become large when 50 ppm of Cr doping exceeds. In other words, the charge transfer resistance in the mesoporous layer by appropriate Cr doping is decreased, leading to a positive influence on the improvement of solar cell performance and that photovoltaic parameters are shown in Table 7.1.

7.4.4.4 Manganese and Cobalt-doped TiO_2

The XRD patterns of manganese (Mn^{2+}) and cobalt (Co^{2+}) doped TiO_2 nanoparticles indicates that the crystal phase of the prepared powders was single anatase phase and were reported by A.E. Shalan and Rashad [74]. The crystalline size was 13 nm for pure nanoparticles, 7 nm for Mn-doped and 15.2 nm for Co-doped TiO_2. It should be noticed that besides the XRD peaks that have arisen from the main phase of anatase TiO_2, there were some additional peaks in XRD patterns of the (Mn, Co) doped samples, which were assigned to $CoTiO_3$ and Mn_2O_3. The optical absorption edges of the (Mn^{2+}, Co^{2+}) doped TiO_2 shifted to a lower energy in the visible light region compared to that of undoped TiO_2. All the produced nanopowders were highly transparent with the doping metals, and the high transparency of the produced powder is attributed to the generation of donor levels accompanying an addition of doping metals. The optical band gap values determined were 4.1 and 4.4 eV for TiO_2 doped with Mn and Co, respectively, and was higher than the ones usually found in the literature [75] and is attributed to the small particle size of the powder characterized in this work. The internal resistances of the three kinds of photoelectrodes (pure TiO_2, Mn–TiO_2, and Co–TiO_2) were studied using Nyquist plot in order to investigate the electron transfer at the TiO_2/dye/electrolyte interface. A high electron accumulation must occur because photogenerated electrons are not extracted at the electrode contact under illumination. The photovoltaic parameters of the Mn^{2+} and Co^{2+} doped TiO_2 nanoparticles are shown in Table 7.1 and its efficiency is 1.85% and 1.06%, respectively.

7.4.4.5 Iron-doped TiO_2

The synthesis of iron-doped TiO_2 nanoparticles using sol–gel technique was prepared by Tae Sung Eom et al. [76]. The XRD pattern of the TiO_2 powder at 500 °C reveal a mixture of anatase and rutile phases. In the Fe_2O_3 phase, the XRD pattern shows the Fe_2O_3-doped TiO_2 surface was sufficient to crystallize the forms. Figure 7.5 shows the Nyquist plots of doped samples. EIS is used to investigate electron transport and recombination in DSSCs. From EIS, the small

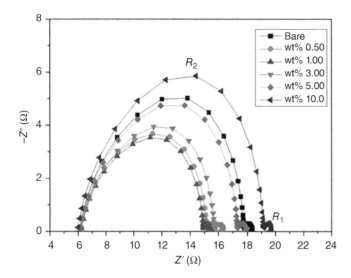

Figure 7.5 Electrochemical impedance spectra of Fe$_2$O$_3$-doped TiO$_2$ films. *Source:* Eom et al. 2014 [76]. Reproduced with the permission of Taylor & Francis.

semicircle is fitted into the charge-transfer resistance (R_1) and a constant phase, and the large semicircle is fitted to the transfer resistance (R_2) and a constant phase. Because R_1 is not affected by Fe, a minimum for the Fe$_2$O$_3$, we can mainly focus on R_2. The first semicircle was (1 wt%), which was related to the charge-transfer resistances of the FTO/TiO$_2$ and the TiO$_2$/electrolyte interfaces (R_2). The decreased R_2 of Fe (1 wt%) indicated a reduction in the electron recombination and increased electron transport efficiency. However, in the case of increased Fe$_2$O$_3$ (10 wt%), R_2 is increased, because the Fe (10 wt%) with more trap sites obstructed the movement of electrons from the nanoporous TiO$_2$ layer to the FTO electrode. It seems that the electron flow improved because the Fe$_2$O$_3$ co-semiconductor reduced the resistance (even at the counter electrode) due to the efficient applied potential transport [76].The absorption spectra of the N-719 dye over the 400–800 nm wavelength range for various TiO$_2$ composite electrode films (0.5–10 wt% Fe$_2$O$_3$). Between 400 and 500 nm wavelength, the absorbance of the sample containing 1.0 wt% Fe$_2$O$_3$ was the highest, and the absorbance of the sample containing 10 wt% Fe$_2$O$_3$ was the lowest. According to the Lambert–Beer's law, a higher absorbance indicates a higher dye concentration. A suitable amount of Fe$_2$O$_3$ in the film could provide a large surface area for dye adsorption. Therefore, the TiO$_2$ layer with the dye serves as a photoactive layer. It is well known that the photocurrent of a DSSC is correlated directly with the number of dye molecules; the increased adsorption of dye molecules leads to an enhanced harvesting of the incident light as well as a larger photocurrent, and photovoltaic parameters are shown in Table 7.1.

7.4.4.6 Nickel-doped TiO$_2$

The nickel (Ni)-doped TiO$_2$ nanoparticles are prepared and reported by T. Sakthivel et al. [77]. The XRD analysis revealed the crystallinity and phase formation of the prepared nickel (Ni) dopant in TiO$_2$ matrix. XRD patterns of pure and Ni-doped TiO$_2$ nanoparticles with different dopant concentrations (1%, 3%, 5% of Ni ion) show the anatase crystalline phase of TiO$_2$ thereby indicating that the phase remains unchanged with the incorporation of Ni^{2+} ion in the TiO$_2$ lattice. Ni-doped TiO$_2$ nanoparticles show peaks with high angle shift indicating the incorporation of Ni dopant in TiO$_2$ nanoparticles. There is no significant peaks-related to rutile phase and oxide of the dopant Ni ions are observed. When the Ni ion is doped into the TiO_2 matrix, it occupies either "substitutional" or "interstitial" site of TiO_2. Also the Ni ion occupies the interstitial sites to some extent and distorts the crystal lattice structure which results in a slight amorphous nature. The UV–Vis absorption measurements of pure and Ni-doped TiO_2 samples are carried out in the wavelength region of 200–800 nm at room temperature which shows absorption maximum in UV region and red shifted when Ni dopant concentration increases. TiO_2 has a tendency to absorb only ultra-violet light as it has a standard band gap of 3.2 eV for bulk. The estimated band gap energy of pure and Ni-doped TiO_2 samples with doping concentration of 1%, 3%, 5% are 3.51, 2.49, 1.82, and 2.7 eV, respectively. It is observed that band gap energy reduced with an increase in dopant concentration up to 3%, beyond that it again gets increased owing to its reduced transparency. It is noteworthy that the band gap can be easily tuned to a specific level with the choice of doping concentration of transition metal-ions into the host TiO_2 as a result shifting of optical absorption edge from ultra-violet to visible region is observed. The Nyquist plots of pure and Ni-doped TiO_2-based DSSC gives double semicircle behavior. The semicircle in the high-frequency region is due to charge transfer resistance (R) at counter electrode/electrolyte interfaces and semicircle in the mid-frequency region is due to charge transfer resistance (R_{ct1}) at metal oxide/dye/electrolyte interfaces. High-frequency intercept on real axis is attributed to the series (R_{ct2}) resistance which contributes to the sheet resistance of FTO [78, 79]. There is no much difference in series resistance. But the charge transport resistance at metal oxide/dye/LiI interface gets decreased with increase in the Ni-dopant concentration (3% of Ni) which implies efficient charge transportation in Ni-doped TiO_2. Here, electron transportation gets increased through Ni metal doping in TiO_2. But in the case of 5% Ni doping, the R_{ct2} again gets increased owing to the random transit motion of electrons by the formation of intermediate trapsites. It increases the charge recombination and reduces the charge transportation at platinum counter electrode/electrolyte interfaces R. In addition, the chemical capacitance gets increased for Ni-doped TiO_2-based DSSCs which is due to efficient charge separation by metal doping which renders the charge recombination process [80]. The photovoltaic parameters of the Ni-doped TiO_2 nanoparticles are shown in Table 7.1, and it is noted that with the highest efficiency of 3% Ni-doping TiO_2 is found to be 4.043%.

7.4.4.7 Copper-doped TiO_2

The synthesis of Cu-doped TiO_2 nanowall by liquid phase deposition method is reported by Dahyunir Dahlan et al. [81]. No diffraction peaks related to the presence of Cu is observed in the Cu-doped TiO_2. It is an indication of the successful substitution of Cu ion into the anatase lattice, instead of interstitially growing up [82]. It is confirmed later by energy dispersive X-ray analysis (EDX) analysis that it shows the existence of Cu in the nanowall and most of the anatase main peaks visibly shifted toward a higher angle, indicating effective lattice stressing (shrinking) in TiO_2 upon the substitution of Cu atoms. Such effective lattice distortion is expected to induce change in the carrier density of state (DOS) at the valence and CB, enhancing the photosensitivity of the TiO_2 nanowall [83]. The energy gap was found to gradually decrease with the increase of dopant concentration. When the TiO_2 nanowall is doped with 3.3 wt% of Cu, the energy gap becomes 3.32 eV. It further declined to 3.25 eV when the dopant concentration further increased to 6.2 wt%. The effective band gap lowering by the doping may eventually lead to Fermi level reducing, improving the photosensitivity of the sample [84]. Thus, the photoactivity, such as photovoltaic, of the doped samples is expected to be enhanced. The ascent of device performance upon being doped with the Cu ion could be due to the following reasons: (i) the improvement of charge mobility and (ii) charge transfer process in the device. From the electrochemical impedance analysis, the sample that is doped with 6.25 mM indicated the lowest in the series resistance and the charge transfer resistance. This may in turn reduce and limit the electrons and holes recombination and enhance the carrier life time in the device, promoting facile and high charge extraction at the counter electrode as verified by the dark-current analysis. The optical band gap of the sample doped with 6.25 mM, Cu atom is much lower compared to the pristine sample. Lowering the optical band gap reflects the widening of the optical responsivity of the sample, improving the excitons generation during the photovoltaic process. Combining such an effect with a facile carrier transfer and transportation in the device is judged by the EIS. However, because of dominant role of carrier transport properties in the device's performance, which is superior over other samples are indicated by the EIS. The photovoltaic parameters are calculated and shown in Table 7.1.

7.4.4.8   Zinc-doped TiO$_2$
Zinc (Zn)-doped TiO$_2$ hollow fiber (HF) was synthesized by a coaxial electrospinning technique, and it was reported by Zainal Arifin et al. [85]. The XRD pattern of TiO$_2$ NP and Zn-doped TiO$_2$ HF shows the crystallographic planes of TiO$_2$. The absorbance of TiO$_2$ NP is higher than that of other photoanodes, and hence, TiO$_2$ NPs can absorb a large quantity of light entering the photoanode layer. However, the transmittance test results show that the TiO$_2$ NPs in the photoanode show higher light absorbance; however, the transmittance also increase. Conversely, a high content of Zn-doped TiO$_2$ HF decreases the light transmittance as well as absorbance. Hence, the scattering of light in Zn-doped TiO$_2$ HF is better than that in TiO$_2$ NP owing to the larger diameter and longer structure than TiO$_2$. In this study, the double-layer structure is equally efficient in dye-loading by TiO$_2$ NP and light scattering by Zn-doped TiO$_2$ HF. Table 7.1 shows the performance of the DSSC in terms of the values of V_{OC}, J_{SC}, FF, and efficiency for a typical dye loading layer composition in the DSSC photoanode. The DSSC efficiency of Zn-doped TiO$_2$ HF is about 8.41 times higher than that of DSSC with conventional TiO$_2$ NP single layer. The increase in the efficiency of the double-layer DSSC is primarily attributed to the increase in the photocurrent density of the DSSC, which is determined by how many photons are converted into the movement of electrons [85].

7.4.4.9   Yttrium-doped TiO$_2$
The yttrium (Y^{3+})-doped TiO$_2$ photoanodes are prepared and reported by Xiaofei Qu et al. [86]. No obvious difference is observed between the peak characteristics of the yttrium-doped and undoped TiO$_2$ photoanodes. The distribution status of the present rare earth ions could also be reduced to three kinds of spatial location: lattice (replacing Ti^{4+} and interstitial void), grain boundary, and particle surface. Because the ion radius of Y^{3+} (0.90 Å) is much larger than that of Ti^{4+} (0.61 Å), it is hard for it to enter the TiO$_2$ lattice. But in the experiment, the used yttrium nitrate solution is easier to penetrate into the porous film, so Y$_2$O$_3$ can cover on TiO$_2$ surfaces or form the subgrains among the grain boundaries during the second annealing process. For Y^{3+}-doped photoanodes, no additional characteristic peaks (such as Y$_2$O$_3$) except anatase TiO$_2$ is found, which may be attributed to the low content and high dispersity of yttrium species [86]. From light absorption studies, it can be seen that in the visible light region (400–700 nm), all the doped samples show an enhanced absorbance, leading to enhanced visible light photoresponse. With the increase of Y^{3+} doping concentration, the absorption edge exhibited an apparent red-shift for TiO$_2$: 0.006 Y^{3+}, and blue-shift for TiO$_2$: 0.015 Y^{3+} and TiO$_2$: 0.025 Y^{3+}. There are various explanations for the change of the optical absorption edge. Generally, the yttrium doping would lead to a red-shift of the absorption edge [87, 88]. In the work of Niu et al. [87], red-shift of absorption edge was attributed to the charge transfer process between electrons from Y^{3+} and conduction or valence band (CB or VB) of TiO$_2$, which lead to a lower optical band gap energy. According to research work of Kumar et al. [89] and Zhang et al. [90],

blue-shift of absorption edge after yttrium doping is observed which is due to occupied states of Ti 3d and the up-shift of Fermi level (i.e.) yttrium doping-created dopant levels near the CB, resulting in red-shift of absorption edge and lower band gap. With the increase of doping concentration, enhanced dopant incorporation may cause gradual movement of CB, leading to an absorption edge blue-shift and larger band gap. From $I-V$ studies, the corresponding values of photovoltaic performance are summarized in Table 7.1.

7.4.4.10 Zirconium-doped TiO$_2$

Zirconium (Zr)-doped TiO$_2$ was prepared using sol–gel, and it is reported by Anastasia Pasche et al. [91]. The peaks in the diffraction patterns correspond to the characteristic Bragg peaks of anatase TiO$_2$ and 10 and 15 mol% Zr shows a slight hint for zirconia formation. The weakness of the peak is attributed to the relatively low zirconia concentration and the well dispersed Zr^{4+} ions within the crystalline n-TiO$_2$ matrix. However, it is assumed that the concentration of a ZrO$_2$ phase will increase, if higher Zr doping concentrations are chosen. Finally, rather than having a mixed crystal system with a secondary phase segregated from n-TiO$_2$, the diffusion of Zr into the n-TiO$_2$ crystal lattice results in one single phase; an outcome beneficial for uninterrupted electron transport during DSSC operation. Traveling electrons run a smaller risk of recombination, as they do not encounter any grain boundaries, which are common places for traps hypothesis that Zr^{4+} is well dispersed within the TiO$_2$ crystal lattice. The crystal structures of ZrO$_2$ and TiO$_2$ are very similar, and Zr^{4+} ions can easily substitute for Ti anatase matrix. However, the lattice structure of TiO$_2$ will be locally deformed close to a Zr exchanged in the crystal structure. The reason for this is the difference in ionic radii of Ti^{4+} and Zr^{4+} (Ti^{4+} = 61 pm, Zr^{4+} = 72 pm) [92, 93]. It can therefore be concluded that Zr-doping causes an expansion of the anatase lattice parameters with respect to the pure anatase phase at locations where Zr ions have replaced Ti, resulting in a slight shift of XRD peaks to smaller diffraction angles. This shift becomes more pronounced with increasing doping concentration, and samples are kept constant at 10 mol%, but the calcination temperature is increased from 400 to 700 °C. These results show that Zr doping suppresses the anatase-to-rutile phase transition [94]. No phase transformation from anatase to rutile was found across the 10 mol% Zr samples, although they were calcined at temperatures expected to cause an anatase-to-rutile phase transition [95]. Instead, the anatase crystal structure was maintained with relative peak intensities increasing with calcination temperature. This stabilization of the anatase phase can be explained by the fact that Zr^{4+} is more electropositive than Ti^{4+}. As such, the electronic cloud in n-TiO$_2$, containing Zr doping, will be more loosely held, thus favoring the formation of the less dense anatase phase over the denser rutile phase [96]. In other words, the tight packing arrangements required for rutile phase formation is fully suppressed by the substitution of Ti electropositivity of Zr^4 by Zr^{4+} in the crystal lattice. Furthermore, the larger size and higher compared to Ti^{4+} allows the lattice to exhibit better bonding properties, and thus higher thermal stability than pure n-TiO$_2$ [96, 97]. The photovoltaic parameters are tabulated in Table 7.1.

7.4.4.11 Molybdenum-doped TiO$_2$

The molybdenum (Mo)-doped TiO$_2$ nanoparticles are prepared by hydrothermal method, and it is reported by Aisha Malik et al. [98]. XRD analysis shows no change in crystal structure of TiO$_2$ after doping with different concentrations of Mo indicating single-phase polycrystalline material. The complex impedance of Mo-doped TiO$_2$ decreases with increase in frequency resulting in an increase in a.c. conductivity. Mo-doped TiO$_2$ possess higher photocurrent due to a higher dye uptake and also reduce the loss of electron by suppressing their recombination, as a result significant increase in the overall power conversion efficiency (PCE) (7.16%) is observed (Table 7.1).

7.4.4.12 Silver-doped TiO$_2$

Jun Luo et al. reported the silver (Ag) ion implanted into TiO$_2$ film on the performance of DSSC [99]. The spectra indicate no presence of Ag or Ag compounds in the Ag-doped titania thin films. However, Ag element mapping clearly evidenced the presence of Ag. The reason may be that the Ag ions were deeply dispersed below the surface of the titania films and the associated electron clouds failed to scatter X-ray coherently to produce a recognizable pattern. The absence of peaks in the diffraction pattern of Ag titania indicates a phase transformation from rutile to anatase has occurred during the Ag-ion implantation. The phase transition is conducive to electron transfer because electron transport in anatase is faster than in rutile [100]. The Ag-ion implantation also influenced the lattice structure of titania. The $J–V$ curves show apparently a beneficial effect of Ag-ion implantation on the photoelectric performance of the photoanodes and is shown in Table 7.1.

7.4.5 Post-Transition Metals

In the periodic table, the posttransition metals are located between the transition metals and the metalloids [9]. As such, posttransition metals have some nonmetal properties, showing covalent bonding effects. The investigated posttransition metals are aluminum and tin.

7.4.5.1 Aluminum-doped TiO$_2$

Aluminum is a group III metal with good optical quality, low resistivity, and high conductance [9]. Aluminum (Al)-doped TiO$_2$ nanoparticle composite is prepared by sol–gel technique and reported by K. Manoharan and P. Venkatachlam [101]. The SEM image of the Al-doped TiO$_2$ nanoparticles/nanowires (TNPWs) had an uneven crystal grain and flat surface morphology, but the crystal grain size of the sample changed with the Al-doped concentration. This is because the superfluous Al atoms form neutral defects such as interstitial Al atoms and some oxidized to form Al, and lead to crystal growth orientation along (101) and (004) planes. A large surface area of the mesoporous electrode and better charge injection resulted in a larger J_{SC}, while the Al$_2$O$_3$ phase acted as a blocking layer, preventing charge recombination [101]. The photovoltaic parameters are shown in Table 7.1.

7.4.5.2 Gallium-doped TiO$_2$

Aravind Kumar Chandiran et al. reported the synthesis of gallium (Ga^{3+})-doped TiO$_2$ nanoparticles which are used as photoanodes in DSSC [102]. Ga^{3+} is a successful incorporation within the anatase lattice of TiO$_2$ and is confirmed notably by powder XRD. Rietveld refinement carried out in the case of the Ga^{3+} sample suggests the latter to substitute Ti^{4+} lattice ions entailing the formation of oxygen vacancies in conjunction with the C101 dye in liquid-electrolyte-based DSSCs. The higher PCE is obtained with the cell of 1% Ga doped with TiO$_2$ (8.1% PCE (Table 7.1)).

7.4.5.3 Indium-doped TiO$_2$

Xiaohua Sun et al. reported that the indium (In)-doped TiO$_2$ film was prepared by a spin coating technique [103]. The electrode material in a DSSC is very important because the quasi-Fermi level of the semiconductor oxide photoelectrode (TiO$_2$) increases linearly with E_{fb} [104, 105]. To confirm the effect of In doping on the flat-band potentials of TiO$_2$ compact layers, the Mott–Schottky plots of the pure TiO$_2$ and In doped TiO$_2$ films are performed at a frequency of 100 Hz at room temperature. Compared with the pure TiO$_2$ compact layer, the E_{fb} of the 6% In-doped TiO$_2$ compact layer shifted negatively about 0.068 V, which suggests that quasi-Fermi level of the In-doped TiO$_2$ compact layer is higher than that of the pure TiO$_2$ compact layer. That is to say, In-doped TiO_2 compact layer creates a potential barrier between FTO substrate and porous TiO_2 film, which can block the electron injection from the CB of porous TiO_2 to FTO, further increasing the electron density in porous TiO_2 film and then enhance its quasi-Fermi level. In DSSCs, open-circuit voltage (V_{OC}) is determined by the difference between the redox potential of the electrolyte and the quasi-Fermi level of the semiconductor oxide photoelectrode [106]. The rise of the quasi-Fermi level of porous TiO_2 caused by In-doped TiO_2 compact layer with higher flat-band potential thus improves the V_{OC}. The photovoltaic performances of prepared photoanode in DSSC shows an efficiency of 7.96% (Table 7.1).

7.4.5.4 Tin-doped TiO_2

Yandong Duan et al. reported that tin (Sn)-doped TiO_2 was prepared by simple hydrothermal process [107]. XRD pattern of prepared samples show that the anatase nanocrystalline structure is retained after doping. The lattice constants values are expanded which are attributed to the larger effect of ionic radius of the Sn^{4+} ionic (0.069 nm) than that of Ti^{4+} (0.061 nm). In order to identify the effect of Ti atoms partially substituted by the Sn atoms to the electronic conductance of $Ti_{1-x}Sn_xO_2$, DFT is carried out to study the DOSs for our materials. The calculated band gap of TiO_2 is lower than the experimental value of 3.2 eV [108]. The photovoltaic parameters are tabulated (Table 7.1) with an efficiency of 8.14%.

7.4.6 Lanthanides-doped TiO_2

Because of their 4f bands, lanthanides provide interesting optic and electronic properties [109] such as up-conversion, photoluminescence, or down-conversion [110, 111], making it possible to harvest photons that are outside the absorption region of most dyes (400–800 nm). Although there are some reports that show effective up- or down-conversion [111–113], these processes are not yet very efficient in DSSCs, and their exact role in improving device efficiencies is unclear. Because many optical conversion studies do not consider the effect of doping on the electron transport rate and lifetime, the exact contribution of conversion is hard to quantify. Lanthanides that have been used to dope TiO_2 are lanthanum, cerium, neodymium, samarium, europium, erbium, thulium, and ytterbium.

7.4.6.1 Lanthanum-doped TiO_2

Shay Yahav et al. show that La treatment of mesoporous TiO_2 films leads to strong improvements of the photocurrent or photovoltage, which depends on the pH of the deposition conditions [114]. At neutral pH, a strong increase in the photovoltage is observed while the photocurrent is reduced. The photovoltaic performance of La-doped TiO_2 shows V_{OC}, J_{SC}, FF, and efficiency (η) are 0.689 V, 15.0 mA/cm^2, 0.68, and 7.0% (Table 7.1). Charge extraction and lifetime measurements indicate that the V_{OC} increase is due to a shell formation around the TiO_2 nanocrystals, generating an interface dipole that shifts the TiO_2 energy bands toward the vacuum level, whereas the reduced photocurrent is caused by less favorable electron injection through the shell. Recent work suggests that the performance of DSSCs can be improved by doping the TiO_2 nanocrystals during synthesis using lanthanum [115, 116]. For La-doping, it is claimed that an increase in the oxygen vacancy density at the particle surface is responsible for improved dye adsorption, resulting in a higher dye load responsible for an increased photocurrent [114].

7.4.6.2 Cerium-doped TiO_2

Effect of cerium (Ce) doping in the TiO_2 photoanodes on transport of DSSCs are reported by Jing Zhang et al. [117]. All peaks are indexed to the anatase phase, and no cerium oxide phase is observed and the cerium did not alter the TiO_2 crystalline type. It is known that Ce induces unoccupied 4f states just under the TiO_2 CB [118]. This may enlarge the driving force for electron injection from the dye into TiO_2. The crystallite size and dye adsorption capability are not significantly changed upon doping. X-ray photoelectron spectroscopy (XPS) showed that both Ce^{3+} and Ce^{4+} are present and the VB is unaffected. The UV–Vis absorption spectrum showed a red-shift, which indicated that Ce indeed induces unoccupied 4f states just under the TiO_2 CB. This is confirmed by cyclic voltammetry. Charge extraction and EIS measurements showed an increase in the electron density and capacitance due to enhanced electron injection, resulting in a higher J_{SC}, but caused a loss in V_{OC} (Table 7.1) [117]. A follow-up study showed that the dopant acted solely as a surface trap state and therefore, it is possible to counteract some of the adverse effects of Ce-doping by a $TiCl_4$ treatment [119].

7.4.6.3 Neodymium-doped TiO₂

The neodymium (Nd)-doped TiO_2 are synthesized by a solid-state technique and explored as a photoanode material in DSSC and this is reported by Lavenna P. D'Souza et al. [120]. From XRD, the peaks indicate the complete retention of anatase phase after Nd doping [121, 122]. A slight shift to lower theta values is in accordance with Bragg's equation, whereas the larger ionic radius of Nd^{3+} substitutes Ti^{4+}, and this effect appeared more prominent at higher angles [65]. The existence of lanthanide induces oxygen vacancies and better dye absorption together with formation, leading to a band gap narrowing, visible light absorption and photoexcitation in DSSC. Favorable alignment of the energy levels, result in a better rate of electron injection and electron mobility, together with the phenomenon of photoexcitation of the semiconductor and leads to an improved electron density and improved J_{SC}. The photovoltaic performance are shown in Table 7.1. Higher negative flat-band potential is V_{OC} observed in doped samples [120].

7.4.6.4 Samarium-doped TiO₂

Meihua Liu reported the enhanced PCE of DSSC with samarium (Sm)-doped TiO_2 nanoparticles prepared by hydrothermal method [123]. From XRD, the crystallite phase is anatase structure, and there is no evidence of samarium oxide which may be attributed to the low concentration in samarium [123]. From UV–Vis spectra of the photoanodes with Sm doping concentration, the absorption wavelength of about 400 nm is blue shifted. Photoluminescence (PL) analysis shows that the ultra-violet part of the sunlight irradiation may be transformed into visible light and reabsorbed by the N719 dye via the down-conversion luminescence effect of Sm^{3+}. This could enhance the light-harvesting ability of the cell and may improve the cell's performance [124]. The influence of samarium doping on the photovoltaic performance of the cells are indicated in Table 7.1 and shows the 6.08% efficiency.

7.4.6.5 Europium-doped TiO₂

Europium (Eu)-doped TiO_2 is by sol–gel synthesis, and the DSSC fabricated using synthesized material is reported by Farheen H. Fouad et al. [125]. The powder X-ray diffraction (PXRD) pattern indicates the formation of a composite phase wherein Ti atoms are probably replaced with Eu atoms in the basic TiO_2 matrix. The experimental results reveal the formation of a complex doped material. The absorption spectra of the prepared samples shows broad absorption peak spread from 240 to 310 nm and indicates the formation of Eu–TiO_2 complex as revealed from XRD analysis. From fluorescence spectra, it is noted that the synthesized material is able to convert irradiated UV and visible photons in to conduction electrons which expectedly increases with increasing concentration of Eu. There are two emission peaks around 384 and 427 nm that indicates that a trapping level has been created with Eu-doping in TiO_2. The photovoltaic parameters are listed in Table 7.1 and shows the 1.01% efficiency of Eu doped with TiO_2 [125].

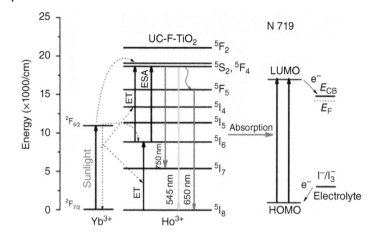

Figure 7.6 Energy transfer mechanism in the upconversion materials UC-F-TiO$_2$. ET, electron transfer; ESA, excited state absorption. *Source:* Yu et al. 2014 [129]. Reproduced with the permission of Elsevier.

7.4.7 Co-doped TiO$_2$

Chi-Hwan Han et al. reported that the Co-doped TiO$_2$, TiO$_2$:[erbium (Er^{3+}), ytterbium (Yb^{3+})] powder was synthesized by novel sol–gel combustion hybrid method [126]. The light scattering layer increased the efficiency of the DSSC by 15.6% [126]. Ghazi M. Abed et al. reported that the Cr–Gd co-doped TiO$_2$ nanoribbons as photoanodes in DSSC with an efficiency of 1.99% [127]. N, La co-doped TiO$_2$ in low-temperature based DSSC show an efficiency of 5.33% due to the enhanced performance that mainly depends on the improvement of J_{SC} and is attributed to the increase of the dye absorption amount, electron lifetimes, and the decrease of the charge transport resistance [128].

7.4.8 Tri-doped TiO$_2$

Jia Yu et al. reported on the Tri (Ho^{3+}, Yb^{3+}, F$^-$)-doped TiO$_2$ and its applications in DSSC with 37% improvement in PCE and the energy transfer mechanism as shown in Figure 7.6. [129]. The improved DSSCs conversion efficiency is as large as 37% (photovoltaic parameters as shown in Table 7.1), which is associated with closer attachment of the in situ upconversion process, enhanced light harvesting, and photogenerated electron–hole pair separation, as well as elevated Fermi level. Although DSSCs have many advantages compared to silicon-based solar cells, particularly in terms of more environmental-friendly manufacture processes, its energy conversion efficiency is still lagging. The near infra-red (NIR)-to-green upconversion fluorescent nanoparticles of Ho^{3+}, Yb^{3+}, F$^-$ tridoped TiO$_2$ combined with other improvements in structure design optimization and the new sensitizers, such as panchromatic sensitizer, will soon make possible DSSCs with higher conversion efficiencies.

7.5 Conclusion

In DSSC, photoanode is one of the components which decide the efficiency of the cell. In order to improve the efficiency by reducing charge recombination and modifying electronic structures, nanostructures of TiO$_2$ and doped TiO$_2$ are tried. The more effective way of modifying electronic properties of TiO$_2$ is by doping with alkaline earth metals, metalloids, nonmetals, transition metals, posttransition metals, lanthanides, Co- and tri-doped with other materials. It is noted that most of the dopants occupy substitutional rather than the interstitial sites. The crystallite size is <20 nm and the dopant did not alter the crystalline phase of TiO_2. The optical band gap of TiO_2 is altered by the dopant due to the creation of low energy levels. DSSCs fabricated using the dopants as photoanode shows an efficiency up to 9.6% for Sc–TiO_2 and 1.01% for Eu–TiO_2. It is concluded that by properly selecting a dopant and multilayered dopants with suitable material, the photovoltaic parameters can be improved and hence the stability and efficiency of DSSC.

References

1 Regan, B.O. and Gratzel, M. (1991). *Nature* 353: 737.
2 Asbury, J.B., Ellingson, R.J., Ghosh, H.N. et al. (1999). *J. Phys. Chem. B* 103: 3110.
3 Ellingson, R.J., Asbury, J.B., Ferrere, S. et al. (1998). *J. Phys. Chem. B* 102: 6455.
4 Van De Lagemaat, V.D.J. and Frank, A.J. (2001). *J. Phys. Chem. B* 105: 11194.
5 Hara, K., Horiuchi, H., Katoh, R. et al. (2002). *J. Phys. Chem. B* 106: 374.
6 Islam, A., Sugihara, H., Singh, L.K. et al. (2001). *Inorg. Chim. Acta* 322: 7.
7 Kubo, W., Kitamura, T., Hanabusa, K. et al. (2002). *Chem. Commun.*: 374.
8 Boschloo, G., Lindstrom, H., Magnusson, E. et al. (2002). *J. Photochem. Photobiol., A* 148: 11.
9 Roose, B., Pathak, S., and Steiner, U. (2015). *Chem. Soc. Rev.* 44: 8326.
10 Jiu, J., Isoda, S., Wang, F., and Adachi, M. (2006). *J. Phys. Chem. B* 110: 2087.
11 Tahcan, Z., Zaban, A., and Rühle, S. (2010). *Sol. Energy Mater. Sol. Cells* 94: 317.
12 Wang, D.A., Yu, B., Zhou, F. et al. (2009). *Mater. Chem. Phys.* 113: 602.
13 Pan, K., Zhang, Q., Wang, Q. et al. (2007). *Thin Solid Films* 515: 4085.
14 Wang, M., Bai, S., Chen, A. et al. (2012). *Electrochim. Acta* 77: 54.
15 Leung, D., Fu, X., Wang, C. et al. (2010). *ChemSusChem* 3: 681.
16 Rauf, M., Meetani, M., and Hisaindee, S. (2011). *Desalination* 276: 13.
17 Park, H., Park, Y., Kim, W., and Choi, W. (2013). *J. Photochem. Photobiol., C* 15: 1.
18 Chen, X. and Mao, S.S. (2007). *Chem. Rev.* 107: 2891.
19 Nah, Y.C., Paramasivam, I., and Schmuki, P. (2010). *ChemPhysChem* 11: 2698.
20 Banerjee, A.N. (2011). *Nanotechnol. Sci. Appl.* 4: 35.
21 Ardakani, H. (1994). *Thin Solid Films* 248: 234.
22 Asahi, R., Taga, Y., Mannstadt, W., and Freeman, A. (2000). *Phys. Rev. B: Condens.* 61: 7459.
23 Meng, S. and Kaxiras, E. (2010). *Nano Lett.* 10: 1238.
24 Ko, K.H., Lee, Y.C., and Jung, Y.J. (2005). *J. Colloid Interface Sci.* 283: 482.
25 Zhang, J., Han, Z., Li, Q. et al. (2011). *J. Phys. Chem. Solids* 72: 1239.
26 Barrett, E.P., Joyner, L.G., and Halenda, P.P. (1951). *J. Am. Chem. Soc.* 73: 373.
27 Barnes, P.R.F., Miettunen, K., Li, X. et al. (2013). *Adv. Mater.* 25: 1881.
28 Sarker, S., Ahammad, A.J.S., Seo, H.W., and Kim, D.M. (2014). *Int. J. Photoenergy* 85: 1705.
29 Subramanian, A., Bow, J.-S., and Wang, H.-W. (2012). *Thin Solid Films* 520: 7011.
30 Cava, R.J., Murphy, D.W., Zahurak, S. et al. (1984). *J. Solid State Chem.* 53: 64.
31 Wagemaker, M., Borghols, W.J.H., and Mulder, F.M. (2007). *J. Am. Chem. Soc.* 129: 4323.
32 Stashans, A., Lunell, S., and Bergstrom, R. (1996). *Phys. Rev. B: Condens.* 53: 159.
33 Kalyanasundaram, K. and Gratzel, M. (1998). *Coord. Chem. Rev.* 77: 347.
34 Lee, K.M., Suryanarayanan, V., and Ho, K.C. (2006). *Sol. Energy Mater. Sol. Cells* 90: 2398.
35 Park, K.H., Jin, E.M., Gu, H.B. et al. (2009). *Mater. Lett.* 63: 2208.

36 Scaife, D.E. (1980). *Sol. Energy* 25: 41.

37 Kapoor, P.N., Uma, S., Rodriguez, S., and Klabunde, K.J. (2005). *J. Mol. Catal. A: Chem.* 229: 145.

38 Iwamoto, S., Kaneko, M., YoheiSazanami, M.I., and Maenosono, A. (2008). *ChemSusChem* 1: 401.

39 Bandaranayake, K.M.P., IndikaSenevirathna, M.K., Prasad We-ligamuwa, P.M.G.M., and Tennakone, K. (2004). *Coord. Chem. Rev.* 248: 1277.

40 Jung, H.S., Lee, J.K., Nastasi, M. et al. (2005). *Langmuir* 21: 10332.

41 Liu, Q., Zhou, Y., Yandongduan, M.W. et al. (2013). *J. Alloys Compd.* 548: 161.

42 Vanmaekelbergh, D. and de Jongh, P.E. (2000). *Phys. Rev. B: Condens.* 61: 4699.

43 Tian, H., Hu, L., Zhang, C. et al. (2011). *J. Mater. Chem.* 21: 863.

44 Finazzi, E., Di Valentin, C., and Pacchioni, G. (2009). *J. Phys. Chem. C* 113: 220.

45 Gopal, N.O., Lo, H.H., and Ke, S.C. (2008). *J. Am. Chem. Soc.* 130: 2760.

46 Nguyen, T.-V., Lee, H.-C., Alam Khan, M., and Yang, O.-B. (2007). *Sol. Energy* 81: 529.

47 Frank, A.J., Kopidakis, N., and Van De Lagemaat, J. (2004). *Coord. Chem. Rev.* 248: 1165.

48 Imahori, H., Kang, S., Hayashi, S. et al. (2006). *Langmuir* 22: 11405.

49 Oekerman, T., Yoshida, T., Zhang, D., and Minoura, H. (2004). *J. Phys. Chem. B* 108: 2227.

50 Chu, D., Yuan, X., Qin, G. et al. (2008). *J. Nanopart. Res.* 10: 357.

51 Nie, X. and Sohlberg, K. (2004). *Materials Research Society Proceedings on Materials and Technology for Hydrogen Economy*, vol. 801, 205. Materials Research Society, Cambridge University Press.

52 Khan, S., Al-Shahry, M., and Ingler, W.B. Jr., (2002). *Science* 297: 243.

53 Kamisaka, H., Adchi, T., and Yamashita, K. (2005). *J. Chem. Phys.* 123: 084704.

54 Gole, J.L., Stout, J.D., Burda, C. et al. (2004). *J. Phys. Chem. B* 108: 1230.

55 Mekprasart, W., Suphankij, S., Tangcharoen, T. et al. (2014). *Phys. Status Solidi A*: 1–7. https://doi.org/10.1002/pssa.201330566.

56 Hou, Y.Q., Zhuang, D.M., Zhang, G. et al. (2003). *Appl. Surf. Sci.* 218: 97.

57 Motlak, M., Akhtar, M.S., Barakat, N.A.M. et al. (2014). *Electrochim. Acta* 115: 493.

58 Yong Neo, C. and Ouyang, J. (2013). *J. Power Sources* 241: 647.

59 Neo, C.Y. and Ouyang, J. (2011). *J. Power Sources* 196: 10538.

60 Zhang, X.T., Liu, H.W., Taguchi, T. et al. (2004). *Sol. Energy Mater. Sol. Cells* 81: 197.

61 Zheng, X., Yang, X., Zhang, J. et al. (2011). *J. Mater. Sci.* 46: 5071.

62 Hou, Q., Zheng, Y., Chen, J.-F. et al. (2011). *J. Mater. Chem.* 21: 3877.

63 Lv, Y.Y., Yu, L.S., Huang, H.Y. et al. (2009). *Appl. Surf. Sci.* 255: 9548.

64 Lv, Y.Y., Yu, L.S., Huang, H.Y. et al. (2009). *J. Alloys Compd.* 488: 314.

65 Lu, X.J., Mou, X.L., Wu, J.J. et al. (2010). *Adv. Funct. Mater.* 20: 509.

66 Cui, J.B., Soo, Y.C., and Chen, T.P. (2008). *J. Phys. Chem. C* 112: 4475.

67 Latini, A., Cavallo, C., Aldibaja, F.K., and Gozzi, D. (2013). *J. Phys. Chem. C* 117: 25276.

68 Liu, J., Duan, Y., Zhou, X., and Lin, Y. (2013). *Appl. Surf. Sci.* 277: 231.

69 Xie, Y., Huang, N., You, S. et al. (2013). *J. Power Sources* 224 (108): 173.

70 Lee, K.M., Suryanarayanan, V., and Ho, K.C. (2007). *Sol. Energy Mater. Sol. Cells* 91: 1416.

71 Wang, Q., Moser, J.E., and Grätzel, M. (2005). *J. Phys. Chem. B* 109: 14945.

72 Huang, N., Liu, Y.M., Peng, T. et al. (2012). *J. Power Sources* 204: 257.

73 Jang, Y.H., Xin, X.K., Byun, M. et al. (2012). *Nano Lett.* 12: 479.

74 Shalan, A.E. and Rashad, M.M. (2013). *Appl. Surf. Sci.* 283: 97.

75 Rashad, M.M. and Shalan, A.E. (2012). *Int. J. Nanopart.* 5: 159.

76 Eom, T.S., Kim, K.H., Bark, C.W., and Choi, H.W. (2014). *Mol. Cryst. Liq. Cryst.* 600: 39.

77 Sakthivel, T., Ashok kumar, K., Senthilselvan, J., and Jagannathan, K. (2018). *J. Mater. Sci. - Mater. Electron.* https://doi.org/10.1007/s10854-017-8137-2.

78 Gao, B., Wang, T., Fan, X. et al. (2017). *Inorg. Chem. Front.* 4: 898–906.

79 Kumar, K.A., Subalakshmi, K., and Senthilselvan, J. (2016). *J. Solid State Electrochem.* 20: 1921.

80 Kumar, K.A., Subalakshmi, K., and Senthilselvan, J. (2016). *AIP Conf. Proc.* 1731: 060017.

81 Dahlan, D., Saad, S.K.M., Berli, A.U. et al. (2017). *Physica E* 91: 185.

82 Mohammad, H.H.J., Noor, R.R., Yahaya, M. et al. (2012). *Adv. Mater. Res.* 364: 485.

83 You, M., Kim, T.G., and Sung, Y.M. (2009). *Cryst. Growth Des.* 10: 983.

84 Saad, S.K.M., Umar, A.A., Rahma, M.Y.A., and Salleh, M.M. (2015). *Appl. Surf. Sci.* 353: 835.

85 Arifin, Z., Suyitno, S., Hadi, S., and Sutanto, B. (2018). *Energies* 11: 2922.

86 Qu, X., Hou, Y., Liu, M. et al. (2016). *Results Phys.* 6: 1051.

87 Niu, X., Sujuan, L., Chu, H., and Zhou, J. (2011). *J. Rare Earths* 29: 225.

88 Zhang, W., Wang, K., Zhu, S. et al. (2009). *Chem. Eng. J.* 155: 83.

89 Kumar, K.S., Song, C.G., Bak, G.M. et al. (2014). *J. Alloys Compd.* 617: 683.

90 Zhang, H., Tan, K., Zheng, H. et al. (2011). *Mater. Chem. Phys.* 125: 15.

91 Pasche, A., Grohe, B., Mittler, S., and Charpentier, P.A. (2017). *Mater. Res. Express* https://doi.org/10.1088/2053-1591/aa742d.

92 Abd El-Lateef, H.M. and Khalaf, M.M. (2015). *Mater. Charact.* 108: 29.

93 Zheleznov, V.V., Sushkov, Y.V., Voit, E.I. et al. (2015). *J. Appl. Spectrosc.* 81: 983.

94 Kim, J., Song, K.C., Foncillas, S., and Pratsinis, S.E. (2001). *J. Eur. Ceram. Soc.* 21: 2863.

95 Hanaor, D.A.H. and Sorrell, C.C. (2011). *J. Mater. Sci.* 46: 855.

96 Venkatachalam, N., Palanichamy, M., Arabindoo, B., and Murugesan, V. (2007). *J. Mol. Catal. A: Chem.* 266: 158.

97 Bendoni, R., Mercadelli, E., Sangiorgi, N. et al. (2015). *Ceram. Int.* 41: 9899.

98 Aisha, M., Hameed, S., Siddiqul, M.J. et al. (2014). *J. Mater. Eng. Perform.* 23: 3184. https://doi.org/10.1007/s11665-014-0954-3.

99 Luo, J., Zhou, J., Guo, H. et al. (2014). *RSC Adv.* 4: 56318.

100 Thavasi, V. and Renugopalakrishnan, V. (2009). *Mater. Sci. Eng., R* 63: 81.

101 Manoharan, K. and Venkachalam, P. (2015). *Mater. Sci. Semicond. Process* 30: 208.

102 Chandiran, A.K., Sauvage, F., Etgar, L., and Graetzel, M. (2011). *J. Phys. Chem. C* 115: 9232.

103 Sun, X., Zhang, Q., Liu, Y. et al. (2014). *Electrochim. Acta* 129: 276.

104 Radecka, M., Wierzbicka, M., Komornicki, S., and Rekas, M. (2004). *Physica B* 348: 160.

105 Bandara, J. and Pradeep, U.W. (2008). *Thin Solid Films* 517: 952.

106 Yang, S.M., Kou, H.Z., Song, S.L. et al. (2009). *Colloids Surf., A* 340: 182.

107 Duan, Y., Fu, N., Liu, Q. et al. (2012). *J. Phys. Chem. C* 116: 8888.

108 Yin, W.J., Chen, S., Yang, J.H. et al. (2010). *Appl. Phys. Lett.* 96: 221901.

109 Strange, P., Svane, A., Temmerman, W.M. et al. (1999). *Nature* 399: 756.

110 Auzel, F. (2004). *Chem. Rev.* 104: 139.

111 Strumpel, M., McCann, G., Beaucarne, V. et al. (2007). *Sol. Energy Mater. Sol. Cells* 91: 238.

112 Wang, J., Ming, T., Jin, Z. et al. (2014). *Nat. Commun.* 5: 5669.

113 Masuda, Y., Yamagishi, M., and Koumoto, K. (2007). *Chem. Mater.* 19: 1002.

114 Yahav, S., Ruhle, S., Greenwald, S. et al. (2011). *J. Phys. Chem. C* 115: 21481.

115 Zhang, J., Zhao, Z., Wang, X. et al. (2010). *J. Phys. Chem. C* 114: 18396.

116 Wu, X.H., Wang, S., Guo, Y. et al. (2008). *Chin. J. Chem.* 26: 1939.

117 Zhang, J., Peng, W., Chen, Z. et al. (2012). *J. Phys. Chem. C* 116: 19182.

118 Chen, S.W., Lee, J.M., Lu, K.T. et al. (2010). *Appl. Phys. Lett.* 97: 012104.

119 Zhang, J., Feng, J., Hong, Y. et al. (2014). *J. Power Sources* 257: 264.

120 D'Souza, L.P., Shwethrani, R., Amoli, V. et al. (2016). *Mater. Des.* 104: 346. https://doi.org/10.1016/j.matdes.2016.05.007.

121 D'Soua, L.P., Shree, S., and Balakrishna, G.R. (2013). *Ind. Eng. Chem. Res.* 52: 16162.

122 Balakrishna, G.R. and Devi, G. (2005). *Pol. J. Chem.* 79: 919.

123 Liu, M., Hou, Y., and Qu, X. (2017). *J. Mater. Res.* 32: 3469. https://doi.org/10.1557/jmr.2017.357.

124 Liu, R., Qiang, L.S., Yang, W.D., and Liu, H.Y. (2013). *J. Power Sources* 223: 254.

125 Fouad, F.H., Ansari, S.G., Khan, A.A., and Ansari, Z.A. (2017). *J. Mater. Sci. - Mater. Electron.* https://doi.org/10.1007/s10854-017-6387-7.

126 Han, C.-H., Hak-Soo, K.-w.L., Han, S.-D., and Singh, I. (2009). *Bull. Korean Chem. Soc.* 30: 219.

127 Abed, G.M., Alsommarraie, A.M.A., and Al-abdaly, B.I. (2017). *Nanosci. Nanometrol.* 3: 27.

128 Jiang, Y.F., Chen, Y.Y., Zhang, B., and Feng, Y.Q. (2016). *J. Electrochem. Soc.* 163: F1133.

129 Jia, Y., Yang, Y., Fan, R. et al. (2014). *Inorg. Chem.* 53: 8045.

8

Binary Semiconductor Metal Oxide as Photoanodes

S.S. Kanmani[1], I. John Peter[2], A. Muthu Kumar[2], P. Nithiananthi[2], C. Raja Mohan[2], and K. Ramachandran[2]

[1] *Dr. N.G.P. Arts and Science College, Coimbatore, Tamilnadu, India*
[2] *Nanostructure Lab, Department of Physics, The Gandhigram Rural Institute-Deemed to be University, Gandhigram, Tamilnadu, India*

8.1   Why Metal Oxide Semiconductors?

As both natural and synthetic metal oxide semiconductors (MOSs) have diverse applications and the properties of MOS can be tailored in many ways, viz., varied choice of morphologies, introducing oxygen vacancies, doping. In photovoltaics, MOSs serve as a scaffold layer for loading dyes in dye-sensitized solar cells (DSSCs) and organic–inorganic hybrid perovskites in perovskite solar cells (PSCs), as well as electron and hole transport layers in DSSCs and organic solar cells (OSCs). The function of scaffold in DSSCs is to facilitate charge separation and charge transport, whereas that of the transport layers is to conduct one type of charge carrier block to the other type. Therefore, tailoring their properties is inevitable to develop high-performing photovoltaic devices using them. On the other hand, the electrochemical properties of the MOS such as band edge energies determine their success as photocatalysts [1].

The wide-band-gap MOSs (e.g. >3 eV) having suitable band position relative to dye (or photosensitizer) have been used for the fabrication of DSSCs. Owing to the wide band gap, the MOSs employed for the fabrication of DSSCs have absorption at the ultraviolet region. Therefore, photosensitizer/dye is responsible for the absorption of light at the visible and near-infrared region. Furthermore, the high surface area of nanoporous MOS increases dye loading; thereby enhancing light absorption leading to improved performance of DSSCs. In addition to the above-mentioned physical characteristics, low cost, natural abundance, and facile synthesis methods of MOS combined with facile solution processibility is another key advantage for the application in DSSCs. Binary MOSs such as TiO_2, ZnO, Fe_2O_3, SnO_2, ZrO_2, Nb_2O_5, Al_2O_3, and CeO_2 are the typical materials that have been well tested now for their use as photoelectrodes in DSSCs [2–8]. Among them TiO_2, ZnO, and Fe_2O_3 are 3d transition metal oxides, ZrO_2 and Nb_2O_5 are 4d transition metal oxides, SnO_2 and Al_2O_3 are p-block metal oxides, and CeO_2 is a f-block metal oxide. In TiO_2, the Ti ions are in a distorted octahedral environment and formally have a $Ti^{4+}(3d_0)$ electronic configuration.

Ternary oxides and inorganic perovskites such as $BaTiO_3$, $SrTiO_3$, $SrSnO_3$ and organic–inorganic hybrid halide lead perovskites such as $CH_3NH_3PbI_3$ (toxicity of lead is being seriously viewed these days) also are contributing to the development of efficient solar cells [9–13]. As the lifetime of free charges and open-circuit voltages in DSSC are determined by the flatband potential and trap states in the semiconductor oxide, core–shell nanostructured photoanodes were developed. Here comes the above binary metal oxides for such designs of core–shell structures.

Interfacial Engineering in Functional Materials for Dye-Sensitized Solar Cells, First Edition.
Edited by Alagarsamy Pandikumar, Kandasamy Jothivenkatachalam and Karuppanapillai B. Bhojanaa.
© 2020 John Wiley & Sons, Inc. Published 2020 by John Wiley & Sons, Inc.

8.2 Development of MOS-Based DSSC

There are lot of reports available in the literature for $ZnO-TiO_2$ core–shell nanostructures-based DSSCs to solve some of the shortcomings of the physical and chemical properties of these individual oxides, however, not much improvement has surfaced. Irrespective of their similar band-gap energies of 3.2 eV, their different electronic structures influence their interfacial energetics. It is established that electron transport times are about two orders of magnitude faster in ZnO than in TiO_2. However, ZnO exhibits poor dye attachment caused by $ZnO/TiO_2/N719$ dye aggregation and dissolution of I^-/I_3^- electrolyte, resulting in reduced electron injection rates and higher dark current than TiO_2 photoanodes, in DSSC. Consequently, different ways are there to employ ZnO as a inner core with TiO_2 as a barrier layer to prevent ZnO dissolution and ZnO/TiO_2/dye aggregation, i.e. ZnO/TiO_2 core–shell nanostructures have been proposed as photoanodes. Manthina et al. reported that ZnO nanorods with a good TiO_2 nanoparticle coverage had a faster electron transport and exhibited higher dye adsorption but provided low short circuit current density (J_{SC}) due to the absence of an energy gradient [14]. Wang et al. also studied ZnO nanorods with 10 nm thickness over layer of TiO_2. The prepared ZnO/TiO_2 core–shell sintered at 450°C and achieved a current density (J_{SC}) of 5.3 mA/cm^2, an open-circuit voltage (V_{OC}) of 704 mV, and fill factor (FF) of 56% [15]. Kanmani and Ramachandran, reversibly employed TiO_2/ZnO core–shell nanoparticles because of the faster electron mobility in ZnO than in TiO_2 that could potentially enhance interfacial electron transfer. However, they achieved a low J_{SC} which might be due to the dissolution of ZnO or higher interfacial charge recombination [16]. Roh et al. have demonstrated power conversion efficiency (PCE) of 4.51% with a 30 nm ZnO layer on TiO_2 DSSC photoanode. Employing ZnO/TiO_2 core–shell nanostructures as photoanodes requires that the conduction band (CB) potential of the shell material (TiO_2) be more negative than the core to successfully act as an energy barrier and thereby reduce recombination of transport electrons with oxidized dye and triiodide species in I^-/I_3^- liquid DSSC. More so, ZnO should be protected from dissolving in I^-/I_3^- electrolyte. The CB edge of bare TiO_2 is shown to be slightly more positive than that of bare ZnO [17].

It is therefore highly desirable to design semiconductors with tunable band edge positions and band gap. In this contribution, we describe such band-gap engineering by means of the synthesis of nanoporous titanium–zirconium mixed oxides. As indicated in Figure 8.1, the band edge positions of $Ti_{1-x}Zr_xO_2$ are expected to change with the content of zirconium, x, between the positions of TiO_2 and ZrO_2, similar to earlier observed band edge movement in mixed oxide systems. An increase of zirconium content should therefore allow for a higher CB edge and, in consequence, for a higher open-circuit voltage when used in DSSC. Indeed, very recently, it

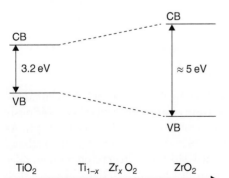

Figure 8.1 Schematics of the dependence of conduction band edge and valence band (VB) edge as a function of zirconium content.

was shown for one fixed value of zirconium content, $x = 5\%$, that an increase in open-circuit voltage can be achieved.

Eguchi et al. prepared various compositions in the TiO_2/Nb_2O_5 system starting from pure TiO_2 to pure Nb_2O_5 and observed that the FF increase steadily with Nb_2O_5 content. Increase in the FF indicates that the loss due to carrier recombination decreased, i.e. increase in the shunt resistance with an increase in the Nb_2O_5 content. The V_{OC} of the composite electrode cell was increased by increasing the Nb_2O_5 content [18].

More recently, Kitiyanan and Yoshikawa prepared TiO_2/ZrO_2 and TiO_2/GeO_2 composite electrodes [6, 19]. The TiO_2/ZrO_2 composite electrodes showed an increase in all the photovoltaic parameters, whereas the FF decreased in the TiO_2/GeO_2 composite compared with cells fabricated using pure TiO_2. Palomares et al. coated layers of ZrO_2, Al_2O_3, and SiO_2 onto 15 nm TiO_2 particles and observed significant increase in the FF and J_{SC} for core/shell electrode compared with the bare TiO_2 electrode [20]. This chapter presents the progress of research in these MOSs-based DSSC in the last two decades, and the possibility for the future development in this area.

8.2.1 TiO$_2$/ZnO Core/Shell Configuration

Other than this, several works have been initiated by various groups to improve the electrical transport property of TiO_2 and ZnO such as preparation of composite electrodes, core-shell structures, and performing effective doping. Band gap engineering of compound semiconductors suggests a new approach to enhance the electron transport property by forming a shell of one metal oxide onto the other MOS as core, which includes TiO_2/ZnO, ZnO/TiO_2, TiO_2/SnO_2, TiO_2/Al_2O_3, etc. [21]. The CB potential of the shell has to be more negative than that of the core in order to establish an energy barrier at the semiconductor/electrolyte interface and confine the electrons inside the core, which results in enhanced fill factor (FF) and short-circuit current density (J_{SC}). The higher CB edge of shell can effectively maximize the open-circuit voltage (V_{OC}), as particularly observed in the case of TiO_2/ZnO core/shell structures.

8.2.2 Preparation of TiO$_2$/ZnO Core/Shell Nanomaterials

Zinc acetate dihydrate ($Zn(C_2H_4O_2)_2 \cdot 2H_2O$), sodium hydroxide (NaOH), and pre-synthesized TiO_2 nanoparticles were used as starting materials for the preparation of TiO_2/ZnO core/shell nanomaterials. In the typical reaction [22], 0.5487 g (10 mM) of $Zn(C_2H_4O_2)_2 \cdot 2H_2O$ was dissolved in 250 ml of deionized water with constant magnetic stirring for 10 minutes; then 0.1997 g of presynthesized TiO_2 nanoparticles (T1) were added to it. To the above stock solution, 250 ml of 0.1 M NaOH aqueous solution was added dropwise and the resulting solution was vigorously stirred for over five hours to get good dispersion. Finally, the white color solution was refluxed at 97 °C for three hours to form ZnO shell on the surface of TiO_2 nanoparticles. By allowing the solution to cool down to room temperature, it was centrifuged, washed with deionized water and acetone to remove any unreacted ZnO_2^{2-} ions and dried at room temperature to obtain TiO_2/ZnO core/shell nanomaterials. Here, the Ti^{2+}/Zn^{2+} ratio was taken as 1 : 1, and identified as T1Z1. Similarly, they were also prepared in the ratio of 1 : 0.50 and 1 : 0.25, which are identified as T1Z0.5 and T1Z0.25, respectively. The final products of all samples were annealed in air at 450 °C for two hours for further characterizations and used as photoanode material in the fabrication of DSSC.

8.2.3 TiO$_2$/ZnO Core/Shell Nanomaterials

The UV–Vis absorption spectra of the synthesized samples are shown in Figure 8.2. All the samples show strong absorption in UV region and high transparency in the visible region. From

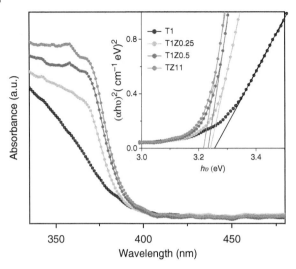

Figure 8.2 UV–Vis absorption spectra of T1, T1Z0.25, T1Z0.5, and T1Z1 samples. Inset corresponds to its α-absorption (Tauc's) plot.

the inset figure (Tauc's plot – extrapolation of the absorption edge onto the energy axis), band gap of sample-T1 is calculated as 3.25 eV and for ZnO-coated TiO$_2$, it is red shifted from T1. It is found that the absorption of T1 occurred at 381 nm, blue shifted from bulk anatase phase (388 nm), increased significantly after ZnO coating on the surface of TiO$_2$. Literatures reported that the wavelength of absorption edge of ZnO (391 nm) is higher than TiO$_2$, and hence, the absorption edge of ZnO modified TiO$_2$ increases from core TiO$_2$. In addition, the absorbance slightly increased with the increase of Zn; this also reveals the crystalline nature of coated ZnO on TiO$_2$, similar to the report of Liao et al. [23] and agreeing with the X-ray diffraction (XRD) results. Brus formula is used then to find the size of the nanoparticles from the band gap, which increased with the increase of Zn content, and confirms the formation of different thickness of ZnO layer on TiO$_2$.

Figure 8.3 shows the room temperature photoluminescence (PL) spectra of synthesized samples, for an excitation wavelength of 280 nm. Inset shows the normalized spectra of actual PL.

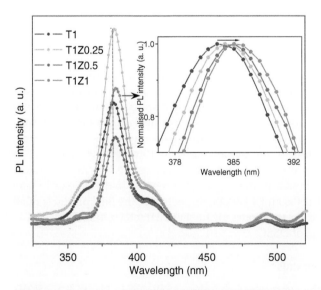

Figure 8.3 PL spectra of T1, T1Z0.25, T1Z0.5, and T1Z1 samples. Inset shows the normalized spectra of actual PL.

The spectra show a strong characteristic UV emission peak at 383.1 nm, red shifted with the increase of Zn content. This observation is quite consistent with the UV–Vis absorption spectral results. The strong UV emission is attributed to radiative annihilation of excitons. Similarly, weak characteristic emission peaks at 362, 412, and 493 nm are also observed. Emission peak that occurred at 412 nm corresponds to the self-trapped excitons localized at TiO_6 octahedral sites and 493 nm corresponds to electron transition mediated by defects levels in the band gap, such as oxygen vacancies formed during sample preparation.

From the comparison of the ratio of UV/Visible emission, it is observed that the ratio of T1Z0.25 are relatively higher, which describes the good quality of core/shell nanomaterials and agreeing with the results of XRD. Similarly, the low value of UV emission of T1Z0.5 suggests the reduced crystallinity that may decrease J_{SC} due to the difficulties occur in diffusion of photogenerated charge carriers. The visible emission also plays a key role in determining the performance of DSSC, here the T1Z0.25 exhibits more defect-related emission compared to all other samples. This result indicates that the crystal orientation and defects have a large effect on the performance of DSSC.

8.2.4 DSSC Performance of TiO$_2$/ZnO Core/Shell Configuration

Figure 8.4 shows the photocurrent-voltage characteristics of the cells based on bare TiO_2 and ZnO-coated TiO_2 systems. Under illumination, the bare TiO_2-based DSSC exhibits higher J_{SC} than that of the ZnO-coated TiO_2, agreeing with the results of Roh et al. [17]. They showed 0.18% solar efficiency for the coating of 300 nm thickness of ZnO on predeposited TiO_2 film with N3 dye in liquid electrolyte medium. Similarly, the reduction may occur due to the precipitation of Zn^{2+}-dye complex aggregate, which limits the electron exchange between dye and electrolyte, leading to lower electron injection efficiency (i.e. low photocurrent). Another fact is that the roughness factor of ZnO-coated film decreases, which can reduce the amount of dye adsorption and leads to low photocurrent. In ZnO-coated TiO_2 DSSC, TZ0.25 exhibits minimum J_{SC}, and this is mainly associated with the higher visible emission observed in PL spectra.

The performance of bare TiO_2 and ZnO-coated TiO_2 DSSCs are presented in Table 8.1. It is seen that the V_{OC} of bare TiO_2 is low, which is mainly attributed to the increased charge

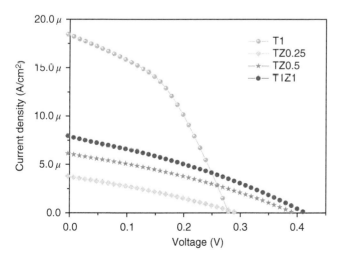

Figure 8.4 Current–voltage characteristics of bare TiO$_2$ and ZnO-coated TiO$_2$ DSSC.

Table 8.1 Performance of DSSC based on bare TiO_2 and ZnO coated TiO_2 systems.

Materials	J_{SC} ($\mu A/cm^2$)	V_{OC} (V)	FF	η % in 10^{-3}
T1	18.4	0.278	0.395	2.0
T1Z0.25	3.64	0.301	0.242	0.3
T1Z0.5	6.16	0.390	0.315	0.8
T1Z1	7.91	0.412	0.306	4.0

recombination with either oxidized dye or tri-iodide electrolyte. But, after ZnO coating, the V_{OC} is increased from 0.278 to 0.412 V, confirming the active role of ZnO in reducing the recombination by acting as an inherent energy barrier and agreeing with the results of Kang et al. [24]. However, the higher V_{OC} also results from an elevation of Fermi level with the addition of ZnO layer [25]. In contrast to V_{OC}, the decrease in FF of ZnO-coated TiO_2 can be explained in terms of the formation of ZnO barrier layer at the TiO_2/indium tin oxide (ITO) interface, which could slightly impede electron tunneling to the collecting layer. On the other hand, the resistance of the polymer electrolyte increases due to the difficulties originated during penetration. This is because of the formation of barrier layer, so the shell layers slightly obstruct the electrolyte penetration by suppressing the easiest electron flow to the electrolyte, i.e. the porous nature of the electrode reduces with ZnO coating.

A lot of work has already been reported with high solar conversion efficiency, by using liquid electrolyte and efficient dyes. In comparison with the results already reported, here J_{SC} values are too small. This is mainly due to the smaller particle size of the synthesized materials, which do not support space charge region (no electric field that can assist the separation of the electrons in the semiconductor), leading to recombination of the photogenerated charge carriers. J_{SC} will be further improved by enhancing the particle size to optimum level and the crystallinity of the synthesized materials by increasing the annealing temperature.

It is to be noted that the synthesis here involved two steps such as (i) the preparation of core materials and (ii) then the shell layer coated on presynthesized core surface. This would (Figure 8.5) offer energy barrier not only at electrode/electrolyte interface but also between individual core nanoparticles, providing pathway for interfacial recombination. This leads to a decrease in J_{SC} for ZnO-coated TiO_2 than bare TiO_2 system. Most of the studies that support this work (DSSC-based on TiO_2/ZnO) are carried out by any one of layer-by-layer deposition which reduces the photocurrent loss (that occurs due to recombination). But Rajkumar et al. [26] reported that eosin (EY) dye-sensitized TiO_2/ZnO nanocomposites photoanode showed increase in both J_{SC} and V_{OC} for the addition of ZnO in TiO_2 matrix when making composite by using liquid electrolyte. Similarly, when EY dye is compared with other efficient ruthenium complex dyes, only one anchoring COOH group is available for fixation on TiO_2 surface, limiting the J_{SC} directly.

Therefore, the changes in TiO_2-based DSSC parameters with ZnO coating are due to the following reasons: (i) the increase in V_{OC} is caused by the negative shift in Fermi level of TiO_2, which retards electron recombination process by forming energy barrier, (ii) the decrease in J_{SC} is due to Zn^{2+}/dye complexes that passivates the transport of dye molecules and increases back reactions due to defect states, (iii) the reduced FF is due to the obstacles observed in the penetration of electron to the electrolyte because of the decreased porosity, and (iv) the enhancement in η is associated with the contribution of all the above parameters and is mainly attributed to the increase in energy barrier to reduce the overall charge recombination. Figure 8.5 schematically

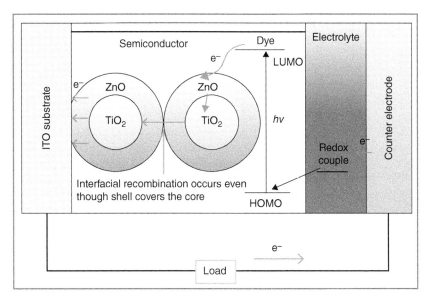

Figure 8.5 Schematic representation of the electron transport and the possibility of interfacial recombination taking place in core–shell model-based DSSC.

represents the electron transport and the possibility of interfacial recombination taking place in core/shell model-based DSSC.

Initially, the work started with TiO_2, since it is the worldwide photovoltaic material with good surface properties. Here an attempt was made with TiO_2 nanoparticles, which can offer high surface area for effective dye adsorption. But the performance was hindered by the trapping/detrapping mechanisms that occur in nanoparticles-based DSSC. Further, DSSCs were assembled with nanowires and achieved the enhancement in J_{SC} values very well up to 1021.8 ($\mu A/cm^2$) (three to fourfold times compared to nanoparticles), which is due to the improved direct electron transport pathway to the collection electrode with fewer particle-particle junction barriers. Additionally, the good ionic conductivity of poly ethylene glycol (PEG)-based electrolytes resulted in improved V_{OC} and remarkable enhancement in solar cell performance up to 0.313%.

Apart from TiO_2, various morphological forms of ZnO were also tried due to their higher electron mobility and reduced recombination rate over TiO_2. A comparative study of DSSC based on ZnO nanoparticles, nanorods, nanoflowers, and chunk nanostructures were performed and found that the ZnO nanorods exhibit maximum conversion efficiency of 0.163%. This increase in J_{SC} is mainly attributed to the high interaction of nanorods with the incident photons without any loss and due to the direct conduction pathway for electron transport without recombination. Next to nanorods, chunk nanostructures comprised of a regular arrangement of nanoparticles show the conversion efficiency of 0.123%. Due to the presence of more branched nanorods with good crystallinity, it is expected that nanoflowers would give better performance, but this is not fulfilled due to the existence of more defect-related states that act as recombination centers.

As discussed here, various attempts were tried with different forms of TiO_2 and ZnO morphologies, but in all cases, one critical factor namely recombination that takes place parallel limits the overall solar conversion efficiencies. One possible way to minimize the recombination rate is to build an energy barrier at the electrode/electrolyte interface, and this can be achieved by using two different materials in core (TiO_2)/shell (ZnO) configuration. Addition of ZnO layer increased V_{OC} from 0.278 to 0.412 V, confirming its active role as energy barrier leading to the reduction in recombination losses. In contrast to V_{OC}, J_{SC}, and FF were decreased due to the increased Zn^{2+}/dye aggregation. The overall conversion efficiency of TiO_2/ZnO (T1Z1) DSSC showed slight enhancement due to reduction in recombination and enhanced V_{OC}.

The conversion efficiency of ZnO decreases due to the instability of the ZnO in the dye solution that leads to Zn^{2+}-dye aggregates. To overcome these difficulties by improving surface stability, energy level positions, and carrier concentrations, different types of dopants were introduced into the host ZnO lattice. Besides spintronics applications, nowadays many attempts are going on to use these materials for DSSC photoelectrodes. Here, this concept was tried with Ti, Mg, and Sn-doped ZnO and found that Sn-doped ZnO showed the maximum efficiency of 0.693% due to the enhanced carrier concentration and appreciable negative shift in Fermi level. All these investigations were carried out with EY dye as sensitizer in DSSC.

The performance of DSSC equally depends on the types of the semiconductors and the sensitizers used. Similar to organic dyes, CdS semiconductor is also tried as an effective sensitizer in sensitizing TiO_2 and ZnO nanomaterials, because the illumination of CdS with single photon may generate multiple excitons. The formation of type-II heterojunction by CdS with TiO_2 or ZnO can promote very fast and easy charge transfer process. Generally, the CdS coating is done with chemical bath deposition (CBD) or successive ion layer adsorption and reaction (SILAR), but both are time-consuming processes and offer poor adsorption of CdS on the precoated MOS films. As a substitute to this, nanocomposites of CdS with TiO_2, and ZnO are prepared and found that the ZnO/CdS systems exhibit higher V_{OC}.

Overall, it is suggested that if all the solar cell parameters maintain a certain balance between them, then only the performance of the cell would improve so as to meet the world's energy requirements.

8.3   Importance of Heterostructures

Tennakone et al. [27] have proposed the advantage of SnO_2/ZnO core-shell and the effect of thin outer coating of ZnO:

i) During the illumination, dye is absorbing the photons and creates the excitons, after that inject electrons into the CB of interconnected SnO_2 layer from dye CB. Since dye molecules are attached to the ZnO, which is of ~1 nm thickness, the injection of electrons to SnO_2 particles could take place through tunneling via this ZnO. Moreover, ZnO layer acts only to block the electrons from SnO_2 CB to electrolyte (assuming complete surface coverage). Hence, as to suppress recombination of these electrons with the electrolyte or oxidized dye.

ii) During the illumination, the dye can absorb and inject electrons to the CB of ZnO which then undergoes downhill transition to SnO_2 CB (Figure 8.6). It is quite likely that there are trap levels in both semiconductors and trap-mediated transfer. In this state, the electrons in both SnO_2 and ZnO could undergo recombination with electrolyte solution and the oxidized dye (if SnO_2 particles are not fully covered by ZnO layer).

In 2013, Ripon Bhattacharjee et al. [28] observed ZnO/SnO_2 (SN-Z) nanocomposite photoanode-based solar cells obtained a high V_{OC} of 670 mV, which is relatively higher than other reported values of SnO_2 nanoparticles (SNP) and SnO_2 nano-flowers (SNF) which is shown in Figure 8.7. The low shift of CB and suppression of charge recombination are reasons for higher value of V_{OC}. In addition, SnO_2-based DSSCs had quite low fill factor (FF) value (44.47%), whereas SN-Z-based DSSC shows considerably higher FF of 68.9%. Moreover, electrochemical impedance spectroscopy (EIS) results show that fast electron transport and longer

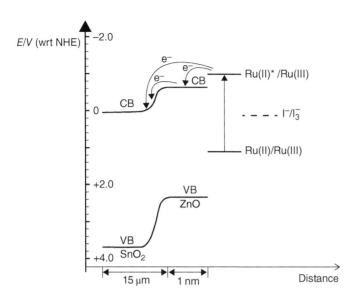

Figure 8.6 Schematic energy level diagram showing relative potentials of SnO_2 and ZnO.

electron lifetime are the major reasons for the high PCE. Finally, the SnO_2 nanocomposite fully covered by an insulating layer of ZnO enhanced the conversion efficiency of SnO_2 DSSCs.

8.4 *I–V* Characteristics

When ZnO is used as the coating material, it is possible that ZnO would act in a different manner to insulating materials. For SnO_2/ZnO DSSCs, injected electrons from the excited dye species attached to ZnO outer layer would fill the CB of ZnO and then would be transferred to the CB of the SnO_2. According to CB position of ZnO (−4.3 eV with respect to vacuum level) and SnO_2 (−5.0 eV with respect to vacuum level), and the results which were obtained for the SnO_2/ZnO composite system, the above assumption seems to be more valid as the electrons try to move toward the lower potential region, according to thermodynamics. Initially, SnO_2/ZnO composite showed lower value of J_{SC} and V_{OC} due to the recombination occurring at uncovered sites of SnO_2. When the amount of ZnO is increased, SnO_2 surface tend to become fully covered with ZnO and thereby the recombination occurring at the interfaces of dye/SnO_2/electrolyte will be reduced. Because of the electrons in the CB, SnO_2 have a lesser possibility to reach the surface traps of the ZnO in order to recombine with triiodide ions or oxidized dye molecules. After the certain point of ZnO amount, J_{SC} and V_{OC} started to decrease (Figure 8.8). This is due to the lower electron tunneling probability through ZnO coating layer as the higher ZnO amount will lead to an increase in the coating layer thickness. Here, two different types of electrolytes were used for making the solar cells and the measured $I–V$ spectrum is shown in Figure 8.9 [29].

8.5 Matching of Bandgaps

Milan et al. [30] observed composite bi-oxide-layered photoanode results in the dramatic improvement of the overall device performances (Figure 8.10). In particular, V_{OC} was 0.60 V,

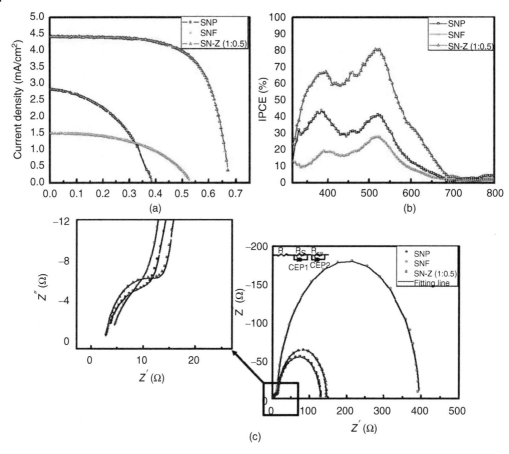

Figure 8.7 (a) I–V characteristic and (b) incident photon conversion efficiency (IPCE) spectra. (c) Nyquist plot of SNP, SNF and SN-Z based DSSCs.

J_{SC} was 10.28 mA/cm^2, and FF was 57%. The PCE was 3.53%, i.e. about three times larger than for pure SnO$_2$ and three and a half times than pure ZnO. These results can be tentatively explained by considering the optimal position of ZnO CB, with respect to N719 LUMO, which guarantees for V_{OC} enhancement, compared to pure SnO$_2$. Increased photocurrent calls for good electron transport, guaranteed by the tightly connected SnO$_2$ network.

The first work exploring the properties of mixed ZnO/SnO$_2$ photoanodes hypothesized that an effect of band gap engineering might be induced by surrounding the SnO$_2$ particles with ZnO species, favoring the charge injection from the N719 LUMO to the ZnO CB and then transferring the photo-generated electrons to the SnO$_2$ CB. This favorable band alignment would result in two relevant improvements, as schematically illustrated in Figure 8.10a: the first advantage would be the possibility to properly inject photo-generated electrons from N719 to SnO$_2$ through the ZnO (which however still presents issues as for injection in itself) and the second relevant advantage would be the elimination of the so-called back recombination between SnO$_2$ CB and the electrolyte redox couple (represented by the dashed grey arrow in Figure 8.10a), as the outer ZnO shell acts as effective tunneling barrier between the SnO$_2$ nanoparticles (NP) and the electrolyte. However, this schematic would be realistic only for very particular ZnO/SnO$_2$ architectures, as those represented in Figure 8.10b, in which SnO$_2$ is partially surrounded by ZnO and dye is exclusively adsorbed on ZnO.

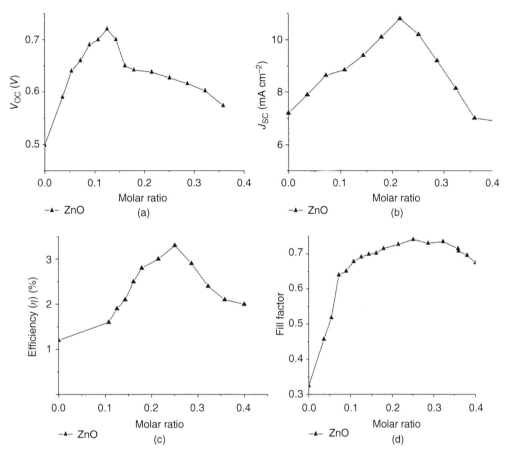

Figure 8.8 Variations of solar cell parameters of DSSCs made using various compositions of ZnO-coated SnO$_2$ on with electrolyte X. Variation of (a) V_{OC}, (b) J_{SC}, (c) efficiency, and (d) fill factor.

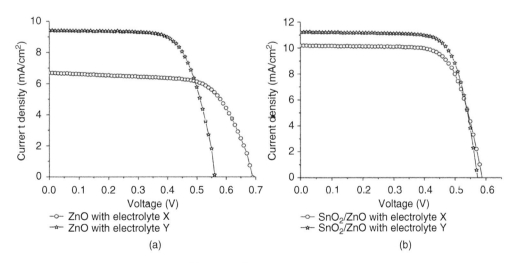

Figure 8.9 *I–V* characteristics curves of the DSSCs fabricated with composite systems: (a) pure ZnO and (b) SnO$_2$/ZnO with electrolytes X and Y.

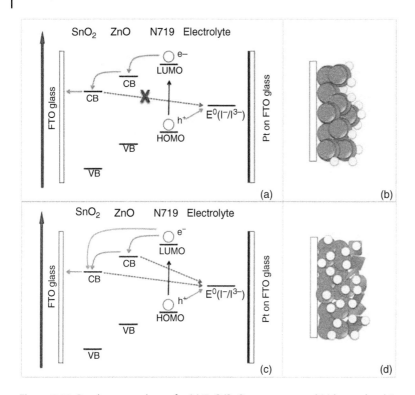

Figure 8.10 Band energy scheme for (a) ZnO/SnO$_2$ structures and (c) layered architecture in this work. (b) and (d) show the two configurations theoretically corresponding to band energy diagrams reported in (a) and (c), respectively.

Additional constraint would be the contact between SnO$_2$ and fluorine doped tin oxide (FTO), where ZnO should not be involved at all, in order to avoid an energy gap between SnO$_2$ CB and ZnO CB, as it is for instance in core-shell SnO$_2$/ZnO systems, through which electron is not allowed to run. In photoanode structures different from those above described, for instance disordered SnO$_2$–ZnO networks or the layered configuration herein proposed (Figure 8.10d), there are several limitation as for charge transport processes that should be taken into account. In those cases, dye would be anchored into both metals and then electron injection would take place in both SnO$_2$ and ZnO, as well as back recombination with electrolyte (as depicted in Figure 8.10c). Also a systematic study of the possible effect that is exerted by the layered oxide configuration by changing the relative number of ZnO/SnO$_2$ layers and modulating the time applied for dye uptake was also studied.

Electrochemical investigation exposed that the enhancement of device performances shown by ZnO/SnO$_2$-based cells is mainly ascribable to the synergistic effect attained by the simultaneous exploitation of these two oxides, which collaborate in optimizing the resistance toward the exciton recombination and the capability to accumulate the photo-generated charges at the oxides. The device exploiting bare SnO$_2$ features low R_{REC} and high C_μ, with corresponding high J_{SC} and low V_{OC}, as compared with a solar cell whose photoanode is based on pure ZnO, which on the contrary presents high R_{REC} and low C_μ and associated higher V_{OC} and lower J_{SC} which is shown in Figure 8.11. The enhanced R_{REC} observed for layered electrode architectures, as compared with bare SnO$_2$, is ascribable to the presence of ZnO, which alone features the best R_{REC} among the analyzed devices and, at the same time, exerts a capping effect on the underlying SnO$_2$.

Figure 8.11 *I–V* curves of DSSCs based on ZnO (triangle), SnO$_2$ (sphere) and a mixed ZnO/SnO$_2$ network composed of 3 ZnO and 3 SnO$_2$ layers (3@3 sample, square).

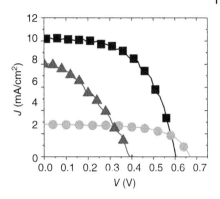

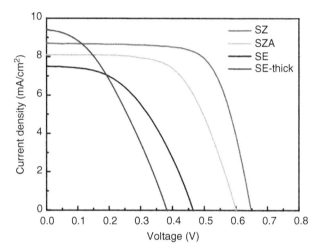

Figure 8.12 *I–V* behaviors of SnO$_2$ the nanoparticles electrodes (SE), the ZnO/SnO$_2$ electrode (SZ – 5 wt%) and the nanostructured SnO$_2$ electrode (SZA) from SZ.

Jung-Hoon Lee et al. [31] reported and achieved the higher PCE of acetic acid-treated ZnO–SnO$_2$ (SZ) electrode compared to bare SnO$_2$ which is presented in Figure 8.12. The bare SnO$_2$ electrode exhibited a low open-circuit voltage V_{OC} of 0.465 V, a fill-factor (FF) of 47.7%, and a moderate current density J_{SC} of 7.49 mA/cm^2. The SnO$_2$/ZnO nano-grain electrode exhibited greatly enhanced photovoltaic performances. V_{OC} and FF greatly increased to 0.648 V and 70.2%, respectively, leading to a conversion efficiency as high as 3.96%. The acetic acid-treated ZnO/SnO$_2$ nano-grain exhibited a remarkable increase in the conversion efficiency up to 2.98%, which corresponded to an 80% increase from the bare SnO$_2$. This difference was associated with the substantial suppression of the recombination of the electrons in the SZ and the SnO$_2$/ZnO (SZA). The small V_{OC} of the conventional SnO$_2$ electrode (SE) was primarily attributed to the low-lying conduction band edge (CBE) of SnO$_2$, which was about 0.4 V lower than TiO$_2$. Additionally, the highly populated surface state of the SE beneath the CBE that originated from the small particle size decreased the V_{OC} more than expected from the CBE position. Therefore, the drastic drop in the V_{OC} for the SE with increase in thickness suggested the existence of deep traps. Actually, coating the surface of SnO$_2$ electrodes with insulating oxides, such as MgO, ZnO, Al$_2$O$_3$, and ZrO$_2$ was reported to significantly increase the V_{OC} by reducing the surface states. The SnO$_2$, SnO$_2$–ZnO, and SnO$_2$–ZnO–acetic acid exhibited maximum incident photon conversion efficiency (IPCE)

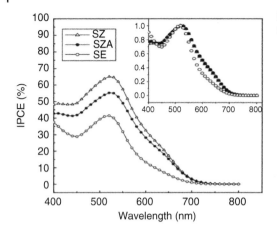

Figure 8.13 IPCE spectra of the SE, SZ, and SZA–DSSCs. The inset shows the normalized IPCE.

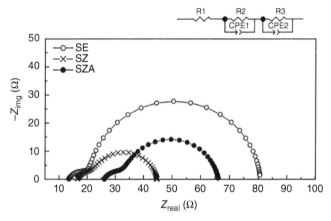

Figure 8.14 AC-impedance spectra of the SE, SZ, and SZA–DSSCs under the one sun illumination condition.

values of 41.4%, 65.0%, and 55.2%, respectively, at 520 nm (Figure 8.13). The internal resistances of the three electrodes were studied using ac-impedance spectroscopy in order to investigate the electron transfer at the SnO_2 (or SnO_2/ZnO)/dye and electrolyte interfaces. The Cole–Cole plots of the three impedance spectra were compared in Figure 8.14. Two semicircles were monitored at 10–100 Hz and in the kHz ranges and a static resistance at higher frequencies for all of the samples (Figure 8.14). The SZA exhibited a larger R1 (26 Ω) than the SE (13 Ω) and the SZ (16 Ω) because the FTO layer was eventually etched by acetic acid. R2 values of around 6–7 Ω were found for all three electrodes, implying that the electrical contact of the FTO and SnO_2 films was quite similar in the three electrodes. For R3, the nano-grainelectrodes exhibited substantially reduced values of 22 (SZ) and 34 Ω (SZA) with respect to the value of 60 Ω for the SE.

The SnO2-based DSSCs show low FF ~ 0.49 and low $V_{OC} \sim 0.49$ V, while high short current density (J_{SC}) is ~ 16.3 mA/cm^2 due to its high mobility. On the other hand, ZnO and TiO2 nanoflowers (NFs) device showed higher V_{OC} due to higher CB compared to SnO2, however, low $J_{SC} \sim 4.8$ mA/cm^2 of ZnO device could be attributed to the formation of Zn^{2+} complexes with the dye ligands, which hinders charge separation, and finally, the dye is not able to inject electrons into the ZnO. However, the composite ZnO–SnO2 composite ZnO/SnO2 (CNFs) photo-anode has shown superior performance, $V_{OC} \sim 0.738$ V, $J_{SC} \sim 13$ mA/cm^2, FF ~ 0.60

yielding PCE ~ 5.6% (Figure 8.15). The PCE of the composite-based device has increased by ~400% than the ZnO-based device and over ~30% than the SnO2-based device [32]. Similarly, for the TiO2–SnO2 DSSCs, the PCE of the CNFs-based device was increased by ~100% and 50% than SnO2 and TiO2-based device, respectively, as reported before. The superior performance of the CNFs devices is expected from the more negatively shifted Vfb and their high electrical conductivity, a synergy has thus been achieved in the CNFs-based devices. The electrochemical characteristics were analyzed through Nyquist plot which is shown in Figure 8.16. Moreover, the EIS results are used to calculate the charge transport resistance, recombination resistance, transit time, electron lifetime (Figure 8.17).

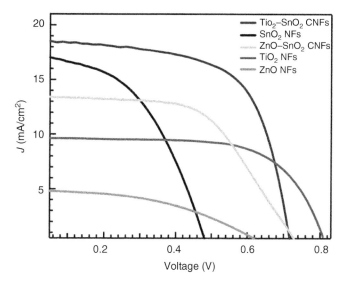

Figure 8.15 *I–V* curves of the DSSCs fabricated using TiO_2–SnO_2 NFs and ZnO–SnO_2 CNFs.

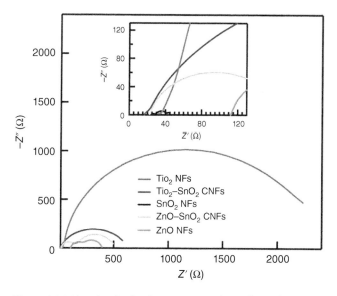

Figure 8.16 Nyquist plot for the composite electrodes and (inset) magnified portion of the Nyquist plot at the high-frequency region showing the differences in the series and transport resistance.

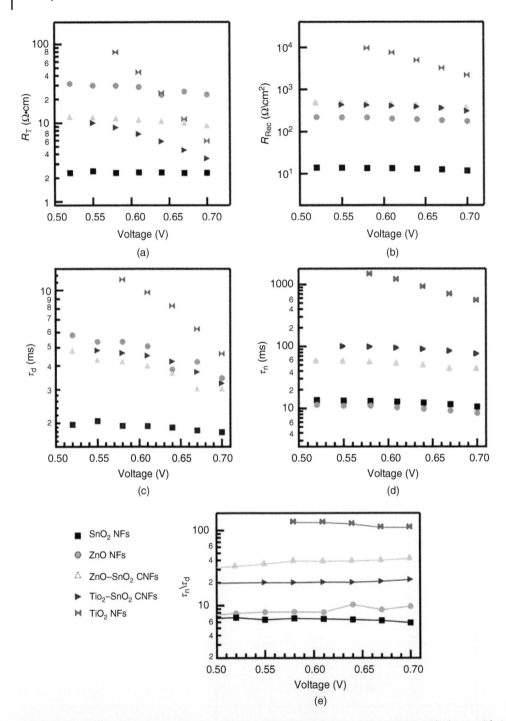

Figure 8.17 (a) Charge transport resistance, (b) recombination resistance, (c) transit time, (d) electron lifetime, (e) ratio between transit and electron lifetime as a function of applied voltage measured under dark conditions for TiO_2–SnO_2 CNFs and ZnO–SnO_2 CNFs photoanodes.

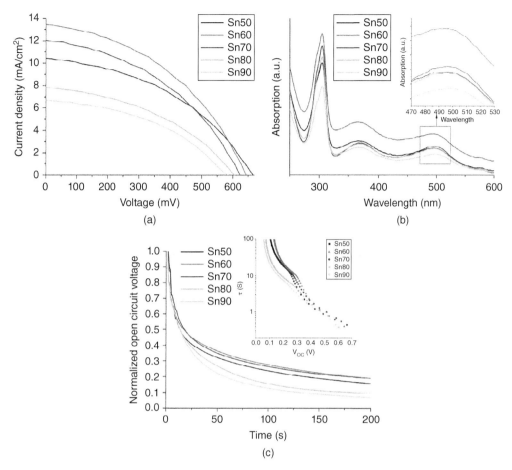

Figure 8.18 (a) $I-V$ characteristics curves of the fabricated DSSCs, (b) UV–Vis absorbance spectra of photoanodes, and (c) OCVD diagrams of the DSSCs.

Masoud Abrari et al. [33] reported and optimized the concentration of Sn on the ZnO–SnO$_2$(SnXX) nanocomposite. The efficiency of the cells fabricated from Sn50, Sn60, Sn70, Sn80, and Sn90 composite nanostructures are 2.90%, 3.64%, 3.10%, 1.92%, and 1.64%, respectively. As it is seen, the cell fabricated from Sn60 material has the highest efficiency (Figure 8.18a). From the $I-V$ spectrum, there is an increase in the efficiency in this situation (Sn60) that can be attributed to the improvement of three parameters: current density, open circuit voltage, and fill factor (FF). The XRD results have previously shown that the crystallite size of Sn60 sample is larger than that of the other samples. Therefore, it is not far from mind that the cell fabricated from this sample enjoys a better electron transport property that enhances the FF and current density. Furthermore, it is clear from the field emission scanning electron microscope (FESEM) images that hexagonal structures are formed in Sn70, Sn60, and Sn50 samples. These hexagonal structures will improve the 2D electron transport and also create light scattering in the cell, so the current density will further increase. On the other hand, by increasing the share of zinc oxide in the composite, the amount of open circuit voltage is increased in comparison with the pure tin oxide due to the relocation of the Fermi levels to higher energies. Moreover, Sn60 cell having the longer lifetime of the electrons is an indicator of lower recombination centers and thus, the Sn60 DSSC will have a better photovoltaic performance. The 2D and hexagonal structures of zinc oxide nanoparticles provide a suitable electron transport property. They can transfer the electrons to the contacts easier and prevent the recombination with the electrolyte. Finally, he exploited the synthesized nanostructures for DSSC fabrication and achieved a 150% growth of efficiency in comparison with the cells prepared from pure SnO2 nanoparticles. UV–Vis spectroscopy is used to acquire the absorbance spectra of these photoanodes, which are shown in Figure 8.18b. In order to compare the dye adsorption of the photoanodes, we can use the peak at 500 nm, which corresponds to the N719 dye. As we can see, Sn60 photoanode has the highest peak at this wavelength and therefore, it has the highest dye adsorption. For a better analysis of the photovoltaic behavior, we calculated the carrier lifetime in the DSSCs fabricated from the nanostructures prepared with different duty cycles at 0.9 Hz frequency through open-circuit voltage decay (OCVD) analysis. As we can see from the Figure 8.18c, the electron lifetime of the Sn60 DSSC is higher than the other samples in most of the graph. The longer lifetime of the electrons in this cell is an indicator of lower recombination centers and thus the Sn60 DSSC will have a better photovoltaic performance.

Jiaxing Song and cowokers [34] prepared ZnO–SnO$_2$ nanocomposite thin films by spin-coating of mixed nanoparticles as electron collection layers (ECLs) to fabricate low-temperature-processed PSCs. The device architecture and energy-level position are shown in Figure 8.19. They observed clear dependence of short-circuit current (J_{SC}) and fill factor (FF) on the component content in the composite films despite almost unchanged open-circuit voltage (V_{OC}). With planar-layered structures, the PSCs exhibited relatively high PCE of 14.3% under the standard AM 1.5 G (100 mW/cm^2) simulated sunlight illumination with an optimal weight ratio of 2 : 1 for ZnO/SnO$_2$ in solution. Most importantly, the CH3NH3PbI3 layer and device based on ZnO–SnO$_2$ nanocomposite thin film exhibited improved thermal stability and device stability in comparison with that based on ZnO, respectively. Figure 8.20 shows the current–voltage (J–V) characteristics measured from the reversed voltage scan of PSCs with each ZnO, SnO$_2$, and ZnO–SnO$_2$ nanocomposites (5 : 1, 2 : 1, 1 : 1, 1 : 2). Interestingly, nonsignificant change in the V_{OC} value was observed, despite the CBE of SnO$_2$ is significantly below that of ZnO. This insensitivity of the V_{OC} value to the CBE level of the ECL could be attributed to both the passivation effect of the remnant PbI$_2$ and the effective charge collection by both ZnO and SnO$_2$ ECL layers. As the result of the varied

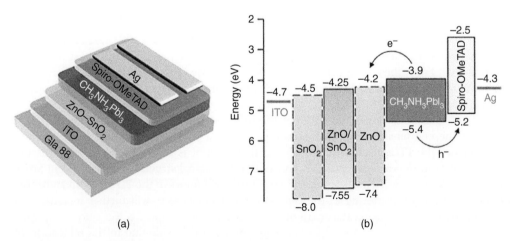

(a)   (b)

Figure 8.19 Device architecture (a) and energy level diagram (b) of the ITO/ECL/CH$_3$NH$_3$PbI$_3$/spiro-OMeTAD/Ag PSCs based on ZnO, SnO$_2$, or ZnO/SnO$_2$ (2 : 1).

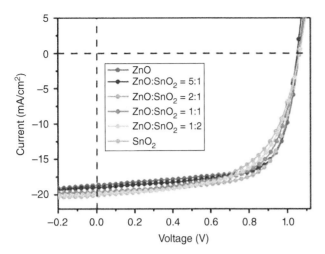

Figure 8.20 *I–V* characteristic for the PSC devices based on ZnO–SnO$_2$ ECL prepared from different mass ratio for ZnO and SnO$_2$ in the solution.

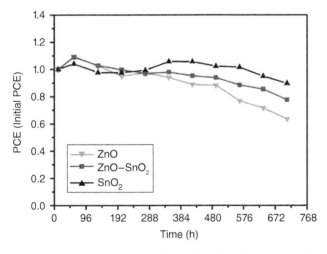

Figure 8.21 Durability of the ZnO-based, ZnO–SnO$_2$-based and SnO$_2$-based PSC devices exposed to ambient air.

J_{SC} and FF values, the highest PCE of 14.3% was determined by a J_{SC} of 19.6 mA/cm^2, a V_{OC} of 1.06 V, and a FF of 0.688 was achieved for the best fabricated device with ZnO/SnO$_2$(2 : 1). Figure 8.21 shows the stability result of PSCs based on ZnO, ZnO/SnO$_2$ (2 : 1), and SnO$_2$ as ECLs during the storage in ambient air environment in room temperature. Moreover, they carried out EIS measurements in the dark at 0 V relative to the open circuit potential on the devices with the configuration of ITO/ECL/CH$_3$NH$_3$PbI$_3$/Spiro-OMeTAD/Ag, in which ECL is SnO$_2$, ZnO/SnO$_2$, and ZnO, respectively. Figure 8.22 shows the Nyquist plots of PSC devices with these ECLs. The Nyquist plot is composed of two irregular arcs, i.e. a small arc at high frequency and a large arc at low frequency. From the result, the SnO$_2$-based device gives the lowest Rct value indicating that charge extraction is most efficient at the SnO$_2$/CH$_3$NH$_3$PbI$_3$. Moreover, the lower Rs and Rct values must jointly contribute to the high J_{SC} value of SnO$_2$-based PSCs.

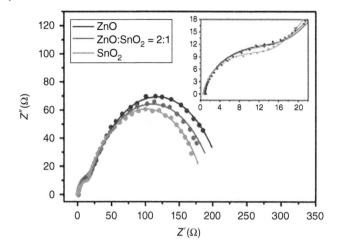

Figure 8.22 (a) Nyquist plots for the perovskite solar cells with SnO_2, ZnO/SnO_2, and ZnO ECLs, respectively. Magnified view of the Nyquist plots at high frequency.

Palomares et al. reported and compared the function of three different metal oxide overcoats Al_2O_3, ZrO_2, and SiO_2, grown on nanocrystalline TiO_2 films. The films of Al_2O_3, ZrO_2, and SiO_2 have the capability to reduce interfacial electron transfer dynamics to increase the performance of DSSCs [20].

In 2006, Dürr et al. explored titanium-zirconium mixed oxides solved by this issue and obtained the shift in the CB edge that leads to an increase of the open-circuit voltage when incorporating the Zr content into TiO_2, also reduced the electron recombination [35]. The schematic representation of the TiO_2–ZrO_2 core–shell formation is shown in Figure 8.23. Zirconia (ZrO_2), is a ceramic oxide and acts as a good photo catalyst that occurs in three different phases: tetragonal, monoclinic, and cubic. Being a wide bandgap metal oxide similar to TiO_2, it has been used in the preparation of DSSCs. As indicated in Figure 8.1, the band edge positions of $Ti_{1-x}Zr_xO_2$ are changed with respect to the content of zirconium, x, between the positions of TiO_2 and ZrO_2, and they optimized the Zr content concentration at $x = 1\%$.

Figure 8.23 Schematic of core–shell working electrode.

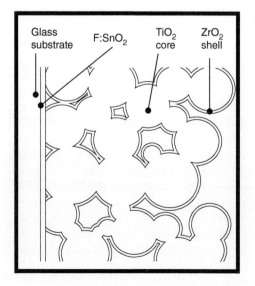

Figure 8.24 Schematics of the dependence of conduction band edge and valence band (VB) edge as a function of zirconium content.

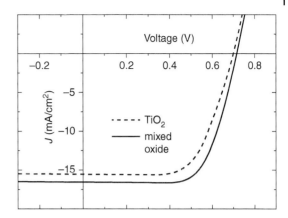

For higher Zr content, we observe a drop in short circuit current density due to the reduced driving force for electron injection into the semiconductor, which cannot be compensated by the higher V_{OC}. The efficiency increased from 7% to 8.1% when incorporating the Zr on TiO_2 which is shown in Figure 8.24.

David B. Menzies et al. [36] reported that ZrO_2 deposited onto TiO_2 and is made in the form of core shell type heterostructure and optimize the shell thickness. The bare TiO_2 as a photo-anode used in this study achieved the relatively low performance due to the opaque nature of the cells. The ZrO_2 coating on the TiO_2 nanoparticle produced a reliable trend in power conversion efficiencies. Here, all ZrO_2-coated TiO_2 electrodes performed to higher efficiency than the bare or uncoated cell. However, 0.4 nm thickness shell achieved the higher performance of PCE 2.27% compared to other thickness shells. As ZrO_2 is a more insulating oxide than TiO_2, the photo-excited electrons can be transformed from the excited dye and tunnelled via the ZrO_2 shell into the TiO_2 semiconductor CB. The electron tunneling is used to assist the fast electron injection processes. Simultaneously, the charge recombination process is slowed by offering an energy barrier from the semiconductor to the oxidized dye or electrolyte. Moreover, when increasing the shell thickness upto a certain level the photo-excited electrons were not properly injected into the TiO_2 CB rather than being recombined from the ZrO_2 to the dye or electrolyte. Thus, this ZrO_2-modified TiO_2 electrodes in DSSCs have enhanced the performance of the inefficient device.

In 2014, K. Manoharan et al. [37] obtained the efficiency enhanced from 5.21% to 8.03% when the ZrO_2 shell was coated on TiO_2 (Figure 8.25). The improvement of the PCE was attributed to the increase in the J_{SC}, V_{OC}, and FF, and these results suggested that electron injection is more efficient through the ZrO_2 shell compared with the uncoated cell. There were two possible mechanisms and the major reasons for the increased V_{OC}, such as

i) blocking effects
ii) surface dipole effects

Here, ZrO_2-coated electrodes have higher CB edge and coating layer act as blocking barrier for the recombination of the injected electrons either with the oxidized dye or with the oxidized redox couple. If the recombination of the injected electrons is reduced, the electron population at the TiO_2 CB will be increased, hence resulting quasi-Fermi level would increase, and as a result V_{OC} is increased. Another possible mechanism is "surface dipole effect" which is not related directly to the electron recombination. The coating layer may induce the positive charges near the surface resulting in an electrostatic field that will be generated, which could increase the conduction edge, hence the V_{OC} is increased. For these reasons, the formation of the mixed metal oxide nanostructure enhanced the performance of the DSSCs. Also,

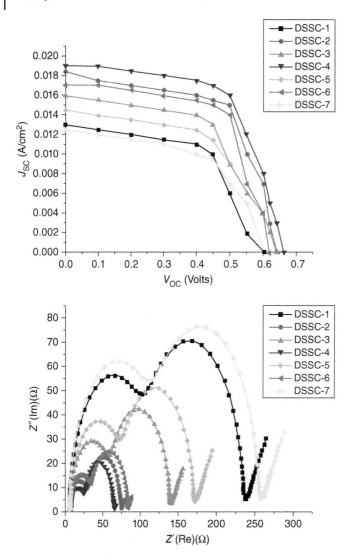

Figure 8.25 *I–V* characteristics of DSSCs and Nyquist plot of DSSCs.

the ZrO_2-coated cell is having a lower charge transfer resistance at the TiO_2/dye/electrolyte interface (Figure 8.25). Moreover, this cell is having higher values of chemical capacitance at the TiO_2/dye/electrolyte interface, τ_{eff}, L_n, D_{eff}, μ, σ, and η_{cc} as compared to other DSSCs. In addition, k_{eff} and concentration of electrons in the TiO_2/dye/electrolyte interface are lower values as compared to that without ZrO_2 DSSCs. In addition, some researchers incorporating the cation dopant on the photo-anode materials (metal oxides) exerts larger dipole moment, that changes the interface energetic leads to the efficient electron transfer. Finally, they conclude, that core–shell structure-based DSSCs are essential for efficient electron injection and thereby increase the overall performance by designing a proper shell material (metal oxide) with suitable band gap and optimal thickness of shell layer.

Natalie O.V. Plank et al. [38] have applied an MgO and a ZrO_2 shell deposition method to control the interface between two indolene-based organic molecular dyes, termed D102 and D149 which are displayed in Figure 8.26. Solar cells were assembled as solid-state DSCs

Energy (eV)

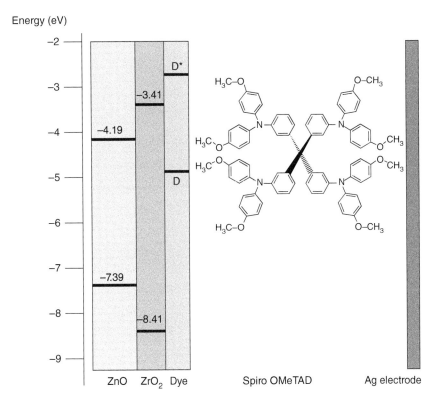

Figure 8.26 Schematic of the energy levels at the photovoltaic heterojunction in the solid-state DSC.

with 2,2′,7,7′-tetrakis (N,N-di-p-methoxyphenylamine)-9,9′-spirobifluorene(spiro-OMeTAD) as the molecular hole-transporter. Surprisingly, these measurements reveal that the deposition of a surface shell of MgO and ZrO_2 predominantly enhances the photo-induced charge generation and does not appear to inhibit the electron hole-recombination. The $I–V$ characteristics under the solar simulator AM 1.5 G for the D102 devices are shown in Figure 8.27. The ZnO/ZrO_2 nanowire (NW) device shows the highest short-circuit current, 2.14 mA/cm^2, and the ZnO NW demonstrates the lowest at 0.72 mA/cm^2. All of the devices prepared are shown to have an average V_{OC} of 0.47 V. The η of the devices has been improved for both the ZnO–MgO NW and the ZnO–ZrO_2 NW devices to 0.155% and 0.283%, respectively. Figure 8.27 shows the $I–V$ curves of the D102 and D149 devices, respectively, one day (a, d), one month (b, e), and two months (c, f) after fabrication. During this time period, the devices were not encapsulated and were stored at room temperature in the dark under ambient atmosphere conditions. As a result of the enhanced photocurrent for both the ZnO–MgO NW and the ZnO–ZrO_2 NW in comparison to the ZnO NW devices, there is a higher overall PCE for all of the coated NWs. During the period of two months, the devices have remained stable. The highest performance of 0.7% was achieved for devices incorporating the ZnO–ZrO_2 NW arrays in conjunction with D149 after one month. To the best of our knowledge, this represents the highest efficiency reported for a ZnO nanowire hybrid solar cell incorporating an organic hole-transporter. The measured photovoltaic parameters are displayed in Table 5.1.

It is clear that the "insulating" and wide band gap shells of MgO and ZrO_2 greatly enhanced the photovoltaic process. The broad intention of using these core-shell materials, in many different systems, is to "passivate" the semiconductor surface, primarily to inhibit electron-hole

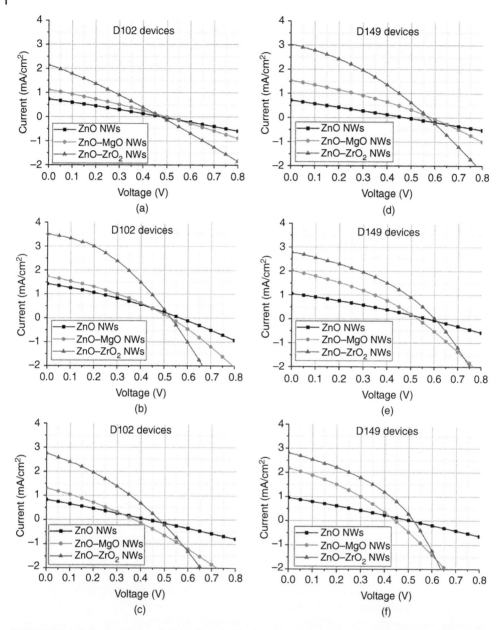

Figure 8.27 *I–V* characteristics of solar cells-sensitized D102 with ZnO NWs, ZnO–MgO NWs, and ZnO–ZrO$_2$ NWs after (a) one day, (b) one month, and (c) two months and *I–V* characteristics of the solar cell with D149 dye with ZnO NW, ZnO–MgO NW, and ZnO–ZrO$_2$ NW devices after (d) one day, (e) one month, and (f) two months.

recombination. To investigate if this is occurring in this system, we have performed transient absorption spectroscopy on complete devices and dye-sensitized NW arrays (Figure 8.28). All samples were pumped with the same excitation influence and exhibited similar optical absorption. Contrary to what we expected, the charge generation is increased on the nanosecond time scale for the MgO-coated NWs and even more so for the ZrO$_2$ coated NWs. This could be due to favorable electron transfer through or to the core-shell materials as compared to bare ZnO.

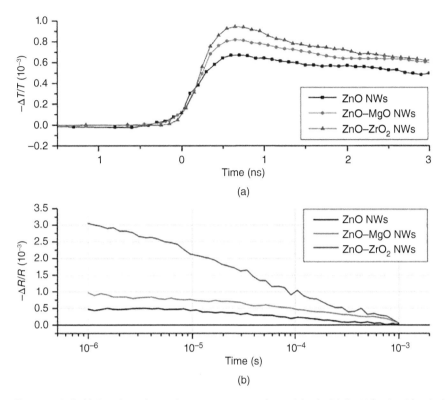

Figure 8.28 (a, b) Transient absorption spectra nanorods sensitized with D102 having identical optical densities.

However, we note that this may also be due to the presence of the ZrO_2 layer reducing early time recombination. Figure 8.29 looks at the absorption spectra of ZnO NW, ZnO–MgO NW, and ZnO–ZrO_2 NW films submerged in D102 for 35 minutes. At these increased dye loading times, we observe a larger signal from the dye for both the ZrO_2 and the MgO barrier layer samples in comparison to bare ZnO. It is therefore sensible to conclude that the increased efficiency of the ZrO_2 and MgO barrier layer devices in comparison to the pristine ZnO can be attributed to an overall increase in dye loading.

Athapol Kitiyanan et al. [6] reported the improvement of a short-circuit photocurrent up to 11%, an open-circuit voltage up to 4%, and a solar energy conversion efficiency up to 17%, for the DSSC fabricated by mesoporous zirconia/titania mixed system when compared to the cell that was fabricated only by nanostructured TiO_2. The cell fabricated by 5 mm thick mixed TiO_2–ZrO_2 electrode gave the short-circuit photocurrent about 13 mA/cm^2, open-circuit voltage about 600 mV, and the conversion efficiency 5.4%. Also they measured the relationship between the current density (J_{SC}), open circuit voltage (V_{OC}), efficiency, and the film thickness of TiO_2 and TiO_2–ZrO_2 electrode, and this was reported in Figure 8.30.

Xinning Luan et al. [39] observed that the I–V characteristics of DSSCs are based on ZrO_2 coated and uncoated TiO_2 electrode film and $TiCl_4$-treated TiO_2 electrode film with the same thickness of 8.8 μm (Figure 8.31). Compared to untreated TiO_2, the DSSC consisting of $TiCl_4$-treated TiO_2 electrode film increases the energy conversion efficiencies by significantly enhancing the surface area for dye loading. In DSSCs, electron recombination can transfer electrons in milliseconds from the CB of TiO_2 to the ground state of a dye or to the electrolyte, which reduces the energy conversion efficiency. The DSSC based on ZrO_2-TiO_2

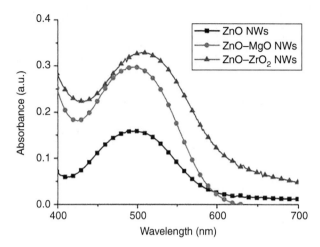

Figure 8.29 Optical absorption in the visible range for ZnO NWs, ZnO–MgO NWs, and ZnO–ZrO2 NWs coated with D102 dye.

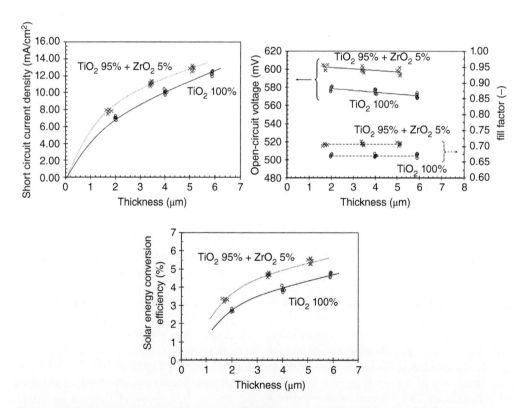

Figure 8.30 Relationship between the short-circuit current, open-circuit voltage, and efficiency against the thickness of electrodes fabricated by TiO_2 100% and mixed metal oxides (ZrO_2 5% + TiO_2 95%).

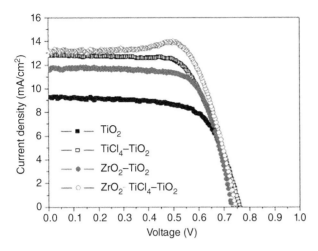

Figure 8.31 *I–V* characteristics of DSSCs based on 8.8 μm thick TiO$_2$ nanosheet electrode film and TiCl$_4$ treated TiO$_2$ nanosheet electrode film coated with and without ZrO$_2$.

(TiO$_2$ electrode film coated with ZrO$_2$) delivers an energy conversion efficiency of 6.12% with V_{OC} of 0.73 V, J_{SC} of 11.60 mA/cm^2, and FF of 0.72, which represents a 28.6% increase in cell efficiency compared to that of the DSSC based on untreated TiO$_2$ NSs (4.76%). The significant enhancements in J_{SC} and PCE can be ascribed to conformed growth of ZrO$_2$ on TiO$_2$ NSs. The highest efficiency of 7.33% is achieved by a DSSC based on TiCl$_4$-ZrO$_2$-TiO$_2$ (TiCl$_4$ treated TiO$_2$ NSs electrode film coated with ZrO$_2$), with V_{OC}, J_{SC}, and FF equal to 0.74 V, 13.20 mA/cm^2, and 0.75, respectively. This 54% increase in cell efficiency compared to that of the DSSC based on untreated TiO$_2$ NS film (4.76%) indicates that the ZrO$_2$ atomic layer deposition (ALD) coating combined with TiCl$_4$ treatment on TiO$_2$ NS film can effectively enhance the efficiency of DSSCs by significantly creating more surface area for dye loading and preventing electron-hole recombination between TiO$_2$ and the dye/electrolyte, respectively.

8.6 Conclusion

The above mentioned results clearly suggest that core-shell combination of metal oxide is more efficient for increasing the PCE. Here, the shell material acts as the electron-blocking layer, hence, it reduces the charge recombination from photo anode CB to dye or electrolyte valence band due to the arrangement of Type II formation. Moreover, the shell acts not only as the blocking layer but it also acts as the protection layer which protect the core from the environment and also some shell material acts as the sensitizer and depends on the band gap. Finally, after the strong discussions, it is concluded that binary metal oxide-based photo anode will play an important role in future.

References

1 Elumalai, N.K., Vijila, C., Jose, R. et al. (2015). Metal oxide semiconducting interfacial layers for photovoltaic and photocatalytic applications. *Mater. Renewable Sustainable Energy* 4: 11.

2 Bach, U., Lupo, D., Comte, P. et al. (1998). Solid-state dye-sensitized mesoporous TiO$_2$ solar cells with high photon-to-electron conversion efficiencies. *Nature* https://doi.org/10.1038/26936.

3 Law, M., Greene, L.E., Johnson, J.C. et al. (2005). Nanowire dye-sensitized solar cells. *Nat. Mater.* https://doi.org/10.1038/nmat1387.

4 Cavas, M., Gupta, R.K., Al-Ghamdi, A.A. et al. (2013). Preparation and characterization of dye sensitized solar cell based on nanostructured Fe_2O_3. *Mater. Lett.* https://doi.org/10.1016/j.matlet.2013.04.053.

5 Hara, K., Horiguchi, T., Kinoshita, T. et al. (2000). Highly efficient photon-to-electron conversion with mercurochrome-sensitized nanoporous oxide semiconductor solar cells. *Sol. Energy Mater. Sol. Cells.* https://doi.org/10.1016/S0927-0248(00)00065-9.

6 Kitiyanan, A., Ngamsinlapasathian, S., Pavasupree, S., and Yoshikawa, S. (2005). The preparation and characterization of nanostructured TiO_2–ZrO_2 mixed oxide electrode for efficient dye-sensitized solar cells. *J. Solid State Chem.* https://doi.org/10.1016/j.jssc.2004.12.043.

7 Tripathi, M. and Chawla, P. (2015). CeO_2-TiO_2 photoanode for solid state natural dye-sensitized solar cell. *Ionics (Kiel)* https://doi.org/10.1007/s11581-014-1172-6.

8 Chen, S.G., Chappel, S., Diamant, Y., and Zaban, A. (2001). Preparation of Nb_2O_5 coated TiO_2 nanoporous electrodes and their application in dye-sensitized solar cells. *Chem. Mater.* https://doi.org/10.1021/cm010343b.

9 Zhang, L., Shi, Y., Peng, S. et al. (2008). Dye-sensitized solar cells made from $BaTiO_3$-coated TiO_2 nanoporous electrodes. *J. Photochem. Photobiol. A Chem.* https://doi.org/10.1016/j.jphotochem.2008.01.002.

10 Guo, E. and Yin, L. (2015). Tailored $SrTiO_3$/TiO_2 heterostructures for dye-sensitized solar cells with enhanced photoelectric conversion performance. *J. Mater. Chem. A.* https://doi.org/10.1039/c5ta02556g.

11 Shin, S.S., Kim, J.S., Suk, J.H. et al. (2013). Improved quantum efficiency of highly efficient perovskite $BaSnO_3$-based dye-sensitized solar cells. *ACS Nano.* https://doi.org/10.1021/nn305341x.

12 Loh, L., Briscoe, J., and Dunn, S. (2016). Bismuth ferrite enhanced ZnO solid state dye-sensitised solar cell. *Procedia Eng.* https://doi.org/10.1016/j.proeng.2015.09.235.

13 Cui, J., Meng, F., Zhang, H. et al. (2014). CH3NH3PbI3-based planar solar cells with magnetron-sputtered nickel oxide. *ACS Appl. Mater. Interfaces* https://doi.org/10.1021/am507108u.

14 Manthina, V., Correa Baena, J.P., Liu, G., and Agrios, A.G. (2012). ZnO-TiO_2 nanocomposite films for high light harvesting efficiency and fast electron transport in dye-sensitized solar cells. *J. Phys. Chem. C* https://doi.org/10.1021/jp304622d.

15 Wang, M., Huang, C., Cao, Y. et al. (2009). Dye-sensitized solar cells based on nanoparticle-decorated ZnO/TiO_2 core/shell nanorod arrays. *J. Phys. D Appl. Phys.* https://doi.org/10.1088/0022-3727/42/15/155104.

16 Kanmani, S.S. and Ramachandran, K. (2012). Synthesis and characterization of TiO_2/ZnO core/shell nanomaterials for solar cell applications. *Renew. Energy* https://doi.org/10.1016/j.renene.2011.12.014.

17 Roh, S.J., Mane, R.S., Min, S.K. et al. (2006). Achievement of 4.51% conversion efficiency using ZnO recombination barrier layer in TiO_2 based dye-sensitized solar cells. *Appl. Phys. Lett.* 89: 253512. https://doi.org/10.1063/1.2410240.

18 Eguchi, K., Koga, H., Sekizawa, K., and Sasaki, K. (2011). Nb_2O_5-based composite electrodes for dye-sensitized solar cells. *J. Ceram. Soc. Jpn.* 108: 1067–1071. https://doi.org/10.2109/jcersj.108.1264_1067.

19 Kitiyanan, A., Kato, T., Suzuki, Y., and Yoshikawa, S. (2006). The use of binary TiO_2-GeO_2 oxide electrodes to enhanced efficiency of dye-sensitized solar cells. *J. Photochem. Photobiol. A Chem.* https://doi.org/10.1016/j.jphotochem.2005.08.002.

20 Palomares, E., Clifford, J.N., Haque, S.A. et al. (2003). Control of charge recombination dynamics in dye sensitized solar cells by the use of conformally deposited metal oxide blocking layers. *J. Am. Chem. Soc.* https://doi.org/10.1021/ja027945w.

21 Diamant, Y., Chen, S.G., Melamed, O., and Zaban, A. (2003). Core–shell nanoporous electrode for dye sensitized solar cells: the effect of the $SrTiO_3$ shell on the electronic properties of the TiO_2 core. *J. Phys. Chem. B* 107: 1977.

22 Zhang, X.L., Kim, Y.H., and Kang, Y.S. (2007). Synthesis and properties of TiO_2/ZnO core/shell nanomaterials. *Solid State Phenom.* 119: 239.

23 Liao, H., Hsu, C.H., and Chen, D.H. (2006). Preparation and properties of amorphous titania-coated zinc oxide nanoparticles. *J. Solid State Chem.* 179: 2020.

24 Wang, K.P. and Teng, H. (2009). Zinc-doping in TiO_2 films to enhance electron transport in dye-sensitized solar cells under low-intensity illumination. *Phys. Chem. Chem. Phys.* 11: 9489.

25 Jose, R., Thavasi, V., and Ramakrishna, S. (2009). Metal oxides for dye-sensitized solar cells. *J. Am. Ceram. Soc.* 92 (2): 289.

26 Rajkumar, N., Kanmani, S.S., and Ramachandran, K. (2011). Performance of dye-sensitized solar cell based on TiO_2:ZnO nanocomposites. *Adv. Sci. Lett.* 4: 627.

27 Tennakone, K., Perera, V.P.S., Kottegoda, I.R.M., and Kumara, G.R.R.A. (1999). Dye-sensitized solid state photovoltaic cell based on composite zinc oxide/tin (IV) oxide films. *J. Phys. D. Appl. Phys.* https://doi.org/10.1088/0022-3727/32/4/004.

28 Bhattacharjee, R. and Hung, I.-M. (2013). A SnO_2 and ZnO nanocomposite photoanodes in dye-sensitized solar cells. *ECS Solid State Lett.* 2: Q101–Q104. https://doi.org/10.1149/2.013311ssl.

29 Wanninayake, W.M.N.M.B., Premaratne, K., and Rajapakse, R.M.G.R. (2016). Enhancing performance of SnO_2-based dye-sensitized solar cells using ZnO passivation layer. *Int. J. Photoenergy* 2016: 1–8. https://doi.org/10.1155/2016/9087478.

30 Milan, R., Selopal, G.S., Epifani, M. et al. (2015). ZnO@SnO_2 engineered composite photoanodes for dye sensitized solar cells. *Sci. Rep.* 5: 14523. https://doi.org/10.1038/srep14523.

31 Lee, J.H., Park, N.G., and Shin, Y.J. (2011). Nano-grain SnO_2 electrodes for high conversion efficiency SnO_2 DSSC. *Sol. Energy Mater. Sol. Cells* https://doi.org/10.1016/j.solmat.2010.04.027.

32 Bakr, Z.H., Wali, Q., Yang, S. et al. (2019). Characteristics of ZnO-SnO_2 composite nanofibers as a photoanode in dye-sensitized solar cells. *Ind. Eng. Chem. Res.* 58: 643–653. https://doi.org/10.1021/acs.iecr.8b03882.

33 Abrari, M., Ahmadi, M., Ghanaatshoar, M. et al. (2019). Fabrication of dye-sensitized solar cells based on SnO_2/ZnO composite nanostructures: a new facile method using dual anodic dissolution. *J. Alloys Compd.* 784: 1036–1046. https://doi.org/10.1016/j.jallcom.2018.12.299.

34 Wang, X.-F., Tian, W., Zheng, E. et al. (2015). Low-temperature-processed ZnO–SnO_2 nanocomposite for efficient planar perovskite solar cells. *Sol. Energy Mater. Sol. Cells.* 144: 623–630. https://doi.org/10.1016/j.solmat.2015.09.054.

35 Dürr, M., Rosselli, S., Yasuda, A., and Nelles, G. (2006). Band-gap engineering of metal oxides for dye-sensitized solar cells. *J. Phys. Chem. B.* https://doi.org/10.1021/jp063857c.

36 Menzies, D.B., Cervini, R., Cheng, Y.B. et al. (2004). Nanostructured ZrO_2-coated TiO_2 electrodes for dye-sensitised solar cells. *J. Sol-Gel Sci. Technol.* https://doi.org/10.1007/s10971-004-5818-0.

37 Manoharan, K., Joby, N.G., and Venkatachalam, P. (2014). A novel TiO_2 nanoparticles/nanowires composite core with ZrO_2 nanoparticles shell coating photoanode for high-performance dye-sensitized solar cell based on different electrolytes. *Ionics (Kiel)* https://doi.org/10.1007/s11581-013-1050-7.

38 Plank, N.O.V., Howard, I., Rao, A. et al. (2009). Efficient ZnO nanowire solid-state dye-sensitized solar cells using organic dyes and core-shell nanostructures. *J. Phys. Chem. C.* https://doi.org/10.1021/jp904919r.

39 Luan, X. and Wang, Y. (2014). Ultrathin exfoliated TiO_2 nanosheets modified with ZrO_2 for dye-sensitized solar cells. *J. Phys. Chem. C* https://doi.org/10.1021/jp5052112.

9

Plasmonic Nanocomposite as Photoanode

Su Pei Lim[1,2]

[1] *Xiamen University Malaysia, School of Energy and Chemical Engineering, Jalan Sunsuria, Bandar Sunsuria, Sepang, Selangor Darul Ehsan, Malaysia*
[2] *Xiamen University, College of Chemistry and Chemical Engineering, Xiamen, China*

9.1   Introduction

Semiconductor oxides used in dye-sensitized solar cell (DSSC) include TiO_2, ZnO, SnO_2, and Nb_2O_5 and the lists go on, which serve as the carrier for the monolayers of the sensitizer using their large surface and electron transfer to the conducting substrate. Nanocrystalline semiconductor films adsorb a large amount of the dye molecules and increase the harvesting efficiency of the solar energy. However, the major drawback associated with the use of large surface area TiO_2 is its random electron transport, which will cause the electron–hole recombination process and hence affect the overall device performance [1, 2]. To overcome this problem, designing a photoanode with an efficient transport pathway from the photoinjected carriers to the current collector seems to be a possible alternative to enhance the performance of DSSCs. With this aim, surface modification with metal, doping, semiconductor coupling, and hybridizing with carbon material have been attempted [3–6]. Modification of metal oxide with plasmonic particles such as gold (Au) [7–9] and silver (Ag) [10, 11] were reported actively in the DSSC application to prevent the recombination of the photogenerated electron–hole pairs and improve the charge transfer efficiency.

The past few years have witnessed a surge of interest in the development of a hybrid metal oxide with plasmonic nanoparticles (NPs) such as Ag and Au by taking advantage of improving the efficiency of a DSSC. The plasmonic nanoparticles play dual roles in the DSSC performance, including the enhancement of the absorption coefficient of the dye and optical absorption due to surface plasmonic resonance [12–14]. Moreover, they act as an electron sink for photo-induced charge carriers, improve the interfacial charge transfer process, and minimize the charge recombination, thereby enhancing the electron transfer process in a DSSC [13–17]. Hence, the performance of a DSSC with metal oxide plasmonic nanocomposite material-modified photoanodes has been actively investigated [15–17] since year 2011.

9.2   Plasmonic Nanocomposite Modified TiO_2 as Photoanode

Among them, Qi et al. [18] and coworkers reported on $Ag@TiO_2$ core–shell nanostructure with the aim to utilize the localized surface plasmon (LSP) to improve the performance of the DSSC. Initially, they synthesized Ag nanoparticles using a modified polyol process and

Interfacial Engineering in Functional Materials for Dye-Sensitized Solar Cells, First Edition.
Edited by Alagarsamy Pandikumar, Kandasamy Jothivenkatachalam and Karuppanapillai B. Bhojanaa.
© 2020 John Wiley & Sons, Inc. Published 2020 by John Wiley & Sons, Inc.

collected it using centrifugal method. Later, the Ag nanoparticles were coated with TiO_2 by adding titanium isopropoxide. They have studied the effect of localized surface plasmon resonance (LSPR) on the performance of DSSC. They found out the LSPR from Ag nanoparticles increased the absorption of the dye molecules that allowed the decreasing of the thickness of the photoanodes and eventually improved the electron collection and device performance. A small amount of Ag nanoparticles improves the efficiency from 7.8% to 9.0%, while the photoanode thickness was decreased by 25% as compared to TiO_2-based DSSC (Figure 9.1a). By decreasing the thickness of photoanode not only can save the material but also can prevent the recombination of the electron (Figure 9.1b,c).

Pascal Nbelayim et al. [19] investigated the wide-range effects of Ag nanoparticles on TiO_2 by preparing different concentrations of Ag. They have fabricated Ag@TiO_2 core–shell doped TiO_2 paste and undoped one for comparison. There are two optimum concentrations that were reported to achieve considerable higher efficiency among all the samples. The first one is with the concentration of 0.1% of Ag and 4.88% efficiency. This is due to the efficient charge transfer by lowering the Fermi level of TiO_2, hence, contributing to the higher J_{SC}. The Fermi level of the photoanode and the redox potential can be described using the V_{OC} value. It can be observed that the sample with 0.1% Ag provides the less V_{OC} value, lowering the Fermi level to the optimum state and promise efficient charge transfer. The other sample is with 1% of Ag; with the improvement of electron sink effect for the better J_{SC} and enhanced efficiency of 5.00%. Similarly, Lim et al. [20] also has carried on an experiment to study the effect of Ag on TiO_2 for the improvement of DSSC efficiency. The plasmonic TiO_2 with various concentrations of Ag (0, 1, 2.5, 10, and 20 wt%) was prepared using one-step chemical reduction method. The transmission electron microscopic (TEM) images show a uniform distribution of Ag nanoparticles with a particle size range of 2–4 nm on the TiO_2 surface (Figure 9.1d), which made DSSC based on Ag/TiO_2 exhibit a better performance. It was observed that the optical properties in the region of 400–500 nm was significantly enhanced after addition of the Ag surface plasmon resonance effect (Figure 9.1e). The sample with 2.5 wt% of Ag gave the best result and show a faster electron transport time form the Nyquist plot. The electron lifetime was significantly increased and survived from the recombination after incorporation of Ag on TiO_2. Therefore, the DSSC fabricated using Ag@TiO_2 show improved J_{SC} compared to TiO_2.

Instead of using Ag nanoparticles, a number of works related to the incorporation of Au into metal oxides have also been reported. Song et al. [21] reported on microwave-assisted hydrothermal-prepared Au@TiO_2 core–shell nanoparticle with the ~40 nm Au core and ~60 nm TiO_2 shell (Figure 9.1f). However, they found that Au nanoparticles were unstable and could dissolve in iodide electrolyte. Therefore, in order to overcome the problem, a hollow TiO_2 (Figure 9.1g) is produced and used as scattering layer on top of the nano-crystalline TiO_2. The scattering layer was produced by selectively etching of as-prepared Au@TiO_2 core–shell NPs. Finally, they managed to obtain 7.40% efficiency with the TiO_2 hollow spheres, compared with 5.21% for the electrode with commercial TiO_2 nanoparticle. Apart from a single-metal source incorporation on metal oxide for the application of photoanode, Lim and the group also carried out [22] a study on the effect of DSSC by adding two metal source, which is Ag and Au. One-step chemical reduction method was performed to co-deposit various concentrations of Ag and Au nanoparticles by adding different compositions of $HAuCl$:$AgNO_3$ (Au0:Ag100, Au25:Ag75, Au50:Ag50, Au75:Ag25, and Au100:Ag0) in the preparation of TiO_2 plasmonic materials. The size of Au decoration (8–20 nm) is larger than the Ag nanoparticles (2–5 nm) on the TiO_2 (Figure 9.1h). By comparing their own fabricated DSSC, it can be observed from the J–V curve that the Au–Ag@TiO_2 was having a significant boost in the short-circuit current up to 23.5 mA/cm^2 compared to the TiO_2 (5.83 mA cm/cm^2), Ag/TiO_2 (7.07 mA cm/cm^2), and Au/TiO_2 (14.11 mA cm/cm^2). The improvement of 7.33% compared

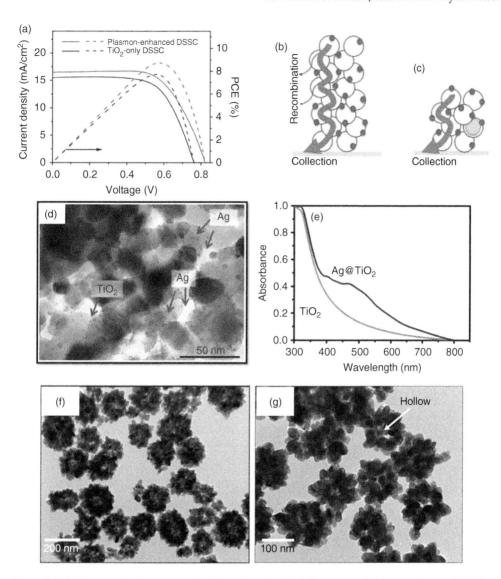

Figure 9.1 (a) Current density and PCE of TiO$_2$ and Ag@TiO$_2$. (b) Illustration of plasmon enhanced DSSCs require thinner film and less material to achieve the same PCE, (c) plasmon enhanced DSSC, (d) TEM images Ag@TiO$_2$, (e) absorption spectra of Ag@TiO$_2$, (f) TEM image of Au@TiO$_2$ core–shell, (g) hollow TiO$_2$ nanoparticles, (h) Au–Ag@TiO$_2$ nanocomposites, and (i) bleach recovery dynamics of Au, TiO$_2$/Au100:Ag0, and TiO$_2$/Au75:Ag25 nanocomposites derived from the transient absorption spectra. (j) Absorbance spectra of Au and Ag nanoparticle, (k) comparison of optical absorption curves for sintered powders, without dye for determination of the energy band gap values, (l) flat band potential (V_{fb}) and V_{OC} value of the samples, and (m) impedance plots for three different DSSCs in frequency range of 1×10^{-2}–1×10^6 Hz. *Source:* (a–c) Qi et al. 2011 [18]. Adapted with permission of The American Chemical Society. (d, e) Lim et al. 2014 [20] https://pubs.rsc.org/en/content/articlelanding/2014/ra/c4ra05689b#!divAbstract. Licensed under CCBY 3.0. https://creativecommons.org/licenses/by/3.0/. Adapted with permission of The Royal Society of Chemistry. (f–i) Adapted from Song et al. 2014 [21] and Lim et al. 2017 [22]. Adapted with permission of The Royal Society of Chemistry. (j–m) Dissanayake et al. 2016 [23]. Adapted with permission of Springer Nature.

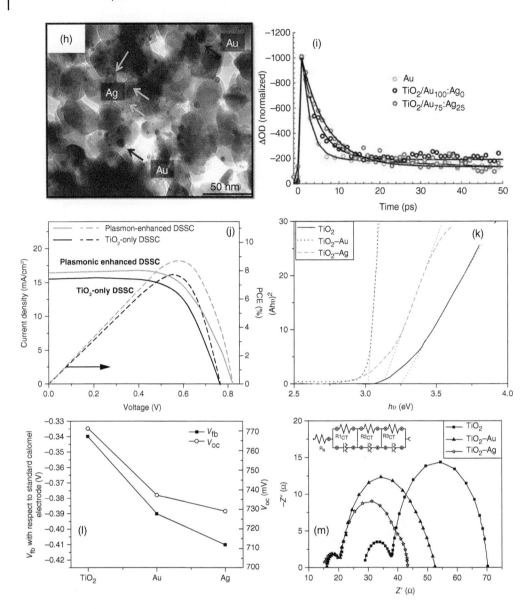

Figure 9.1 (*Continued*)

to unmodified TiO_2 (2.22%) was mainly due to the synergistic effect shown by codeposition of plasmonic Ag and Au on the TiO_2 surface. From the recovery dynamics of bleach band spectra (Figure 9.1i), the time constant is getting longer for Au–Ag@TiO_2. The elongation in time indicates better electron–hole separation that is attributed to the formation of a Schottky barrier at the interface between TiO_2 and the plasmonic metal which acts as an electron sink. The incorporation of Ag and Au not only improves the optical absorption but also minimizes the charge recombination process.

A similar strategy by engineering bi-metal approach onto TiO_2 for DSSC has also been reported by Dissanayake and coworkers [23]. The image of colloidal Au and Ag nanoparticles is shown in Figure 9.1j. Briefly, 1% of trisodium citrate was added to the hydrogen tetrachloroaurate(III) and silver nitrate solution, respectively, to obtain gold and silver nanoparticles. For the preparation of gold nanoparticles, the solution will change the solution form blue to deep red while for the silver nanoparticle, and it will turn the solution from colorless to yellow. The nanoparticles with various concentrations were then added into the TiO_2 and ground for 15 minutes before being fabricated into the photoanode using doctor blade method. Figure 9.1k shows the absorption spectra of the Au and Ag colloids and exhibits a surface plasmon absorption at the region 520 and 420 nm. It was observed that the efficiency increased with the addition of Au and Ag nanoparticles until it reached maximum amount where the excess nanoparticles will start a competing mechanism and eventually decreases the photocurrent and the efficiency of DSSC. The energy band gaps determined from UV–Vis spectra (Figure 9.1l) provided an evidence of the band gap narrowing after addition of the nanoparticles in TiO_2. Their studies on the flat band potential measurements also show that the Fermi level has shifted to more negative values for two nanoparticles incorporated into TiO_2 and reduced of V_{OC} values. Therefore, among all the samples, the Ag–TiO_2 show the best efficiency, as evidenced form Nyquist plots (Figure 9.1m). The lower value of series resistance and charge transfer resistance favors the electron collection efficiency at the counter electrode and efficient electron transport TiO_2/electrolyte. Tables 9.1 and 9.2 show the summary of selected plasmonic-based TiO_2 for DSSC application.

9.3 Plasmonic Nanocomposite Modified ZnO as Photoanode

Similar to the development of plasmonic nanoparticles modified TiO_2 for the DSSC application, researchers also applied the same approach on ZnO to improve the DSSC efficiency. For example, Iwantono et al. [75] have carried out an experiment to investigate the photoactivity properties of Ag on ZnO. Their Ag–ZnO was synthesized using two steps. First, the ZnO nanorod (NRs)was prepared using seed and growth method, followed by spin coating of ethanolic solution of 0.01 M zinc acetate hydrate. It was then immersing the samples in $AgNO_3$ to obtain Ag–ZnO with the addition of $NaBH_4$ as a reducing agent. The FESEM image (Figure 9.2a–c) show that there is no Ag nanoparticles growth on the ZnO nanorod, and they suggested that the Ag might effectively substitute into the ZnO host lattices. The defect found on the hexagonal-like structure is caused by the successful substitutes of Ag into the ZnO lattices, where it interfered with the normal growth of the ZnO nanorod. By comparing their DSSC efficiency, it was found that the efficiency of their Ag–ZnO was enhanced from 0.46% to 1.12%. This enhancement is attributed to the increase in the photoactivity properties of the sample after addition of Ag and the improvement of the facile carrier transportation in the device. In another study, Mangesh Lanjewar et al. [76] fabricated Ag–ZnO by utilizing sol–gel spin coating. Their studies reported that there is 3.67 times enhancement in J_{SC} after doped ZnO with the Ag nanoparticles. The improvement in the device efficiency is attributed to the reduced band gap from 3.28 eV for pure ZnO film to 2.65 eV for Ag–ZnO film, where the photoanode could absorb the visible light range to the greatest extent.

Apart from that, bilayer TiO_2:Ag/ZnO:Ag (TZO:Ag) architecture was presented by Tripathi et al. [77] via a combined chemical reaction followed by doctor blade technique. The first step is preparing the TiO_2:Ag and ZnO:Ag solution separately by using titanium butoxide, zinc acetate dehydrate, and silver nitrate as the precursor. The following step is the deposition of the ZnO:Ag on the flourine doped tin oxide (FTO) substrates using doctor blade technique and sintering it at 550 °C in air. The bilayer is prepared by doctor blade method of the TiO_2:Ag onto the ZnO:Ag film. They also investigated the effect of different dyes in this report. The eosin-Y dye

Table 9.1 Comparison of the photovoltaic performance of some Ag–TiO$_2$ and Au–TiO$_2$ photoanode.

Precursor of Ag	Ag preparation method	Sensitizer	J_{SC} (mA/cm^2)	V_{OC} (V)	η (%)	References
Ag–TiO$_2$ nanocomposite photoanode						
AgNO$_3$	Chemical reduction	*cis*-[(dcbH2) 2Ru(SCN)2]	14.73	0.77	7.84	[24]
AgNO$_3$	Modified polyol	N719	13.68	0.75	7.05	[25]
AgNO$_3$	UV light reduction	N719	13.55	0.74	6.86	[26]
AgNO$_3$	UV light reduction	N719	6.67	0.69	1.83	[27]
AgNO$_3$	Reduction using syzygium extract	N719	4.4	0.78	4.13	[28]
Ag nonowires	N/A	C106	16.83	0.71	8.84	[29]
AgNO$_3$	Stirring	N719	16.01	0.72	8.19	[30]
Ag	Deposited using e-beam evaporation	N719	9.81	0.75	5.14	[31]
AgNO$_3$	Garlic extract reduction	N/A	12.30	0.40	4.17	[32]
AgNO$_3$	Dip coating	N3	8.40	0.64	3.60	[33]
AgNO$_3$	Sonication	N719	1.95	0.87	3.60	[34]
AgNO$_3$	Pulse-current electrodeposition	N719	4.37	0.79	1.68	[35]
AgNO$_3$	Trisodium citrate reduction	N719	11.66	0.70	5.19	[36]
AgNO$_3$	Assisted polyol reduction	N719	11.28	0.72	5.56	[37]
AgNO$_3$	Heating	N719	5.65	0.67	2.40	[38]
AgNO$_3$	Trisodium citrate reduction	N535	13.86	0.73	6.51	[23]
AgNO$_3$	Reflux	N719	1.07	0.72	0.40	[39]
AgNO$_3$	Stirring	N719	18.7	0.67	7.5	[40]
AgNO$_3$	UV light reduction	N719	7.11	0.74	2.83	[41]
[Ag(NH$_3$)]$_2$OH	Glucose reduction	N719	17.69	0.70	1.80	[42]
AgNO$_3$	Sodium borohydride reduction	N3	5.75	0.56	1.54	[43]
AgNO$_3$	Papaya leaf extract reduction	N719	10.67	0.75	5.06	[44]
AgNO$_3$	Solvothermal	N719	6.97	0.86	5.03	[45]
AgNO$_3$	Electrodeposition	Rose bengal	0.13	0.32	N/A	[46]
Ag nanoparticle	Mixing	N719	13.02	0.68	5.74	[47]
AgNO$_3$	One pot synthesis	N719	8.09	0.80	4.58	[48]
Ag paste	Screen printing	N719	17.75	0.76	9.45	[49]
AgNO$_3$	Hydrothermal	N719	26.90	N/A	10.90	[50]
AgNO$_3$	Stirring	N719	3.51	0.61	1.5	[51]

Table 9.1 (Continued)

Precursor of Au	Au preparation method	Sensitizer	J_{sc} (mA/cm^2)	V_{OC} (V)	η (%)	References
Au−TiO$_2$ nanocomposite photoanode						
HAuCl$_4$	Sodium tris-citrate reduction	N719	~6	~0.7	3.3	[52]
HAuCl$_4$	Hydrothermal	N719	13.3	0.74	6.0	[9]
HAuCl$_4$	Addition and stirring with NaOH	N719	4.49	0.60	1.19	[53]
HAuCl$_4$	Impregnation method	N719	7.02	0.77	3.33	[54]
HAuCl$_4$	Sodium tris-citrate reduction	N719	16.7	0.64	7.35	[55]
HAuCl$_4$	Sodium tris-citrate reduction	N719	16.22	0.64	7.13	[56]
HAuCl$_4$	Sodium borohydride reduction	N719	18.81	0.72	5.61	[57]
Au nanoparticles	Self-assembly	N719	5.15	0.76	2.26	[58]
HAuCl$_4$	Modified colloidal wet-chemical method	N719	17.25	0.78	9.09	[59]
Gold plate	Lase ablation	[RuL2(NCS)2]	11.32	0.73	5.36	[60]
HAuCl$_4$	Sol−gel reaction	N719	12.53	0.67	5.62	[61]
HAuCl$_4$	Sodium borohydride reduction	N719	2.79	0.80	1.24	[62]
HAuCl$_4$	Seed and growth method	N719	17.2	0.78	8.45	[63]
HAuCl$_4$	Sodium tris-citrate reduction	N719	12	0.74	6.23	[23]
Au nanoparticles	Commercial obtain	N749	6.42	0.71	3.12	[64]
HAuCl$_4$	Solvothermal	N719	14.56	0.76	7.37	[65]
HAuCl$_4$	Seed and growth method	N719	20.9	0.67	4.69	[66]
HAuCl$_4$	Sodium tris-citrate reduction	N719	10.8	0.63	5	[67]
HAuCl$_4$	Stirring with (3-aminopropyl) trimethoxysilane and ethanol	N719	20.8	0.7	8.31	[68]
HAuCl$_4$	Ionic adsorption and photoreduction	N719	11.71	0.71	5.63	[69]
HAuCl$_4$	Stirring overnight	N719	12.33	0.72	6.40	[70]
HAuCl$_4$	Ionic adsorption and photoreduction	N719	10.35	0.70	4.63	[71]

(Continued)

Table 9.2 Comparison of the photovoltaic performance of some bi-metal-TiO$_2$ photoanode.

Fabrication of Au (precursor, preparation method)	Fabrication of Ag (precursor, preparation method)	Sensitizer	J_{SC} (mA/cm^2)	V_{OC} (V)	η (%)	References
HAuCl$_4$, seed mediated method	AgNO$_3$, ascorbic acid reduction	N719	16.53	0.73	8.43	[72]
HAuCl$_4$, UV light reduction	AgNO$_3$, UV light reduction	N719	15.2	0.74	7.51	[73]
HAuCl$_4$, sodium borohydride reduction	AgNO$_3$, sodium borohydride reduction	N719	23.50	0.76	7.33	[74]
HAuCl$_4$, galvanic displacement	AgNO$_3$, grow on zinc plate	Black dye	19.24	0.65	9.2	[22]

produced a better efficiency compared to cocktail. By comparing all the samples, ZnO:Ag-based photoanode gave the best efficiency of 0.773% compared to pure ZnO (0.11%), and TZO:Ag gave 0.158%. From here, we can observe the significant role of Ag nanoparticle in boosting the DSSC performance outweigh the bilayer approach. Figure 9.2e illustrated the charge transfer process in TZO:Ag-based DSSC. The presence of Ag nanoparticles will raise the Fermi level to more negative potentials until they match with the Fermi level of the ZnO. This process will ease the electron transfer from the Ag to the ZnO. Additionally, the charge recombination can be prevented by the formation of Schottky barrier formed at the ZnO/Ag interface. These factors contribute to the increase of the J_{SC} and the overall efficiency.

Not only that ZnO also has been extensively decorated with Au nanoparticles to enhance the DSSC performance. Marwa Abd-Ellah et al. [78] fabricated Au–ZnO nanorod using electrodeposition technique. First, ZnO nanorod was directly electrodeposited from 0.5 mM ZnCl$_2$ primary electrolyte. Then, the Au nanoparticles are electrodeposited on the ZnO nanorod in the aqueous solution of 1.5 and 10 mm AuCl$_3$ with NaClO$_4$ and KCl as supporting electrolyte. The resulting ZnO nanorod is 1–1.5 μm long and 200–500 nm in diameter (Figure 9.2f–k). Their studies show that different gold concentration will produce different nanoparticles size. The nanoparticles size increase from 12, 20–50 to more than 50 nm when the concentration of Au increase from 1 to 10 m. The increase of the electrodeposition time also increased the density of Au nanoparticles on the ZnO nanorod. However, the light absorbance in the visible region is found to be decrease when the electrodeposition time is prolonged due the aggregation of the nanoparticles. The UV–Vis absorption spectra show 25% higher photon absorption after addition of Au (Figure 9.2l). The higher photon absorption due to the plasmonic effect of the Au nanoparticles will contribute to the improvement of the DSSC performance as shown in the (Figure 9.2m). Figure 9.2n show the photon-to-electron conversion mechanism. It explained the shifting of the Fermi level toward ZnO and formation of the Schottky barrier would benefit the electron transfer and hence enhance the DSSC efficiency. Similarly, ZnO nanorods decorated with Au nanoparticles have also been synthesized by Tanujjal Bora et al. [79] and fabricated it into DSSC. They prepared ZnO nanorod onto FTO via hydrothermal method by using Zinc acetate dehydrate, zinc acetate hexahydrate, and hexamethylenetetramine as the precursors. It was then followed by dipping the ZnO film into the gold chloride hydrate solution to obtain ZnO–Au. The resulting ZnO–Au photoanode was found to exhibit the following features, i.e. having higher optical absorption near 520 nm, which enhance the J_{SC} by ~35% compared to pure ZnO photoanode. Besides that, the time-correlated single-photon

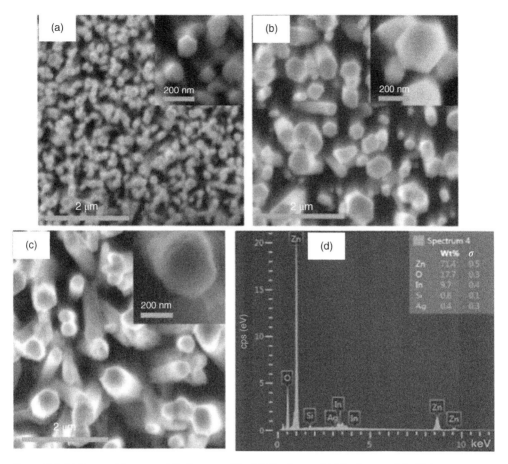

Figure 9.2 (a) Pure and Ag-treated ZnO nanorod for (b) 10 minutes, (c) 35 minutes while (d) is energy-dispersive X-ray spectroscopy (EDX) analysis of sample (b). (e) Schematic representation of electron transfer process involved in DSSC based on TZO:Ag particles (1) photo excitation of electron from highest occupied molecular orbital (HOMO) to lowest unoccupied molecular orbital (LUMO) of dye, (2) injection of electrons into Ag nanoparticles, (3) injection of electron from HOMO to TiO_2 CB directly, (4) transport of electron from Ag nanoparticles to CB of TiO_2, (5) transfer of electron from TiO_2 CB to ZnO CB, (6) subsequent electron transfer from ZnO to FTO by diffusion process, and (7) injection from FTO to electrolyte through platinum from where it goes to HOMO of dye. (f) SEM images of pristine ZnO-NT, and Au/ZnO-NT obtained by Au electrodeposition in (g) 1 mM, (h, i) 5 mM, and (j, k) 10 mM $AuCl_3$ electrolytes, all mixed with 0.1 M $NaClO_4$ and 0.1 M KCl supporting electrolyte, at room temperature. (l) UV–Vis spectra of ZnO and Au–ZnO, (m) current density versus voltage curve, (n) schematic diagrams of photon to electron conversion mechanisms, (o) HRTEM image ZnO–Au nanoparticles, (p) absorption spectra of bare ZnO and ZnO–Au, (q) excitation spectra of ZnO NPs and ZnO–Au NCs monitored at 368 and 550 nm, (r) steady state emission spectra of ZnO NPs and ZnO–Au NCs are shown (excitation at 320 and 375 nm). The inset shows that the defect related green emission is composed of two bands. (s) Dependence of the incident photon conversion efficiency on the incident wavelength for ZnO NP and ZnO–Au NC films cast on an FTO plate. (t) Photocurrent–voltage (*J–V*) characteristics of ZnO–Au NC and ZnO NP (inset) based DSSC. *Source:* (a–d) Lanjewar and Gohel 2017 [76]. Adapted with permission of Taylor & Francis. (e) Tripathi et al. 2015 [77]. Adapted with permission of Elsevier. (f–n) Abd-Ellah et al. 2016 [78]. Adapted with permission of Royal Society of Chemistry. (o–t) Liu et al. 2018 [82]. Adapted with permission of Elsevier.

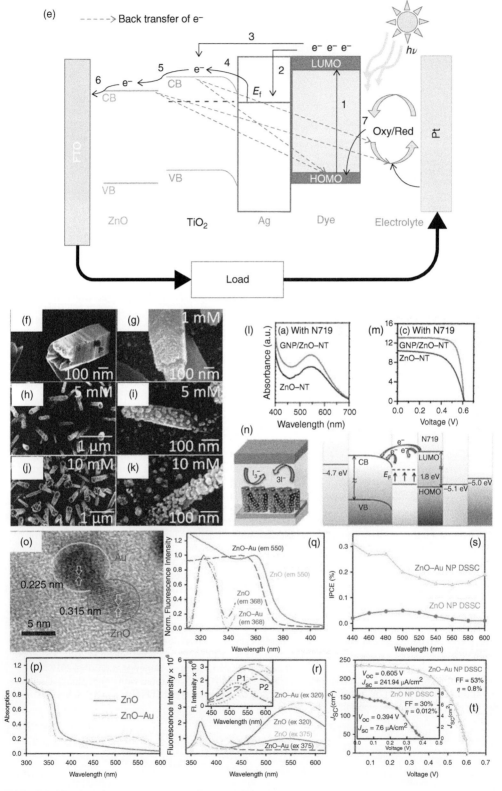

Figure 9.2 (*Continued*)

count spectroscopic show a longer decay time observed in the presence of Au nanoparticles, indicating the formation of the Schottky barrier at the ZnO/Au interface that have effectively hindered the electron recombination.

From most of the reported works, exploitation of plasmonic nanoparticles opens new opportunities for the remarkable advancement of the DSSC efficiency. However, structural disadvantages such as agglomeration of Au nanoparticles on the ZnO surface will limit the DSSC performance. For example, Peh et al. [80] fabricated DSSC based on ZnO nanorods by employing zinc acetate dehydrate as the precursor in the alcoholic solution. The Au nanoparticles with various amounts were then added into the ZnO suspension before being fabricated into the photoanode. Their prepared ZnO nanorod device gave better performance of 5.2% as compared to the Au–ZnO which is only 2.5%. The thick layer of Au nanoparticles aggregation could lead to the distortion of the plasmonic effect. From their report, the further addition of Au will further decrease the cell performance because the aggregated Au nanoparticles will cover the ZnO surface. It prevents direct contact of ZnO with the dye and limit the uptake of the dye. Therefore, it is of utmost importance to control the amount of plasmonic nanoparticles on the metal oxide surface. Only the optimum amount can produce well the distribution of plasmonic nanoparticles and contribute to the enhancement of cell performance.

Other than the modification of ZnO nanorod with Au nanoparticles, there are a few studies reported on the modification on the ZnO nanoparticles. For example, Soumik Sarkar et al. [81] prepared the Au@ZnO nanoparticles for DSSC through a chemical reaction and in situ reduction of Au. The dumbbell-like ZnO–Au was obtained with the Au nanoparticles attached on the ZnO surface. The ZnO and Au nanoparticles having the average diameter of 6 and 8 nm, respectively (Figure 9.2o). With the incorporation of Au, the optical absorption is appeared at 525 nm (Figure 9.2p). It can be seen from the photoluminescence spectra (Figure 9.2q,r), the bare ZnO and ZnO–Au having one broad emission band upon excitation below the band gap ($\lambda_{ex} = 375$ nm) and two emission bands upon excitation above the band edge ($\lambda_{ex} = 320$ nm). When excitation below 375 nm was used, the defect-related emission suppressed the presence of the Au nanoparticles. The incident photon-to-current conversion efficiency (IPCE) (Figure 9.2s) and short-circuit current (Figure 9.2t) are significantly improved in the presence of the Au nanoparticles which attributed to the better electron mobility of ZnO–Au semiconductor. Another work reported by Liu et al. [82] also show a successful fabrication of Au@ZnO nanoparticles coupled with TiO_2 for use of the photoanode. It was found that the conversion efficiency increased with the amount of Au@ZnO added to TiO_2. The highest efficiency of 8.91% was obtained after introducing optimum amount of Au@ZnO (1.93%) to TiO_2, exhibiting better performance compared to bare TiO_2 (7.50%).

9.4 Plasmonic Nanocomposite Modified with Less Common Metal Oxide as Photoanode

Although the most commonly reported metal oxides, such as TiO_2 and ZnO, are widely used in the plasmonic-enhanced DSSC, less common metal oxides have been also reported for the investigation as a promising material for DSSC. A gold coated on silica (Au@SiO_2) nanotubes have been reported as the photoanode by Holly F. Zarick et al. [83]. The synthesis of the Au@SiO_2 was made through a two-step method whereby the Au nanotube was synthesized using seed growth method. Following this, the Au nanotubes with various concentrations were first functionalized before added with the sodium silicate. The Au@SiO_2 was finally obtained after four hours stirring at 50 $^{\circ}$C in the oil bath as shown in Figure 9.3a. The morphological structure of cube shape was obtained at 45–60 nm size range. Based on their previous work, this size was chosen because it gave the optimum absorption to light scattering ratios. In order to enhance the overall optical absorption of the devices, the Au@SiO$_2$ is integrated with TiO$_2$. As in the optical absorption spectrum (Figure 9.3b), the absorption of TiO$_2$ sensitized with N719 is significantly enhanced after incorporated Au@SiO$_2$. In terms of the DSSC use, the Au@SiO$_2$ was added into the TiO$_2$ paste and mixed until homogenous before being coated onto TiCl$_4$-treated FTO using doctor blade method. The schematic representation of photoanode is shown in (Figure 9.3c). The photocurrent spectra (Figure 9.3d) and IPCE spectra (Figure 9.3f) show various photoanodes used to investigate for the cell performance. The best efficiency of 7.8% was obtained for the DSSC with 1.8% of the Au@SiO$_2$. Based on this work, the optical absorption (Figure 9.3d) gradually increases with the increasing of nanocube concentration across the entire spectrum, indicating a stronger coupling between nanocube and dye molecules. The enhanced light absorption integrated with nanocubes will ultimately provide an alternative to design a thin-film solar cell with better stability and cost.

A similar approach has been reported by Zheng et al. for the incorporation of Au@SiO$_2$ into TiO$_2$ as the photoanode [84]. At first, the Au nanorods were prepared using HAuCl$_4$ as a precursor via a seed-mediated method. In order to control the nanorod's aspect ratio of Au nanorods, AgNO$_3$ is added into the solution (Figure 9.3g). After obtaining the Au nanorods through the centrifugation, SiO$_2$ is coated on the Au nanorod using single-step coating method to prevent corroding by the electrolyte. For the fabrication of photoanode, the Au@SiO$_2$ was incorporated into TiO$_2$ paste before being screen printed onto FTO. They found out it is important to control the length to diameter aspect ratio of Au nanorods because it can exhibits different plasmon

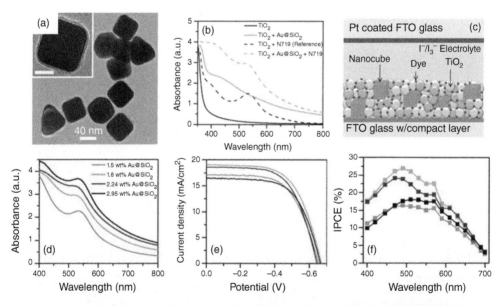

Figure 9.3 (a) Low-magnification TEM micrograph of Au@SiO$_2$ nanocubes and high-magnification image provided in the inset. (b) Optical absorption spectra, (c) schematic representation of plasmon-enhanced DSSCs showing nanocubes embedded within the N719-sensitized mesoporous TiO$_2$ layer with I$^-$/I$_3^-$ liquid electrolyte. (d) Optical absorption spectra of N719-sensitized mesoporous TiO$_2$ with varied particle density of Au@SiO$_2$ nanocubes embedded in the photoanodes. (e) Corresponding current density spectra of the devices. (f) IPCE (%) of the same devices as a function of excitation wavelength. (g) Schematic illustration of the synthesis process of Au NRs and Au NRs@SiO$_2$ for a DSSC, (h) device structure of tandem DSSC. TEM images of (i) Au NRs-1 and (j) Au NRs-2. UV–Vis absorption spectra of (k) Au NRs-1 and (l) Au NRs-2. TEOS, Tetraethyl orthosilicate *Source:* (a–f) Zarick et al. 2014 [83]. Adapted with permission of American Chemistry Society. Zheng et al. 2018 [84]. Adapted with permission of Elsevier.

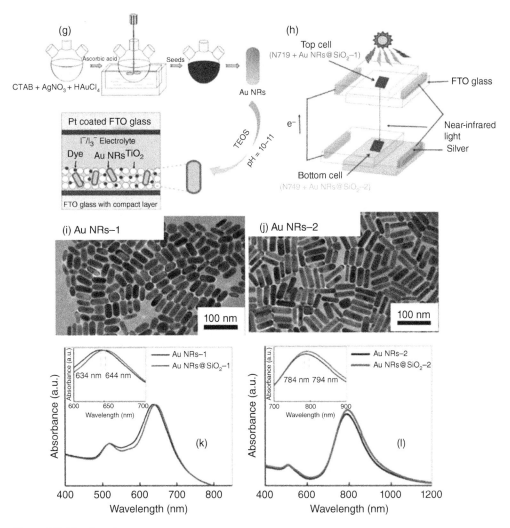

Figure 9.3 *(Continued)*

wavelength which is essential for the dye absorption. During the synthesis, Ag ion plays a key role in determining the aspect ratio of Au NRs. High concentrations of Ag ion will induce the growth of large-aspect ratio Au nanorod (AuNRs-1), whereas low concentrated Ag ion leads to small-aspect ratio one (AuNRs-2). The AuNRs-1 exhibits average dimensions of 38 ± 2 nm in length and 15 ± 1 nm in diameter corresponding to an aspect ratio of ~2.5) (Figure 9.3i). For Au NRs-2, average dimensions of 55 ± 3 nm in length together with 14 ± 1 nm in diameter are obtained, corresponding to an aspect ratio of ~3.9 (Figure 9.3j). From Figure 9.3k,l, an absorption peak (black lines) centered at about 520 nm can be observed for the both Au NRs-cored nanostructures, corresponding to the transverse plasmon absorption of Au NRs. The longitudinal plasmon absorption peaks (black lines) are centered at ~634 nm for Au NRs-1 and ~784 nm for Au NRs-2. They prepared N719 and N749 as the sensitizer and constructed both into top and bottom sub-cells (Figure 9.3h) for the device testing so that the cells are capable of harvesting the wide visible and near-infrared (NIR) light. They managed power conversion efficiency (PCE) of 10.73% compared to 9.02% of the pure TiO_2 photoanode. The enhancement is due to the incorporation of $Au@SiO_2$ that facilitated the reduction of charge recombination and efficient transport path through the 1D Au nanorod.

Apart from that, a $Ag@Nb_2O_5$ architecture as a blocking layer in DSSC was presented by Suresh et al. [85] via a grounding method for six hours to achieve homogenous mixing of samples followed by RF magnetron sputtering onto FTO glass. The cell with Ag addition achieved 9.24% compared to reference which is 7.6% due to the plasmonic effect of Ag. However, the cell performance is highly dependent on the amount of Ag added. They found out the cell efficiency increases with the increasing amount of Ag until it reached maximum at 3 wt% of Ag where the efficiency will deteriorate since then. The larger amount of Ag not only increase the light reflection but also reduce the Nb_2O_5 crystallinity and then diminish the cell performance additionally.

9.5 Conclusion

All in all, the performance of DSSC is still highly dependent on the metal oxide used. Till now, among the materials studied for use in DSSC, nanocrystalline TiO_2 have been most commonly used as metal oxide material in high-efficiency DSSC. The exploitation of modifying metal oxides with plasmonic nanoparticles have opened new opportunities for the remarkable advancement of the development, structural, electronic, and optical properties of metal oxides toward the application of DSSC. Nevertheless, not all the incorporation of plasmonic nanoparticles can enhance the efficiency of the DSSC. High amount of plasmonic nanoparticles will lead to aggregation and cover the surface of metal oxide and result in the deterioration of the DSSC. Therefore, controlling the amount of plasmonic nanoparticles added in the metal oxides is crucial to develop an efficient DSSC.

References

1 Kopidakis, N., Neale, N.R., Zhu, K. et al. (2005). Spatial location of transport-limiting traps in TiO_2 nanoparticle films in dye-sensitized solar cells. *Appl. Phys. Lett.* 87: 202106.

2 van de Lagemaat, J., Park, N.G., and Frank, A.J. (2000). Influence of electrical potential distribution, charge transport, and recombination on the photopotential and photocurrent conversion efficiency of dye-sensitized nanocrystalline TiO_2 solar cells: a study by electrical impedance and optical modulation techniques. *J. Phys. Chem. B* 104: 2044–2052.

3 Lai, Y., Zhuang, H., Xie, K. et al. (2010). Fabrication of uniform Ag/TiO_2 nanotube array structures with enhanced photoelectrochemical performance. *New J. Chem.* 34: 1335–1340.

4 Macak, J.M., Schmidt-Stein, F., and Schmuki, P. (2007). Efficient oxygen reduction on layers of ordered TiO_2 nanotubes loaded with Au nanoparticles. *Electrochem. Commun.* 9: 1783–1787.

5 Yang, L., He, D., Cai, Q., and Grimes, C.A. (2007). Fabrication and catalytic properties of Co−Ag−Pt nanoparticle-decorated titania nanotube arrays. *J. Phys. Chem. C* 111: 8214–8217.

6 Zhao, H., Chen, Y., Quan, X., and Ruan, X. (2007). Preparation of Zn-doped TiO_2 nanotubes electrode and its application in pentachlorophenol photoelectrocatalytic degradation. *Chin. Sci. Bull.* 52: 1456–1461.

7 Du, J., Qi, J., Wang, D., and Tang, Z. (2012). Facile synthesis of $Au@TiO_2$ core−shell hollow spheres for dye-sensitized solar cells with remarkably improved efficiency. *Energy Environ. Sci.* 5: 6914–6918.

8 Jang, Y.H., Jang, Y.J., Kochuveedu, S.T. et al. (2014). Plasmonic dye-sensitized solar cells incorporated with Au–TiO$_2$ nanostructures with tailored configurations. *Nanoscale* 6: 1823–1832.

9 Muduli, S., Game, O., Dhas, V. et al. (2012). TiO$_2$–Au plasmonic nanocomposite for enhanced dye-sensitized solar cell (DSSC) performance. *Sol. Energy* 86: 1428–1434.

10 Gao, Y., Fang, P., Chen, F. et al. (2013). Enhancement of stability of N-doped TiO$_2$ photocatalysts with Ag loading. *Appl. Surf. Sci.* 265: 796–801.

11 Jiang, W., Liu, H., Yin, L., and Ding, Y. (2013). Fabrication of well-arrayed plasmonic mesoporous TiO$_2$/Ag films for dye-sensitized solar cells by multiple-step nanoimprint lithography. *J. Mater. Chem. A* 1: 6433–6440.

12 Eagen, C.F. (1981). Nature of the enhanced optical absorption of dye-coated Ag island films. *Appl. Opt.* 20: 3035–3042.

13 Mock, J.J., Barbic, M., Smith, D.R. et al. (2002). Shape effects in plasmon resonance of individual colloidal silver nanoparticles. *J. Chem. Phys.* 116: 6755–6759.

14 Schaadt, D.M., Feng, B., and Yu, E.T. (2005). Enhanced semiconductor optical absorption via surface plasmon excitation in metal nanoparticles. *Appl. Phys. Lett.* 86: 063106.

15 Lee, K.-C., Lin, S.-J., Lin, C.-H. et al. (2008). Size effect of Ag nanoparticles on surface plasmon resonance. *Surf. Coat. Technol.* 202: 5339–5342.

16 Wen, C., Ishikawa, K., Kishima, M., and Yamada, K. (2000). Effects of silver particles on the photovoltaic properties of dye-sensitized TiO$_2$ thin films. *Sol. Energy Mater. Sol. Cells* 61: 339–351.

17 Zhao, G., Kozuka, H., and Yoko, T. (1997). Effects of the incorporation of silver and gold nanoparticles on the photoanodic properties of rose bengal sensitized TiO$_2$ film electrodes prepared by sol–gel method. *Sol. Energy Mater. Sol. Cells* 46: 219–231.

18 Qi, J., Dang, X., Hammond, P.T., and Belcher, A.M. (2011). Highly efficient plasmon-enhanced dye-sensitized solar cells through metal@oxide core–shell nanostructure. *ACS Nano* 5: 7108–7116.

19 Nbelayim, P., Kawamura, G., Tan, W.K. et al. (2017). Systematic characterization of the effect of Ag@TiO$_2$ nanoparticles on the performance of plasmonic dye-sensitized solar cells. *Sci. Rep.* 7: 15690.

20 Lim, S.P., Pandikumar, A., Huang, N.M., and Lim, H.N. (2014). Enhanced photovoltaic performance of silver@titania plasmonic photoanode in dye-sensitized solar cells. *RSC Adv.* 4: 38111–38118.

21 Song, M.-K., Rai, P., Ko, K.-J. et al. (2014). Synthesis of TiO$_2$ hollow spheres by selective etching of Au@TiO$_2$ core–shell nanoparticles for dye sensitized solar cell applications. *RSC Adv.* 4: 3529–3535.

22 Lim, S.P., Lim, Y.S., Pandikumar, A. et al. (2017). Gold–silver@TiO$_2$ nanocomposite-modified plasmonic photoanodes for higher efficiency dye-sensitized solar cells. *Phys. Chem. Chem. Phys.* 19: 1395–1407.

23 Dissanayake, M., Kumari, J., Senadeera, G., and Thotawatthage, C. (2016). Efficiency enhancement in plasmonic dye-sensitized solar cells with TiO$_2$ photoanodes incorporating gold and silver nanoparticles. *J. Appl. Electrochem.* 46: 47–58.

24 Lan, Z., Wu, J., Lin, J., and Huang, M. (2012). Bi-functional TiO$_2$ cemented Ag grid under layer for enhancing the photovoltaic performance of a large-area dye-sensitized solar cell. *Electrochim. Acta* 62: 313–318.

25 Sebo, B., Huang, N., Liu, Y. et al. (2013). Dye-sensitized solar cells enhanced by optical absorption, mediated by TiO$_2$ nanofibers and plasmonics Ag nanoparticles. *Electrochim. Acta* 112: 458–464.

26 Peng, W., Zeng, Y., Gong, H. et al. (2013). Silver-coated TiO_2 electrodes for high performance dye-sensitized solar cells. *Solid-State Electron.* 89: 116–119.

27 Zhang, X., Liu, J., Li, S. et al. (2013). Bioinspired synthesis of $Ag@TiO_2$ plasmonic nanocomposites to enhance the light harvesting of dye-sensitized solar cells. *RSC Adv.* 3: 18587–18595.

28 Tian, Z., Wang, L., Jia, L. et al. (2013). A novel biomass coated Ag-TiO_2 composite as a photoanode for enhanced photocurrent in dye-sensitized solar cells. *RSC Adv.* 3: 6369–6376.

29 Dong, H., Wu, Z., Lu, F. et al. (2014). Optics–electrics highways: plasmonic silver nanowires@TiO_2 core–shell nanocomposites for enhanced dye-sensitized solar cells performance. *Nano Energy* 10: 181–191.

30 Dong, H., Wu, Z., Gao, Y. et al. (2014). Silver-loaded anatase nanotubes dispersed plasmonic composite photoanode for dye-sensitized solar cells. *Org. Electron.* 15: 2847–2854.

31 Huang, H.-H., Chang, H., Liu, H.-W. et al. (2014). Plasma-etched nanoporous TiO_2 using Ag nanoparticle masks: application for photoanodes of dye-sensitized solar cells. *Mater. Res. Express* 1: 025505.

32 Wang, L., Jia, L., and Li, Q. (2014). A novel sulfur source for biosynthesis of (Ag, S)-modified TiO_2 photoanodes in DSSC. *Mater. Lett.* 123: 83–86.

33 Berginc, M., Opara Krašvec, U., and Marko, T. (2014). Solution processed silver nanoparticles in dye-sensitized solar sells. *J. Nanomater.* 2014: 11.

34 Lim, S.P., Huang, N.M., Lim, H.N., and Mazhar, M. (2014). Surface modification of aerosol-assisted CVD produced TiO_2 thin film for dye sensitised solar cell. *Int. J. Photoenergy* 2014: 1–12.

35 Luan, X. and Wang, Y. (2014). Plasmon-enhanced performance of dye-sensitized solar cells based on wlectrodeposited Ag nanoparticles. *J. Mater. Sci. Technol.* 30: 1–7.

36 Chandrasekhar, P., Chander, N., Anjaneyulu, O., and Komarala, V.K. (2015). Plasmonic effect of $Ag@TiO_2$ core–shell nanocubes on dye-sensitized solar cell performance based on reduced graphene oxide–TiO_2 nanotube composite. *Thin Solid Films* 594: 45–55.

37 Jang, I., Kang, T., Cho, W. et al. (2015). Preparation of silver nanowires coated with TiO_2 using chemical binder and their applications as photoanodes in dye sensitized solar cell. *J. Phys. Chem. Solids* 86: 122–130.

38 Rahnejat, B. (2015). Synthesis and characterization of Ag-doped TiO_2 nanostructure and investigation of its application as dye-sensitized solar cell. *J. Nanoanal.* 2: 39–45.

39 Gupta, A.K., Srivastava, P., and Bahadur, L. (2016). Improved performance of Ag-doped TiO_2 synthesized by modified sol–gel method as photoanode of dye-sensitized solar cell. *Appl. Phys. A* 122: 724.

40 Sharma, G. (2017). Electrophoretic deposition of plasmonic nanocomposite for the fabrication of dye-sensitized solar cells. *Indian J. Pure Appl. Phys.* 55: 73–80.

41 Hu, H., Shen, J., Cao, X. et al. (2017). Photo-assisted deposition of Ag nanoparticles on branched TiO_2 nanorod arrays for dye-sensitized solar cells with enhanced efficiency. *J. Alloys Compd.* 694: 653–661.

42 Xu, Y., Zhang, H., Li, X. et al. (2017). Ag-encapsulated single-crystalline anatase TiO_2 nanoparticle photoanodes for enhanced dye-sensitized solar cell performance. *J. Alloys Compd.* 695: 1104–1111.

43 Sakthivel, T., Kumar, K.A., Ramanathan, R. et al. (2017). Silver doped TiO_2 nano crystallites for dye-sensitized solar cell (DSSC) applications. *Mater. Res. Express* 4: 126310.

44 Solaiyammal, T., Muniyappan, S., Keerthana, B.G.T. et al. (2017). Green synthesis of Ag and the effect of Ag on the efficiency of TiO_2 based dye sensitized solar cell. *J. Mater. Sci. - Mater. Electron.* 28: 15423–15434.

45 Nbelayim, P., Kawamura, G., Tan, W.K. et al. (2018). Ag@TiO$_2$ nanowires-loaded dye-sensitized solar cells and their effect on the various performance parameters of DSSCs. *J. Electrochem. Soc.* 165: H500–H509.

46 Parveen, F., Sannakki, B., Jagtap, C.V. et al. (2018). *AIP Conference Proceedings*, vol. 1989, 030015. AIP Publishing.

47 Ran, H., Fan, J., Zhang, X. et al. (2018). Enhanced performances of dye-sensitized solar cells based on Au–TiO$_2$ and Ag–TiO$_2$ plasmonic hybrid nanocomposites. *Appl. Surf. Sci.* 430: 415–423.

48 Nbelayim, P., Kawamura, G., Abdel-Galeil, M.M. et al. (2018). Effects of multi-sized and -shaped Ag@TiO$_2$ nanoparticles on the performance of plasmonic dye-sensitized solar cells. *J. Ceram. Soc. Jpn.* 126: 139–151.

49 Wu, M.-S. and Yang, R.-S. (2018). Post-treatment of porous titanium dioxide film with plasmonic compact layer as a photoanode for enhanced dye-sensitized solar cells. *J. Alloys Compd.* 740: 695–702.

50 Bhardwaj, S., Pal, A., Chatterjee, K. et al. (2018). Fabrication of efficient dye-sensitized solar cells with photoanode containing TiO$_2$–Au and TiO$_2$–Ag plasmonic nanocomposites. *J. Mater. Sci. - Mater. Electron.* 29: 18209–18220.

51 Rajbongshi, B.M. and Verma, A. (2018). Plasmonic noble metal coupled biphasic TiO$_2$ electrode for dye-sensitized solar cell. *Mater. Lett.* 232: 220–223.

52 Nahm, C., Choi, H., Kim, J. et al. (2011). The effects of 100 nm-diameter Au nanoparticles on dye-sensitized solar cells. *Appl. Phys. Lett.* 99: 253107.

53 Pandikumar, A. and Ramaraj, R. (2013). TiO$_2$–Au nanocomposite materials modified photoanode with dual sensitizer for solid-state dye-sensitized solar cell. *J. Renewable Sustainable Energy* 5: 043101.

54 Ninsonti, H., Chomkitichai, W., Baba, A. et al. (2014). Au-loaded titanium dioxide nanoparticles synthesized by modified sol–gel/impregnation methods and their application to dye-sensitized solar cells. *Int. J. Photoenergy* 2014: 1–8.

55 Chander, N., Khan, A., Thouti, E. et al. (2014). Size and concentration effects of gold nanoparticles on optical and electrical properties of plasmonic dye sensitized solar cells. *Sol. Energy* 109: 11–23.

56 Jeong, N.C., Prasittichai, C., and Hupp, J.T. (2011). Photocurrent enhancement by surface plasmon resonance of silver nanoparticles in highly porous dye-sensitized solar cells. *Langmuir* 27: 14609–14614.

57 Lim, S.P., Pandikumar, A., Huang, N.M., and Lim, H.N. (2015). Facile synthesis of Au@TiO$_2$ nanocomposite and its application as a photoanode in dye-sensitized solar cells. *RSC Adv.* 5: 44398–44407.

58 Chou, H.-T., Wu, J.-L., Wu, T.-M. et al. (2015). *2015 IEEE International Conference on Electron Devices and Solid-State Circuits (EDSSC)*, 293–296. IEEE.

59 Bai, Y., Butburee, T., Yu, H. et al. (2014). Controllable synthesis of concave cubic gold core–shell nanoparticles for plasmon-enhanced photon harvesting. *J. Colloid Interface Sci.* 449: 246–251.

60 Al-Azawi, M.A., Bidin, N., Ali, A.K., and Bououdina, M. (2015). The effects of gold colloid concentration on photoanode electrodes to enhance plasmonic dye-sensitized solar cells performance. *J. Mater. Sci. - Mater. Electron.* 26: 6276–6284.

61 Chen, H.-W., Hong, C.-Y., Kung, C.-W. et al. (2015). A gold surface plasmon enhanced mesoporous titanium dioxide photoelectrode for the plastic-based flexible dye-sensitized solar cells. *J. Power Sources* 288: 221–228.

62 Pandikumar, A., Suresh, S., Murugesan, S., and Ramaraj, R. (2015). Dual functional TiO$_2$–Au nanocomposite material for solid-state dye-sensitized solar cells. *J. Nanosci. Nanotechnol.* 15: 6965–6972.

63 Elbohy, H., Kim, M.R., Dubey, A. et al. (2016). Incorporation of plasmonic Au nanostars into photoanodes for high efficiency dye-sensitized solar cells. *J. Mater. Chem. A* 4: 545–551.

64 Careem, M. and Arof, A.K. (2015). Plasmonic effects of quantum size gold nanoparticles on dye-sensitized solar cell. *Mater. Today Proc.* 3: S73–S79.

65 Bai, L., Liu, X., Li, M. et al. (2016). Plasmonic enhancement of the performance of dye-sensitized solar cells by incorporating hierarchical TiO$_2$ spheres decorated with Au nanoparticles. *Electrochim. Acta* 190: 605–611.

66 Shah, A.A., Umar, A.A., and Salleh, M.M. (2016). Efficient quantum capacitance enhancement in DSSC by gold nanoparticles plasmonic effect. *Electrochim. Acta* 195: 134–142.

67 Mayumi, S., Ikeguchi, Y., Nakane, D. et al. (2017). Effect of gold nanoparticle distribution in TiO$_2$ on the optical and electrical characteristics of dye-sensitized solar cells. *Nanoscale Res. Lett.* 12: 513.

68 Li, Y.-Y., Wang, J.-G., Liu, X.-R. et al. (2017). Au/TiO$_2$ hollow spheres with synergistic effect of plasmonic enhancement and light scattering for improved dye-sensitized solar cells. *ACS Appl. Mater. Interfaces* 9: 31691–31698.

69 Guo, M., Chen, J., Zhang, J. et al. (2018). Coupling plasmonic nanoparticles with TiO$_2$ nanotube photonic crystals for enhanced dye-sensitized solar cells performance. *Electrochim. Acta* 263: 373–381.

70 Liu, C., Liang, M., and Khaw, C. (2018). Effect of gold nanoparticles on the performances of TiO$_2$ dye-sensitised solar cell. *Ceram. Int.* 44: 5926–5931.

71 Chen, J., Guo, M., Su, H. et al. (2018). Improving the efficiency of dye-sensitized solar cell via tuning the Au plasmons inlaid TiO$_2$ nanotube array photoanode. *J. Appl. Electrochem.* 48: 1139–1149.

72 Dong, H., Wu, Z., El-Shafei, A. et al. (2015). Ag-encapsulated Au plasmonic nanorods for enhanced dye-sensitized solar cell performance. *J. Mater. Chem. A* 3: 4659–4668.

73 Wang, Y., Zhai, J., and Song, Y. (2015). Plasmonic cooperation effect of metal nanomaterials at Au–TiO$_2$–Ag interface to enhance photovoltaic performance for dye-sensitized solar cells. *RSC Adv.* 5: 210–214.

74 Amiri, O., Salavati-Niasari, M., Mir, N. et al. (2018). Plasmonic enhancement of dye-sensitized solar cells by using Au-decorated Ag dendrites as a morphology-engineered. *Renewable Energy* 125: 590–598.

75 Iwantono, I., Anggelina, F., Saad, M. et al. (2017). Influence of Ag ion adsorption on the photoactivity of ZnO nanorods for dye-sensitized solar cell application. *Mater. Express* 7: 312–318.

76 Lanjewar, M. and Gohel, J.V. (2017). Enhanced performance of Ag-doped ZnO and pure ZnO thin films DSSCs prepared by sol-gel spin coating. *Inorg. Nano Metal Chem.* 47: 1090–1096.

77 Tripathi, S., Rani, M., and Singh, N. (2015). ZnO: Ag and TZO: Ag plasmonic nanocomposite for enhanced dye sensitized solar cell performance. *Electrochim. Acta* 167: 179–186.

78 Abd-Ellah, M., Moghimi, N., Zhang, L. et al. (2016). Plasmonic gold nanoparticles for ZnO-nanotube photoanodes in dye-sensitized solar cell application. *Nanoscale* 8: 1658–1664.

79 Bora, T., Kyaw, H.H., Sarkar, S. et al. (2011). Highly efficient ZnO/Au Schottky barrier dye-sensitized solar cells: role of gold nanoparticles on the charge-transfer process. *Beilstein J. Nanotechnol.* 2: 681.

80 Peh, C., Ke, L., and Ho, G. (2010). Modification of ZnO nanorods through Au nanoparticles surface coating for dye-sensitized solar cells applications. *Mater. Lett.* 64: 1372–1375.

81 Sarkar, S., Makhal, A., Bora, T. et al. (2011). Photoselective excited state dynamics in ZnO–Au nanocomposites and their implications in photocatalysis and dye-sensitized solar cells. *Phys. Chem. Chem. Phys.* 13: 12488–12496.

82 Liu, Q., Wei, Y., Shahid, M.Z. et al. (2018). Spectrum-enhanced Au@ZnO plasmonic nanoparticles for boosting dye-sensitized solar cell performance. *J. Power Sources* 380: 142–148.

83 Zarick, H.F., Hurd, O., Webb, J.A. et al. (2014). Enhanced efficiency in dye-sensitized solar cells with shape-controlled plasmonic nanostructures. *ACS Photon.* 1: 806–811.

84 Zheng, Y.-Z., Tao, X., Zhang, J.-W. et al. (2018). Plasmonic enhancement of light-harvesting efficiency in tandem dye-sensitized solar cells using multiplexed gold core/silica shell nanorods. *J. Power Sources* 376: 26–32.

85 Suresh, S., Unni, G.E., Satyanarayana, M. et al. (2018). Ag@Nb$_2$O$_5$ plasmonic blocking layer for higher efficiency dye-sensitized solar cells. *Dalton Trans.* 47: 4685–4700.

10

Carbon Nanotubes-Based Nanocomposite as Photoanode

Giovana R. Cagnani[1], Nirav Joshi[1], and Flavio M. Shimizu[2]

[1] *São Carlos Institute of Physics, University of São Paulo, Department of Physics, São Paulo, Brazil*
[2] *Brazilian Nanotechnology National Laboratory (LNNano), Brazilian Center for Research in Energy and Materials (CNPEM), Campinas, São Paulo, Brazil*

10.1 Introduction

Over the past few decades, energy is the backbone of technology and economic development. In addition to man, machine, and money, energy is now fourth factor of production. Without energy, no machine will run, electricity is needed for everything. Hence, our energy requirements have increased dramatically in the years following the industrial revolution. Readily accessible fossil fuels, such as coal, natural gas, and oils, are the major energy sources used to meet our current need. However, these sources are nonrenewable and have led to serious environmental issues, global warming, and air pollution, and their increasing consumption rate has accelerated fossil fuel depletion; the search for alternative energy source has become vital.

Among all the renewable energy sources, solar energy is the ultimate solution to growing energy needs due to its abundance and cleanliness for power generation and also promising alternative to fossil fuels. Importantly, the global energy demand can be met by simply covering 0.1% of earth's crust with solar cell having efficiency of 10% by using photovoltaics [1]. Thus, photovoltaic devices, particularly solar cell is crucial for future energy generation. Despite the advantages of solar cells, it requires to be competent and cost-effective in comparison to conventional energy resources. Since significant breakthroughs in 1991 spurred researchers O'Regan and Grätzel [2]. Dye-sensitized solar cells (DSSCs) have accumulated more and more research attention in the last 20 years. Most commercial solar cells are based on silicon, which exhibits high efficiency of more than 20%, but relatively high cost and limits their large-scale application [3].

There are different types of solar cells, such as amorphous silicon, organic solar cells, CdTe solar cells, and copper indium gallium diselenide, which possess low conversion efficiency and toxic properties. In recent time, organic–inorganic lead halide perovskites-based solar cells, have been widely used because of low cost, facile fabrication, and high conversion efficiency; however, the efficiency has slowed down due to the presence of toxic lead in perovskites, poor reproducibility, and moisture instability [4]. Despite of these, DSSCs have been considered and studied due to good conversion efficiency, excellent stability, and low toxicity. The highest certified conversion efficiency of DSSCs achieved is more than 11% as we can see in Figure 10.1b.

Figure 10.2 displays the basic structure of first fabricated DSSC by O'Regan and Grätzel in 1991. An actual DSSC normally comprises of five components (i) a transparent conductive substrate, (ii) a semiconductor film, usually based on metal oxide nanomaterials (TiO$_2$ or

Interfacial Engineering in Functional Materials for Dye-Sensitized Solar Cells, First Edition.
Edited by Alagarsamy Pandikumar, Kandasamy Jothivenkatachalam and Karuppanapillai B. Bhojanaa.
© 2020 John Wiley & Sons, Inc. Published 2020 by John Wiley & Sons, Inc.

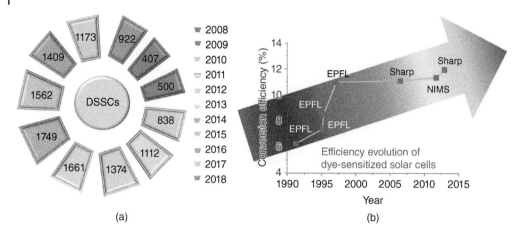

(a)   (b)

Figure 10.1 (a) Number of publications on DSSC in past 10 years. From a Web of Science search for "Sensitized solar cell" with 23800 results – 27 November 2018 (Data source, ISI Web of Science) and (b) efficiency of certified dye-sensitized solar cells. EPFL, E'cole Polytechnique Fédérale de Lausanne, Switzerland; NIMS, National Institute for Materials Science, Japan; Sharp, Sharp Corporation, Japan. *Source:* Fan et al. 2017 [10]. Reproduced with permission of Royal Society of Chemistry. http://www.nrel.gov/ncpv/images/efficiency_chart.jpg.

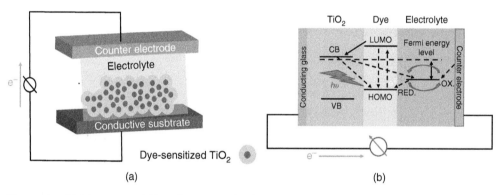

(a)   (b)

Figure 10.2 (a) Schematic and (b) working principle of dye-sensitized solar cells. CB, conduction band; VB, valence band; LUMO, lowest unoccupied molecular orbital; HOMO, highest occupied molecular orbital; RED., reduced; OX., oxidized.

ZnO), (iii) a sensitizer/dye, (iv) an electrolyte containing a redox mediator, and (v) a counter electrode (CE). Technically, the DSSCs performance can be improved by modifying any single or multiple components. Figure 10.2b shows the basic working principle of DSSCs using TiO_2 semiconductor materials. Now, when the sensitizer (dye) absorbs the light energy (photon) and excites an electron from the highest occupied molecular orbital (HOMO) to the lowest unoccupied molecular orbital (LUMO), these generated electrons are entering the conduction band of metal oxide films (TiO_2) and flow through it and are transferred to conductive substrate during the hopping process. After the circulation, the electrons then flow via the external load to the counter electrode to reduce the iodide redox process, thereby regenerating the dye and completing the whole circuit. To understand the performance of DSSCs, Photocurrent density–voltage ($J–V$) and current conversion efficiency (PCE) are used to investigate and short-circuit photocurrent density (J_{SC}), open-circuit voltage (V_{OC}), and fill factor (FF) can be determined from the Photocurrent density–voltage ($J–V$) curves.

Nowadays, researchers have developed various methods to enhance the DSSCs component such as semiconductor films, electrolytes, dyes, and counter electrodes, leading to enhanced current conversion efficiency. Because of the relatively enormous number of reports, this chapter focuses only on carbon nanotubes (CNTs)-based nanocomposite as a photoanodes and their DSSCs performance.

10.2 Recent Advances on DSSC Photoanodes

Generally, a DSSC is composed of a counter electrode, an electrolyte containing an iodide/tri-iodide (I^-/I^{3-}) redox pair, and a photoanode sensitized with a dye [5]. In the photoanode is where the separation and migration of charge carriers take place, which is an important process in the photocatalytic reaction in DSSC. During the operation, the electrons are injected into the conducting band of the anode by photoexcitation of the dye molecules [6]. The electrons are then transported by diffusion through the anode layer and collected on the conducting electrode [6]. In this trend efforts have been focused on strategies to enhance the photoconversion efficiency (PCE) by modifying each component individually, as stated by some reviews [7–11].

An ideal photoanode should cover the following requirements: (i) high surface area to improve the dye adsorption; (ii) be a good electron acceptor; (iii) separating the carriers generated in the dye and facilitating the transport of the electrons from dye; (iv) be stable and resistant to photo corrosion; (v) absorb and scatter sunlight efficiently; and (vi) have a good interface between the dye and the conductive electrode [11]. Therefore, the choice of material, the shape, and size of the photoanode can be an important step to establish a good efficiency of these devices.

The photoanode is a mesoporous n-type semiconductor, most commonly TiO_2 as proposed by O'Regan and Grätzel [2], which should present fast transporting electrons with high capacity to load dye sensitizer. Among the strategies to enhance TiO_2 photoanode performance we can summarize as follows:

Nanostructured semiconductor: Alternative semiconductor materials [12] have been proposed to build photoanodes; however, the best efficiency observed remains with TiO_2. In view thereof, researchers have explored *oxide nanocomposites* as the mixture of CdS and $TiO_{2(A+R)}$ (cadmium sulfide/titanium oxide anatase and rutile) proposed by Alkuam et al. [13], or the $(WO_3)_n–TiO_2$ (tungsten oxide with three different mass ratios, $n = 1\%$, 3%, and 5%,/titanium oxide) reported as a composite that increased from 0.97% to 1.71% studied by Younas et al. [14].

Carbon-based nanocomposites by the addition of semiconductor on graphene [15–18] or carbon nanotubes/nanofibers [17–21].

Metallic nanocomposites such as gold [22–24], silver [25, 26], zinc [27] taking advantage of plasmonic effect [28]. It has been demonstrated that the TiO_2 nanostructure [29] (nanoparticle [30], nanorod (NR) [31], nanotube [32], nano-opal [33], nanofiber [34]), and their combination [35–37] also lead to an increase on PCE values, which is straightly related with the film preparation methodology (sol–gel, hydrothermal, electrospinning [38]).

Light-scattering layer: According to Wang et al. [39], the addition of a transparent scattering layer [40–42] on photoanode is more important than the dye adsorption capacity to increase light distribution and harvesting demonstrated through the use of a scatting layers (PCE = 7.52%). Later Hu et al. [43] reported a double-layer film of $TiO_2/La(OH)_3$: Yb^{3+}/Er^{3+} nanoparticles and porous-hollow TiO_2 microsphere with PCE = 8.89%. The addition of 3D dome-structure revealed an increment at the same level in which Sim et al. [44] obtained a PCE of 7.2%. Pahm et al. [45] incorporated sub-micrometer cavities in the TiO_2 photoanode; however, a low PCE (4.73%) was obtained.

Doping ions: TiO_2 may be doped with different metals [46–48] (Cr, Fe, Ni, Co, Zn), nonmetals [49] (S, N), and rare earth [50–54] (Eu, Ce, Yb, La) to enhance binding of dyes to TiO_2 surface [55].

Dye sensitizer: For this purpose, the molecular designing has been developed with the aid of theoretical/simulation theories [56–58] to accelerate the rational choice of materials or molecular functionalization to increase the dye loading on photoanode.

Among aforesaid strategies carbon-based materials have attracted attention in the last decade due to triple functionality possibility; which means the use of carbon nanotubes in the three main components of DSSCs such as photoanode, counter electrode (CE) [59–64], and electrolyte additive [65–67]. In Section 10.3, we will present the carbon nanotubes and their properties.

10.3 Structure and Properties of Carbon Nanotubes

Carbon nanotubes belong to the fullerene family, which is an allotropic form of carbon. Its atoms are bound by sp^2 hybrid orbitals in a honeycomb lattice that have been rolled up to a cylinder, that is, graphite sheet rolled to form a tube. Due to the characteristic curvature of this structure, the hybridization of molecular orbitals causes σ bonds to be displaced out of the plane of symmetry. To compensate this displacement, the π orbital is moved out of the tube [68, 69]. This makes the CNTs that have remarkable electronic, optical, and mechanical properties, summarized in Table 10.1, ideal for application in DSSCs.

However, the properties of the CNTs can vary according to the number of walls, type of defects, length, methods of synthesis, and concentration. The carbon nanotubes can be found in the form of single-walled carbon nanotube (SWCNT) that are formed by a single rolled sheet, double-walled carbon nanotube (DWCNT) that are composed of a sheet of graphite inside

Table 10.1 CNT properties.

Property	CNTs	Remark
Density (g/cm^3)	1.3–1.4	Low
Surface area (m^2/g)	1500	Very high
Elastic modulus (TPa)	1	Sevenfold higher than steel
Tensile strength (GPa)	100	100 times stronger than steel
Thermal stability	2800 °C in vacuum/750 °C in air	More stable than metal wires in microchips
Thermal conductivity (W/m K)	2000–6000	Twice higher than diamond
Electron mobility (cm^2/V s)	1×10^5	70-fold higher than iron at room temperature
Electrical conductivity (S/cm)	4×10^5	4 times higher
Maximum current density (A/cm^2)	1×10^9 to 10×10^9	More than 1000 times greater than cooper

Source: Adapted from Batmunkh et al. 2015 [70] and Volder et al. 2013 [71].

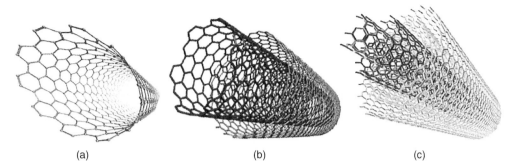

Figure 10.3 Structure of carbon nanotubes. (a) SWCNT, (b) DWCNT, and (c) MWCNT. *Source:* Rafique et al. 2016 [72]. Copyright 2015, Adapted with permission of Taylor & Francis.

another, and multiple walled carbon nanotube (MWCNT) that consist of multiple layers of graphite sheets forming a concentric cylinder, Figure 10.3.

SWCNTs usually have a diameter between 0.8 and 2 nm, while those with multiple walls have diameters between 5 and 20 nm [71]. In addition to length, the direction in which the graphite sheet is rolled interferes with the electronic properties of the carbon tube [69, 70]. This direction is described by its chiral vector (**C**) given by

$$\mathbf{C} = n a_1 + m a_2 \tag{10.1}$$

where the integers (n, m) are the number of steps along the unit vectors (a_1, a_2) of the graphene lattice [69]. The angle between the chiral vector and the indices $(n, 0)$ is described as the chiral angle given by [69]:

$$\theta = \tan^{-1}(m\sqrt{3}/m + 2n) \tag{10.2}$$

In terms of rolling up a graphene sheet, the chiral angle determines the amount of "twist" in the nanotube and ranges from 0° to 30° [73]. If the tube is rolled to form a chiral angle $\theta = 0°$, then we have a carbon nanotube with a zig-zag structure. However, if the tube has a chiral angle $\theta = 30°$ has a carbon nanotube named armchair. All other conformations with angles between $0° < \theta < 30°$ are known as chiral carbon nanotubes [74] (see Figure 10.4).

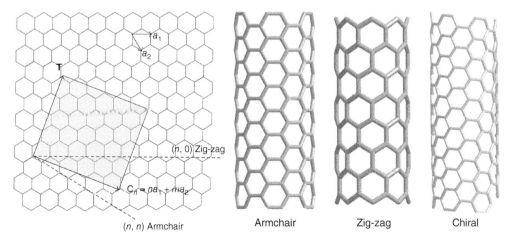

Figure 10.4 Scheme of the angle between the chiral vector and the indices $(n, 0)$ that determine the construction of the carbon nanotube in armchair, zig-zag, and chiral. *Source:* Daniel et al. 2012 [73]. Copyright 2012, Adapted with permission of John Wiley & Sons, Inc.

The electronic properties of the carbon nanotubes are strongly dependent on the chiral angle. In SWCNTs, by slightly changing the chiral angle, it is possible to switch between metallic, low band gap, and high band gap semiconducting carbon nanotubes [73, 74]. If SWCNTs have a chiral index such that $n - m = 3j$ (where $j = 0$) exhibit metallic properties; those with $n - m = 3j$ (where $j \neq 0$) are considered small gap semiconductors or semi-metallic CNTs. When $n - m = 3j \pm 1$, they are large-gap semiconducting CNTs with a band gap around 1 eV for 0.7 nm of diameter [75].

Given above, armchair SWCNTs and approximately 33% of all zig-zag CNTs are metallic (no band gap) at ambient temperatures, whereas the remaining zig-zag SWCNTs and all chiral tubes are considered semiconducting.

MWCNTs are more complex structures due to the interaction of several concentrically arranged tubes. The spacing between the tubes (0.34 nm) is similar to the interplanar spacing of the graphite (0.335 nm). This difference is due to the curvature of the nanotubes as well as the van der Waals forces. The configuration of the concentric tubes imposes restrictions on their diameter, but not on their chiral angle, that can alternate between conductors and metal semiconductors layers [69]. Given this, many studies try to understand how the electronic transport in MWCNTs are since results indicate that the outer layers give the transport [76, 77]. In terms of developing photovoltaic devices as DSSC, the presented characteristics make the CNT excellent materials for application as photoanode.

10.4 CNT-Based Photoanode Material

Since its discovery by Iijima [78] in 1991, carbon nanotubes have been subject of intensive research for a wide range of applications [79] due to their unique electrical, mechanical, and thermal properties [80]. Furthermore, it is well known that an electrode with a fast electron transfer and reduced charge recombination is necessary to obtain high PCE in DSSCs, which motivated the introduction of carbon nanotubes in photoanodes [81]. Later, it was observed that CNTs increase the surface roughness, which may contribute to the increase of both capacity dye loading and light scattering. Usually, the addition of CNTs increases around 35–50% the PCE in comparison to bare semiconductor photoanode as demonstrated by Hu et al. [82] with a triple-layer photoanode architecture composed by ZnO and ZnO/MWCNT ($PCE_{ZnO/MWCNT/ZnO} = 6.25\%$ and $PCE_{ZnO} = 4.61\%$), and by Mehmood et al. [83] that obtained a PCE = 5.25% with a TiO_2 photoanode containing 0.06% MWCNTs, which is 46% greater than unmodified one. As an alternative to the conventional high-cost Pt-based CE, Kilic et al. used iron pyrite (FeS_2) [84] as counter electrode to increase the efficiency to 7.27%.

On the way of low-cost alternatives, Anjidani et al. [85] developed a MWCNT/TiO_2 photoanode via layer-by-layer assembly technique, as schematically depicted in Figure 10.5a, yielding a 7.53% enhancement. FESEM images of the bare TiO_2 and TiO_2/[MWCNT/TiO_2]$_{40}$ multilayer photoanodes are shown in Figure 10.5b,c, with magnified cross section of the photoanodes at interface with fluorine-doped tin oxide (FTO) layer (Figure 10.5d,e) and schematics draw suggesting that TiO_2/[MWCNT/TiO_2]$_{40}$ multilayer act as blocking layer which could decrease electron back reaction at FTO/electrolyte interface, Figure 10.5f,g.

Based on nanostructures combination strategy, Hwang et al. [86] combined TiO_2 nanofibers (TNFs) and MWCNTs to TiO_2 nanoparticles (TNPs) paste to prepare the photoanode. They noticed that adding TNFs, the power conversion efficiency increases, on the other hand CNTs cause an opposite effect by decreasing the PCE if added to the composite. Finally, with 15 wt% of TNF they yield a 4.79% efficiency. Similar observance was reported by Li et al. [87] which unprecedented functionalized CNT with different amounts of urea named C_3N_4

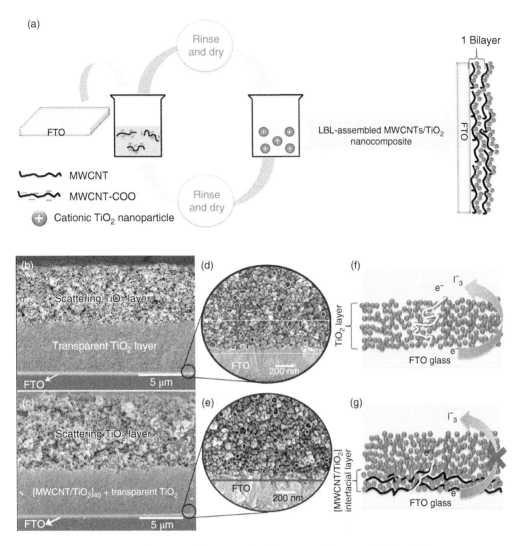

Figure 10.5 (a) Schematic of LbL assembly process for fabrication of [MWCNT/TiO$_2$] multilayer nanocomposite. FESEM images of (b, c) cross section of the bare TiO$_2$ photoanode and the photoanode with a [MWCNT/TiO2]$_{40}$ multilayer (area between lines with thickness about 280 nm), (d and e) magnified cross section of the photoanodes at interface with FTO layer, (f and g) schematics of the bare TiO$_2$ photoanode and the photoanode with a [MWCNT/TiO2]$_{40}$ multilayer between FTO and transparent TiO$_2$ layer, respectively. *Source:* Anjidani et al. 2017 [85]. Copyright 2017, Adapted and reproduced with permission of Elsevier.

powder (CNP), synthesized on a framework of a melamine sponge, achieving a photoelectric conversion efficiency of 6.3% (raw TiO$_2$), 7.1 (TiO$_2$/CNP), and 7.4 (TiO$_2$/CNT). This result indicates that the amount of dye adsorption on photoanodes decreases after adding CNP or CNT. Mehmood et al. [88] reported on carbon allotropes (MWCNTs and/or graphene) mixed with TiO$_2$ paste to form nanocomposites. Results demonstrate that nanocomposite graphene/TiO$_2$ has higher power conversion efficiency (5.25%) than MWCNTs/TiO$_2$ (4.20%). These authors attributed such decrease to diminishing amount of TiO$_2$ in the composite as demonstrated in Figure 10.6.

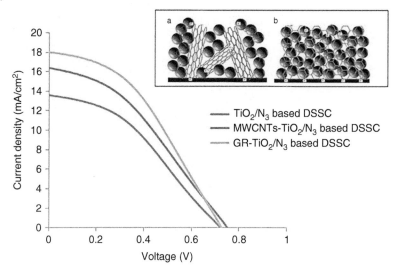

Figure 10.6 The effect of carbon allotrope insertion in TiO_2 nanocomposite on the DSSC performance. *Source:* Mehmood et al. 2018 [88]. Copyright 2018, Adapted and reproduced with permission of Elsevier.

Then an interesting approach was proposed by Davis et al. [89] who employed a modified size exclusion gel chromatography achieving up to 93% purity of single chiralities of SWCNTs to build the photoanode. Mixed and single chirality SWCNTs films tuned the Schottky barrier height at the TiO_2/SWCNT/FTO interface that prevented trap-mediated electron/electrolyte recombination and promoted the charge injection from the dye LUMO to the TiO_2 conduction band. Besides that, energy conversion efficiency has been limited by cells' low shunt resistance. Yoon et al. [90] prepared a metal-free carbon-based photoanode using purified p-type semiconducting SWCNTs (>99.8%) capped by a graphene layer that showed a superior performance to metallic or unsorted CNTs.

Husmann et al. [91] proposed the use of prussian blue (PB) as an innovative alternative of dye sensitizer. Usually, the loading of dye over the TiO_2 surface occurs at high temperature treatments or requires a processing step to increase the surface porosity which limits application in flexible devices. Then they optimized this whole process to a single-step electrodeposition of PB on SWCNT/TiO_2 at room temperature in aqueous environment. Still concerned about flexible devices, Song et al. [19] prepared flexible photoanodes through electrospray deposition and hot-compression procedure. In this study, TiO_2 nanorods (TNRs), Mg^{2+}-doped TiO_2 (Mg^{2+}/TNRs), and MWCNTs/TNRs were assessed in terms of PCE which provided values of 1.7, 2.2, and 2.3, respectively. Despite the low-value observance, they are greatly higher than that of flexible dye-sensitized solar cells (FDSSCs) employing the single-layer NRs film.

The thermal stability of DSSC was a concern explored by Agarwal et al. [92] which studied the thermal aging of DSSC devices. It was observed that the presence of SWCNTs reduced the thermal aging effects on photoanodes because of fast heat transfer rates of CNTs. Later, Mohammadnezhad et al. [93] conducted experiments with MWCNT, and they observed that exposing the devices at 80°C for 240 hours, a minimal loss 20% on PCE was observed for MWCNT/TiO_2 nanocomposite and substantial 59% of loss was determined to raw TiO_2.

Grissom et al. [94] presented a nonmetallic, flexible three-dimensional fiber-type dye sensitized solar cells using a carbon nanotube yarn (CNTY) coated with TiO_2 nano- and microporous layers then decorated with CdS and CdSe quantum dots to prepare both the working and a thread-like counter electrode which reached an optimized efficiency of 7.6%, as schematically

Figure 10.7 Schematic image of a flexible three-dimensional fiber-type dye sensitized solar cells. *Source:* Grissom et al. 2018 [94]. Copyright 2018, Adapted and reproduced with permission of Elsevier.

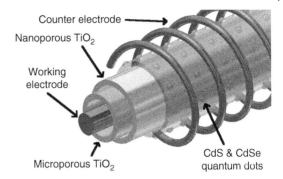

depicted in Figure 10.7. They also demonstrated the feasibility to connect multiple cells in series and parallel arrangements.

10.5 Effect of the Morphology and Interface of the CNT Photoanodes on the Efficiency of the DSSC

The most widely used photoanode is titanium dioxide; however, this is far from an ideal photoanode. The TiO$_2$ layer is usually composed of mesoporous spheres with diameters ranging from 20 to 30 nm, since in order to increase the performance of the DSSC, the best relation between surface area and light scattering has been sought. Large particles increase absorption through light scattering, in contrast to the amount of dye in the particle decreases, while smaller particles increase the amount of dye adsorbed on TiO$_2$ by increasing the contact surface between them (which is advantageous since the dye which provides the electrons for TiO$_2$), but do not promote light scattering [95].

In addition to the morphology of TiO$_2$ particles, other factors influence the performance of DSSC, such as recombination of electrons in photoanode with acceptors in electrolyte. This phenomenon may be in competition with the transport of electrons through the oxide, being in the same time interval. Thus, because the collection of photoinjected electrons competes with recombination, a high efficiency of charge collection requires that the transport be significantly faster than recombination [96, 97]. This can be achieved by using nanostructured photoanodes of composite materials, such as thin films of carbon nanotubes embedded in TiO$_2$ electrodes to serve as electronic transmission supports, increasing the diffusion length and consequently the collection of electrons [98]

Recent studies report the incorporation of carbon nanotubes into particles of TiO$_2$, working electrode of DSSC, with improvement in the transport of the charges. Li et al. verified a 17% increase in the efficiency of the DSSC with the incorporation of CNTs, when compared with DSSC using only TiO$_2$ as photoanode. In these cells, short-circuit current density (J_{SC}) was 17.6 mA/cm^2 and open-circuit voltage (V_{OC}) was 0.69 mV, while the DSSC with only TiO$_2$ photoanode showed short-circuit current density of 15.2 mA/cm^2 and open-circuit voltage was 0.67 V. In this study, they attributed increased efficiency to improved separation and load transport [87].

However, this ideal situation in which CNTs act as a charge carrier to move the photogenerated electrons from one TiO$_2$ nanoparticle to another or to the current collector is challenged by the recombination of the electrons at the carbon with the electrolyte acceptors where they are in touch. Therefore, the improvement in the collection of charges is observed only in photoanodes with small amounts of carbon in TiO$_2$ [99]. Dembele et al. confirmed this through photoelectric

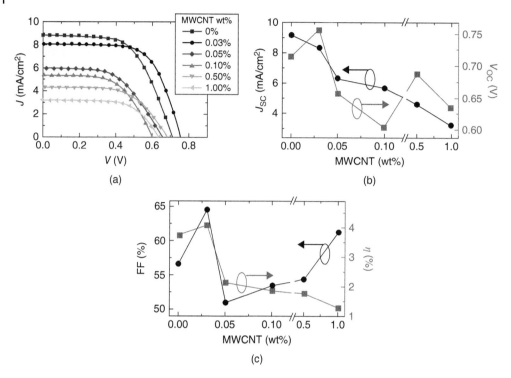

Figure 10.8 DSSCs with different concentrations of MWCNT. (a) *J–V* curves, (b) J_{SC} and V_{OC} values, and (c) FF and η values. *Source:* Dembele et al. 2013 [76]. Copyright 2013, Adapted and reproduced with permission of Elsevier.

measurements performed in DSSC with TiO_2 photoanodes and different amounts of MWCNTs [76]. Among the concentrations studied, ranging from 0.03% to 1.0% by weight of CNTs, the 0.03% solar cell reached the highest efficiency (η) (Figure 10.8a), reaching 4.1%, as presented in Figure 10.8c. The values of short-circuit current density (J_{SC}) and open-circuit voltage (V_{OC}) were 8.2 mA/cm^2 and 0.76 V, respectively (Figure 10.8b).

Dembele et al. [76] also observed that from 0.03 wt% of MWCNT the solar cell begins to lose efficiency drastically (Figure 10.8c), obtaining even lower values than solar cells with pure TiO_2 photoanode. This is because, in this case, the photocurrent density was negatively affected by the addition of MWCNTs due to their competitive light absorption with the dye-sensitized, which causes partial loss of the visible radiation. Another factor is that films with high CNT concentration cause cracks or holes in the TiO_2 compact layer, which impairs the formation of close-packed TiO_2 and exposes the CNT to the electrolyte (causing the electron recombination with acceptor in dye-sensitized).

Another mechanism acting on DSSC is the effect of the Schottky junction on the CNT/TiO_2 interface. According to Moro et al., the formation of a Schottky barrier can serve as an efficient electron trap avoiding the recombination of electron–holes, as a consequence increased efficiency [100]. Song et al. demonstrated the increase in the overall conversion efficiency of DSSCs and attributed to the increase of the photocurrent density due to the formation of the Schottky barrier at the SWCNT/TiO_2 junction [101]. In this study, they used the concentration of 0.1 mg/ml SWCNT and reached efficiency of 6.06%, J_{SC} of 18.2 mA/cm^2 and V_{OC} of 0.57 V. However, this improvement in photocurrent density is only observed in extremely low concentrations of CNTs.

As shown previously in Figure 10.8b, increasing the CNT concentration in the photoanode of TiO_2 occurs with the continuous decrease of the photocurrent density, as proposed by Dembele et al. This seems somewhat contradictory, since the larger the area of CNT/TiO_2 contact, the greater the probability of there being electron traps that prevent recombination. On the other hand, the justification based on competition with the dye in the light harvesting process seems quite satisfactory.

10.6 Summary and Future Prospect

Many studies have been conducted with the goal of developing photoanodes for DSSC. In general, most photoanodes use nanostructured oxides such as TiO_2, producing devices with good efficiency. However, better efficiency results are achieved when the photoanodes are manufactured by combining at least two nanostructures such as spheres or nanorods of TiO_2 and carbon nanotubes. The introduction of CNT to photoanode improves electron transfer and decreases recombination. In addition, they promote the formation of the Schottky junction that act as electron traps. All of these factors mentioned above are responsible for the increased performance of DSSCs. Undeniably, the excellent electronic properties of CNTs promoted its application as photoanode in DSSCs.

Although the addition of CNT has increased the performance of DSSCs, they are still lower in energy efficiency compared to silicon solar cells. To change this scenario, efforts are expected in the development of photoanodes with greater surface area to facilitate the transfer of electrons from the dye to the photoanode and from the photoanode to the conducting electrode. In addition, improving other parts of the DSSCs such as sensitizers, electrolytes, and counter electrodes can also bring an increase in the energy efficiency of these devices.

Acknowledgment

The authors are thankful and acknowledges to Brazilian funding agency FAPESP (2014/23546-1, 2016/23474-6).

References

1 Yeoh, M.-E. and Chan, K.-Y. (2017). Recent advances in photo-anode for dye-sensitized solar cells: a review. *Int. J. Energy Res.* 41 (15): 2446–2467.
2 O'Regan, B. and Grätzel, M. (1991). A low-cost, high-efficiency solar cell based on dye-sensitized colloidal TiO_2 films. *Nature* 353: 737.
3 Blakers, A., Zin, N., McIntosh, K.R., and Fong, K. (2013). High efficiency silicon solar cells. *Energy Procedia* 33: 1–10.
4 Ye, M., Wen, X., Wang, M. et al. (2015). Recent advances in dye-sensitized solar cells: from photoanodes, sensitizers and electrolytes to counter electrodes. *Mater. Today* 18 (3): 155–162.
5 Sreekala, C.S.N.O.A., Indiramma, J., Kumar, K.B.S.P. et al. (2013). Functionalized multi-walled carbon nanotubes for enhanced photocurrent in dye-sensitized solar cells. *J. Nanostruct. Chem.* 3 (1): 19.
6 Brennan, L.J., Byrne, M.T., Bari, M., and Gun'ko, Y.K. (2011). Carbon nanomaterials for dye-sensitized solar cell applications: a bright future. *Adv. Energy Mater.* 1 (4): 472–485.

7 Khan, M.Z.H., Al-Mamun, M.R., Halder, P.K., and Aziz, M.A. (2017). Performance improvement of modified dye-sensitized solar cells. *Renewable Sustainable Energy Rev.* 71: 602–617.

8 Carella, A., Borbone, F., Centore, R. et al. (2018). *Front. Chem.* 6: 481.

9 Sharma, S., Bulkesh, S., Ghoshal, S.K., and Mohan, D. (2017). Dye sensitized solar cells: from genesis to recent drifts. *Renewable Sustainable Energy Rev.* 70: 529–537.

10 Fan, K., Yu, J., and Ho, W. (2017). Improving photoanodes to obtain highly efficient dye-sensitized solar cells: a brief review. *Mater. Horiz.* 4 (3): 319–344.

11 Shakeel Ahmad, M., Pandey, A.K., and Abd Rahim, N. (2017). Advancements in the development of TiO_2 photoanodes and its fabrication methods for dye sensitized solar cell (DSSC) applications. a review. *Renewable Sustainable Energy Rev.* 77: 89–108.

12 Ramzan Parra, M., Pandey, P., Siddiqui, H. et al. (2019). Evolution of ZnO nanostructures as hexagonal disk: implementation as photoanode material and efficiency enhancement in Al: ZnO based dye sensitized solar cells. *Appl. Surf. Sci.* 470: 1130–1138.

13 Alkuam, E., Badradeen, E., and Guisbiers, G. (2018). Influence of CdS morphology on the efficiency of dye-sensitized solar cells. *ACS Omega* 3 (10): 13433–13441.

14 Younas, M., Gondal, M.A., Dastageer, M.A., and Baig, U. (2019). Fabrication of cost effective and efficient dye sensitized solar cells with WO_3-TiO_2 nanocomposites as photoanode and MWCNT as Pt-free counter electrode. *Ceram. Int.* 45 (1): 936–947.

15 Akilimali, R., Selopal, G.S., Benetti, D. et al. (2018). Hybrid TiO_2-graphene nanoribbon photoanodes to improve the photoconversion efficiency of dye sensitized solar cells. *J. Power Sources* 396: 566–573.

16 Mohamed, I.M.A., Dao, V.-D., Yasin, A.S. et al. (2017). Design of an efficient photoanode for dye-sensitized solar cells using electrospun one-dimensional GO/N-doped nanocomposite SnO_2/TiO_2. *Appl. Surf. Sci.* 400: 355–364.

17 Sasikumar, R., Chen, T.-W., Chen, S.-M. et al. (2018). Developing the photovoltaic performance of dye-sensitized solar cells (DSSCs) using a SnO_2-doped graphene oxide hybrid nanocomposite as a photo-anode. *Opt. Mater.* 79: 345–352.

18 Lu, D., Qin, L., Liu, D. et al. (2018). High-efficiency dye-sensitized solar cells based on bilayer structured photoanode consisting of carbon nanofiber/TiO_2 composites and Ag@TiO_2 core-shell spheres. *Electrochim. Acta* 292: 180–189.

19 Song, L., Yin, X., Xie, X. et al. (2017). Highly flexible TiO_2/C nanofibrous film for flexible dye-sensitized solar cells as a platinum- and transparent conducting oxide-free flexible counter electrode. *Electrochim. Acta* 255: 256–265.

20 Kim, H.-S., Chun, M.-H., Suh, J. et al. (2017). Dual functionalized freestanding TiO_2 nanotube arrays coated with Ag nanoparticles and carbon materials for dye-sensitized solar cells. *Appl. Sci.* 7 (6): 576.

21 Rho, W.-Y., Kim, H.-S., Kim, H.-M. et al. (2017). Carbon-doped freestanding TiO_2 nanotube arrays in dye-sensitized solar cells. *New J. Chem.* 41 (1): 285–289.

22 Chandrasekhar, P.S., Parashar, P.K., Swami, S.K. et al. (2018). Enhancement of Y123 dye-sensitized solar cell performance using plasmonic gold nanorods. *Phys. Chem. Chem. Phys.* 20 (14): 9651–9658.

23 Ran, H., Fan, J., Zhang, X. et al. (2018). Enhanced performances of dye-sensitized solar cells based on Au–TiO_2 and Ag–TiO_2 plasmonic hybrid nanocomposites. *Appl. Surf. Sci.* 430: 415–423.

24 Bhardwaj, S., Pal, A., Chatterjee, K. et al. (2018). Fabrication of efficient dye-sensitized solar cells with photoanode containing TiO_2–Au and TiO_2–Ag plasmonic nanocomposites. *J. Mater. Sci. Mater. Electron.* 29 (21): 18209–18220.

25 Wu, W.-Y., Hsu, C.-F., Wu, M.-J. et al. (2017). Ag–TiO$_2$ composite photoelectrode for dye-sensitized solar cell. *Appl. Phys. A* 123 (5): 357.

26 Nbelayim, P., Kawamura, G., Kian Tan, W. et al. (2017). Systematic characterization of the effect of Ag@TiO$_2$ nanoparticles on the performance of plasmonic dye-sensitized solar cells. *Sci. Rep.* 7 (1): 15690.

27 Ahmad, M.S., Pandey, A.K., Rahim, N.A. et al. (2018). Chemical sintering of TiO$_2$ based photoanode for efficient dye sensitized solar cells using Zn nanoparticles. *Ceram. Int.* 44 (15): 18444–18449.

28 Rho, W.-Y., Song, D.H., Yang, H.-Y. et al. (2018). Recent advances in plasmonic dye-sensitized solar cells. *J. Solid State Chem.* 258: 271–282.

29 Kartikay, P., Nemala, S.S., and Mallick, S. (2017). One-dimensional TiO$_2$ nanostructured photoanode for dye-sensitized solar cells by hydrothermal synthesis. *J. Mater. Sci. Mater. Electron.* 28 (15): 11528–11533.

30 Maurya, I.C., Senapati, S., Singh, S. et al. (2018). Effect of particle size on the performance of TiO$_2$ based dye-sensitized solar cells. *ChemistrySelect* 3 (34): 9872–9880.

31 Sriharan, N., Ganesan, N.M., Kang, M. et al. (2019). Improved photoelectrical performance of single crystalline rutile TiO$_2$ nanorod arrays incorporating α-alumina for high efficiency dye-sensitized solar cells. *Mater. Lett.* 237: 204–208.

32 Sun, Q., Hong, Y., Zang, T. et al. (2018). The application of heterostructured SrTiO$_3$–TiO$_2$ nanotube arrays in dye-sensitized solar cells. *J. Electrochem. Soc.* 165 (4): H3069–H3075.

33 Xu, L., Aumaitre, C., Kervella, Y. et al. (2018). Increasing the efficiency of organic dye-sensitized solar cells over 10.3% using locally ordered inverse opal nanostructures in the photoelectrode. *Adv. Funct. Mater.* 28 (15): 1706291.

34 Bakr, Z.H., Wali, Q., Ismail, J. et al. (2018). Synergistic combination of electronic and electrical properties of SnO$_2$ and TiO$_2$ in a single SnO$_2$–TiO$_2$ composite nanofiber for dye-sensitized solar cells. *Electrochim. Acta* 263: 524–532.

35 Liu, Y.-Y., Ye, X.-Y., An, Q.-Q. et al. (2018). A novel synthesis of the bottom-straight and top-bent dual TiO$_2$ nanowires for dye-sensitized solar cells. *Adv. Powder Technol.* 29 (6): 1455–1462.

36 Suriani, A.B., Muqoyyanah, Mohamed, A. et al. (2018). Improving the photovoltaic performance of DSSCs using a combination of mixed-phase TiO$_2$ nanostructure photoanode and agglomerated free reduced graphene oxide counter electrode assisted with hyperbranched surfactant. *Optik* 158: 522–534.

37 Xu, L., Xu, J., Hu, H. et al. (2019). Hierarchical submicroflowers assembled from ultrathin anatase TiO$_2$ nanosheets as light scattering centers in TiO$_2$ photoanodes for dye-sensitized solar cells. *J. Alloys Compd.* 776: 1002–1008.

38 Arifin, Z., Suyitno, S., Hadi, S., and Sutanto, B. (2018). Improved performance of dye-sensitized solar cells with TiO$_2$ nanoparticles/Zn-doped TiO$_2$ hollow fiber photoanodes. *Energies* 11 (11): 2922.

39 Wang, W., Yuan, H., Xie, J. et al. (2018). Enhanced efficiency of large-area dye-sensitized solar cells by light-scattering effect using multilayer TiO$_2$ photoanodes. *Mater. Res. Bull.* 100: 434–439.

40 Nursam, N.M., Hidayat, J., Shobih et al. (2018). A comparative study between titania and zirconia as material for scattering layer in dye-sensitized solar cells. *J. Phys. Conf. Ser.* 1011 (1): 012003.

41 Muhammad, N., Zanoni, K.P.S., Iha, N.Y.M., and Ahmed, S. (2018). The use of rutile- and anatase-titania layers towards back light scattering in dye-sensitized solar cells. *ChemistrySelect* 3 (37): 10475–10482.

42 Zhang, W., Gu, J., Yao, S., and Wang, H. (2018). The synthesis and application of TiO_2 microspheres as scattering layer in dye-sensitized solar cells. *J. Mater. Sci. Mater. Electron.* 29 (9): 7356–7363.

43 Hu, Z., Zhao, L., Guo, H. et al. (2018). Novel double-layered photoanodes based on porous-hollow TiO_2 microspheres and $La(OH)_3$:Yb^{3+}/Er^{3+} for highly efficient dye-sensitized solar cells. *J. Mater. Sci. Mater. Electron.* 30 (1): 212–220.

44 Sim, Y.H., Yun, M.J., Cha, S.I. et al. (2018). Improvement in energy conversion efficiency by modification of photon distribution within the photoanode of dye-sensitized solar cells. *ACS Omega* 3 (1): 698–705.

45 Pham, T.T.T., Mathews, N., Lam, Y.-M., and Mhaisalkar, S. (2018). Influence of size and shape of sub-micrometer light scattering centers in ZnO-assisted TiO_2 photoanode for dye-sensitized solar cells. *Physica B* 532: 225–229.

46 Sakthivel, T., Kumar, K.A., Senthilselvan, J., and Jagannathan, K. (2018). Effect of Ni dopant in TiO_2 matrix on its interfacial charge transportation and efficiency of DSSCs. *J. Mater. Sci. Mater. Electron.* 29 (3): 2228–2235.

47 Shakir, S., Abd-ur-Rehman, H.M., Yunus, K. et al. (2018). Fabrication of un-doped and magnesium doped TiO_2 films by aerosol assisted chemical vapor deposition for dye sensitized solar cells. *J. Alloys Compd.* 737: 740–747.

48 Nguyen, H.H., Gyawali, G., Hoon, J.S. et al. (2018). Cr-doped TiO_2 nanotubes with a double-layer model: An effective way to improve the efficiency of dye-sensitized solar cells. *Appl. Surf. Sci.* 458: 523–528.

49 Mahmoud, M.S., Akhtar, M.S., Mohamed, I.M.A. et al. (2018). Demonstrated photons to electron activity of S-doped TiO_2 nanofibers as photoanode in the DSSC. *Mater. Lett.* 225: 77–81.

50 Fonseca, A.F.V.D., Siqueira, R.L., Landers, R. et al. (2018). A theoretical and experimental investigation of Eu-doped ZnO nanorods and its application on dye sensitized solar cells. *J. Alloys Compd.* 739: 939–947.

51 Xing, G., Zhang, Z., Qi, S. et al. (2018). Effect of cerium ion modifications on the photoelectrochemical properties of TiO_2-based dye-sensitized solar cells. *Opt. Mater.* 75: 102–108.

52 Kharel, P.L., Zamborini, F.P., and Alphenaar, B.W. (2018). Enhancing the photovoltaic performance of dye-sensitized solar cells with rare-earth metal oxide nanoparticles. *J. Electrochem. Soc.* 165 (3): H52–H56.

53 Kumar, V., Pandey, A., Swami, S.K. et al. (2018). Synthesis and characterization of Er^{3+}–Yb^{3+} doped ZnO upconversion nanoparticles for solar cell application. *J. Alloys Compd.* 766: 429–435.

54 Areerob, Y., Cho, K.-Y., and Oh, W.-C. (2018). Strategy to improve photovoltaic performance of DSSC sensitized by using novel nanostructured La dopped TiO_2-graphene electrodes. *J. Mater. Sci. Mater. Electron.* 29 (4): 3437–3448.

55 Ünlü, B., Çakar, S., and Özacar, M. (2018). The effects of metal doped TiO_2 and dithizone-metal complexes on DSSCs performance. *Sol. Energy* 166: 441–449.

56 Shahroosvand, H., Abbasi, P., and Bideh, B.N. (2018). Dye-sensitized solar cell based on novel star-shaped ruthenium polypyridyl sensitizer: new insight into the relationship between molecular designing and its outstanding charge carrier dynamics. *ChemistrySelect* 3 (24): 6821–6829.

57 Pounraj, P., Mohankumar, V., Pandian, M.S., and Ramasamy, P. (2018). Donor functionalized quinoline based organic sensitizers for dye sensitized solar cell (DSSC) applications: DFT and TD-DFT investigations. *J. Mol. Model.* 24 (12): 343.

58 Roy, J.K., Kar, S., and Leszczynski, J. (2018). Insight into the optoelectronic properties of designed solar cells efficient tetrahydroquinoline dye-sensitizers on $TiO_2(101)$ surface: first principles approach. *Sci. Rep.* 8 (1): 10997.

59 Dayan, S., Özdemir, N., and Özpozan, N.K. (2018). Enhanced performance of organic/inorganic hybrid nanomaterials bearing impregnated $[PdL_2]$ complexes as counter-electrode catalyst for dye-sensitized solar cells. *Appl. Organomet. Chem.* 0 (0): e4710.

60 Yu, F., Shi, Y., Yao, W. et al. (2019). *J. Power Sources* 412: 366–373.

61 Chew, J.W., Khanmirzaei, M.H., Numan, A. et al. (2018). Performance studies of ZnO and multi walled carbon nanotubes-based counter electrodes with gel polymer electrolyte for dye-sensitized solar cell. *Mater. Sci. Semicond. Process.* 83: 144–149.

62 Yilmaz, M., Hsu, S.-H., Raina, S. et al. (2018). Dye-sensitized solar cells using carbon nanotube-based counter electrodes in planar and micro-array patterned configurations. *J. Renewable Sustainable Energy* 10 (6): 063501.

63 Monreal-Bernal, A., Vilatela, J.J., and Costa, R.D. (2019). CNT fibres as dual counter-electrode/current-collector in highly efficient and stable dye-sensitized solar cells. *Carbon* 141: 488–496.

64 Chuang, T.-K., Anuratha, K.S., Lin, J.-Y. et al. (2018). Low temperature growth of carbon nanotubes using chemical bath deposited $Ni(OH)_2$ – an efficient Pt-free counter electrodes for dye-sensitized solar cells. *Surf. Coat. Technol.* 344: 534–540.

65 Sakali, S.M., Khanmirzaei, M.H., Lu, S.C. et al. (2018). Investigation on gel polymer electrolyte-based dye-sensitized solar cells using carbon nanotube. *Ionics.*

66 Xianhua, C., Khanmirzaei, M.H., Omar, F.S. et al. (2018). The effect of incorporation of multi-walled carbon nanotube into poly(ethylene oxide) gel electrolyte on the photovoltaic performance of dye-sensitized solar cell. *Polym. Plast. Technol. Eng.*: 1–8.

67 Khannam, M. and Dolui, S.K. (2017). Cerium doped TiO_2 photoanode for an efficient quasi-solid state dye sensitized solar cells based on polyethylene oxide/multiwalled carbon nanotube/polyaniline gel electrolyte. *Sol. Energy* 150: 55–65.

68 Dresselhaus, M.S., Dresselhaus, G., and Eklund, P.C. (1996). Structure of fullerenes, Chapter 3. In: *Science of Fullerenes and Carbon Nanotubes* (eds. M.S. Dresselhaus, G. Dresselhaus and P.C. Eklund), 60–79. San Diego, CA: Academic Press.

69 Dresselhaus, M., Dresselhaus, G., and Eklund, P. (1996). *Science of Fullerenes and Carbon Nanotubes*. San Diego: Elsevier Science & Technology Books.

70 Batmunkh, M., Biggs, M.J., and Shapter, J.G. (2015). Carbon nanotubes for dye-sensitized solar cells. *Small*: 2963–2989.

71 Volder, M.F.L.D., Tawfick, S.H., Hart, A.J., and Baughman, R.H. (2013). Carbon nanotubes: present and future commercial applications. *Science*: 535–539.

72 Rafique, I., Kausar, A., Anwar, Z., and Muhammad, B. (2016). Exploration of epoxy resins, hardening systems, and epoxy/carbon nanotube composite designed for high performance materials: a review. *Polym. Plast. Technol. Eng.* 55 (3): 312–333.

73 Tune, D.D., Flavel, B.S., Krupke, R., and Shapter, J.G. (2012). Carbon nanotube-silicon solar cells. *Adv. Energy Mater.*: 1043–1055.

74 Dresselhaus, M.S., Jorio, A., and Dresselhaus, G. (2004). Unusual properties and structure of carbon nanotubes. *Annu. Rev. Mater. Res.*: 247–278.

75 Belin, T. and Epron, F. (2005). Characterization methods of carbon nanotubes: a review. *Mater. Sci. Eng., B*: 105–118.

76 Dembele, K.T., Nechache, R., Nikolova, L. et al. (2013). Effect of multi-walled carbon nanotubes on the stability of dye sensitized solar cells. *J. Power Sources*: 93–97.

77 Delekar, S.D., Dhodamani, A.G., More, K.V. et al. (2018). Structural and optical properties of nanocrystalline TiO_2 with multiwalled carbon nanotubes and its photovoltaic studies using Ru(II) sensitizers. *ACS Omega*: 2743–2756.

78 Iijima, S. (1991). Helical microtubules of graphitic carbon. *Nature* 354: 56.

79 Davis, F., Shimizu, F.M., and Altintas, Z. (2018). Smart Nanomaterials. In: *Biosensors and Nanotechnology* (ed. Z. Altintas). United States: Wiley.

80 Joshi, N., Hayasaka, T., Liu, Y. et al. (2018). A review on chemiresistive room temperature gas sensors based on metal oxide nanostructures, graphene and 2D transition metal dichalcogenides. *Microchim. Acta* 185 (4): 213.

81 Lee, T.Y., Alegaonkar, P.S., and Yoo, J.-B. (2007). Fabrication of dye sensitized solar cell using TiO_2 coated carbon nanotubes. *Thin Solid Films* 515 (12): 5131–5135.

82 Hu, J., Xie, Y., Bai, T. et al. (2015). A novel triple-layer zinc oxide/carbon nanotube architecture for dye-sensitized solar cells with excellent power conversion efficiency. *J. Power Sources* 286: 175–181.

83 Mehmood, U., Hussein, I.A., Al-Ahmed, A., and Ahmed, S. (2016). Enhancing power conversion efficiency of dye-sensitized solar cell using TiO_2-MWCNT composite photoanodes. *IEEE J. Photovoltaics* 6 (2): 486–490.

84 Kilic, B., Turkdogan, S., Astam, A. et al. (2016). Preparation of carbon nanotube/TiO_2 mesoporous hybrid photoanode with iron pyrite (FeS_2) thin films counter electrodes for dye-sensitized solar cell. *Sci. Rep.* 6: 27052.

85 Anjidani, M., Milani Moghaddam, H., and Ojani, R. (2017). Binder-free MWCNT/TiO_2 multilayer nanocomposite as an efficient thin interfacial layer for photoanode of dye sensitized solar cell. *Mater. Sci. Semicond. Process.* 71: 20–28.

86 Hwang, T.-H., Kim, W.-T., and Choi, W.-Y. (2017). Mixed dimensionality with a TiO_2 nanostructure and carbon nanotubes for the photoelectrode in dye-sensitized solar cells. *J. Nanosci. Nanotechnol.* 17 (7): 4812–4816.

87 Li, X., Pan, K., Qu, Y., and Wang, G. (2018). One-dimension carbon self-doping g-C_3N_4 nanotubes: synthesis and application in dye-sensitized solar cells. *Nano Res.* 11 (3): 1322–1330.

88 Mehmood, U., Ul Haq Khan, A., Ali Qaiser, A. et al. (2018). Nanocomposites of carbon allotropes with TiO_2 as effective photoanodes for efficient dye-sensitized solar cells. *Mater. Lett.* 228: 125–128.

89 Davis, V.L., Quaranta, S., Cavallo, C. et al. (2017). Effect of single-chirality single-walled carbon nanotubes in dye sensitized solar cells photoanodes. *Sol. Energy Mater. Sol. Cells* 167: 162–172.

90 Yoon, K., Lee, J.-H., Kang, J. et al. (2016). Metal-free carbon-based nanomaterial coatings protect silicon photoanodes in solar water-splitting. *Nano Lett.* 16 (12): 7370–7375.

91 Husmann, S., Lima, L.F., Roman, L.S., and Zarbin, A.J.G. (2018). Photoanode for aqueous dye-sensitized solar cells based on a novel multicomponent thin film. *ChemSusChem* 11 (7): 1238–1245.

92 Agarwal, R., Sahoo, S., Venkateswara Rao, C., and Katiyar, R.S. (2017). SWCNT-TiO_2 nanocomposite photo-anode for improved thermal performances of dye-sensitized solar cells. *ECS Trans.* 75 (25): 11–25.

93 Mohammadnezhad, M., Selopal, G.S., Wang, Z.M. et al. (2018). Towards long-term thermal stability of dye-sensitized solar cells using multiwalled carbon nanotubes. *ChemPlusChem* 83 (7): 682–690.

94 Grissom, G., Jaksik, J., McEntee, M. et al. (2018). Three-dimensional carbon nanotube yarn based solid state solar cells with multiple sensitizers exhibit high energy conversion efficiency. *Sol. Energy* 171: 16–22.

95 Wang, Z.-S., Kawauchi, H., Kashima, T., and Arakawa, H. (2004). Significant influence of TiO_2 photoelectrode morphology on the energy conversion efficiency of N719 dye-sensitized solar cell. *Coord. Chem. Rev.* 248 (13): 1381–1389.

96 Sacco, A. (2017). Eletrochemical impedance spectroscopy: fundamental and application in dye-sensitized solar cells. *Renewable Sustainable Energy Rev.*: 814–829.

97 Kai Zhu, N.R.N., Miedaner, A., and Frank, A.J. (2007). Enhanced charge-collection efficiencies and light scattering in dye-sensitized and light scattering in dye-sensitized nanotubes arrays. *Nano Lett.*: 69–74.

98 Nath, N.C.D., Sarker, S., Ahammada, A.J.S., and Lee, J.-J. (2012). Spatial arrangement of carbon nanotubes in TiO_2 photoelectrodes to enhance the efficiency of dye-sensitized solar cells. *Phys. Chem. Chem. Phys.*: 4333–4338.

99 Kavan, L. (2013). Exploiting nanocarbons in dye-sensitized solar cells. *Top. Curr. Chem.* 348: 59–93.

100 Piera Moro, S.S., Donzello, M.P., Fierro, G., and Moretti, G. (2015). A comparison of the photocatalytic activity between commercial and synthesized mesoporous and nanocrystalline titanium dioxide for 4-nitrophenol degradation: Effect of phase composition, particle size, and addition of carbon nanotubes. *Appl. Surf. Sci.*: 293–305.

101 Junling Song, Z.Y., Yang, Z., Amaladass, P. et al. (2011). Enhancement of photogenerated electron transport in dye-sensitized solar cells with introduction of a reduced graphene oxide–TiO_2 junction. *Chem. Eur. J.*: 10832–10837.

11

Graphene-Based Nanocomposite as Photoanode

Subhendu K. Panda, G. Murugadoss, and R. Thangamuthu

CSIR – Central Electrochemical Research Institute, Karaikudi, India

11.1 Introduction

There are several ways of modifying dye-sensitized solar cell (DSSC) in order to improve the power conversion efficiency (PCE). Photoanode modification is one of the important methods in addition to alteration in the counterelectrode. Since the photoanode in DSSC acts as a light harvester as well as a charge carrier, the modification of the photoanode hence plays a crucial role in enhancing the PCE of the cell. Photoanode is the backbone of DSSC that is why its material, shape, and size have to be carefully selected and optimized for the better performance of the cell. For an ideal photoanode, following should be the key features: (i) high surface area for improving dye loading, (ii) fast electron transfer, (iii) high resistance to photo-corrosion, (iv) good transparency, and (v) it should have ability to absorb/scatter the sun light. In general, mesoporous TiO_2 has been widely used most effective metal oxide for photoanode in DSSCs due to its low cost, high stability, abundant in nature, attractive optical and electronic properties, and environment friendliness. The most efficient solar cell developed with TiO_2 delivered approximately 12–14% photo-conversion efficiency [1, 2]. Various other metal oxides such as ZnO [3], SnO_2 [4], Nb_2O_5 [5], $SrTiO_3$ [6], Zn_2SnO_4 [7], and WO_3 [8] have been investigated. Among the other metal oxides, hexagonal structured ZnO show promising results compared to conventional TiO_2. Though, ZnO possess better electron mobility than TiO_2, the efficiency does not exceed the TiO_2-based DSSCs, which is due to low dye pickup and less stability of ZnO in acidic environment compared to TiO_2. In general, it is more important for an ideal photoanode to have high surface area along with high light scattering capability. Commercially available TiO_2 nanoparticles have high surface area, however, particle diameter is less compared to the wavelength of the incident light so unable to scatter the sun-light. Surface area of the nanoparticles and light scattering ability are completely opposite phenomena. Hence, improving scattering with high surface area, double layer and multilayers of mesoporous TiO_2 or nanocomposites are used [9]. A single material cannot provide all the mentioned properties of the photoanode, hence, multiple combinations of materials and their hybrid nanostructures are developed to optimize with other components. In this context, graphene-based materials have been incorporated into each part of DSSCs and interesting results were obtained.

Graphene, a novel two-dimensional carbon nanomaterial discovered by Novoslov et al. in 2004 has attracted tremendous interest in various applications such as photovoltaic devices, aerospace, electronics. Graphene is an allotrope of carbon that consists of flat monolayers of carbon atoms tightly packed into honeycomb crystal lattices [10–13]. Graphene is a 2D material having a one-atom thick network like structure which exhibits high thermal conductivity at

Interfacial Engineering in Functional Materials for Dye-Sensitized Solar Cells, First Edition.
Edited by Alagarsamy Pandikumar, Kandasamy Jothivenkatachalam and Karuppanapillai B. Bhojanaa.
© 2020 John Wiley & Sons, Inc. Published 2020 by John Wiley & Sons, Inc.

room temperature ($\sim 5 \times 10^3$ W/m/K), high charge/hole mobility (2×105 cm^2/V/s), and charge carrier concentration (10^{13} cm^{-2}). A single layer of graphene shows an optical transparency of 97.7% and a three-layered graphene stack shows around 90.8% optical transparency. The addition of each layer corresponds to a 2.3% decrease in optical transparency, while retaining 90% optical transparency. In a double-layer graphene, the Fermi energy levels changes which also changes the optical properties of the material. Also tuning the external electrical gating of graphene, the spectral absorption properties can be modified. So far the good optical absorption related to the Fermi energy level changes it could be used as an efficient anode material in solar cell devices. Apart from the high surface area of the graphene thin layer, doping and decorating of the suitable metal ions on the surface via various chemical and physical methods has been attractive research topic recently, especially in the electrochemistry. For example, nitrogen-doped graphene oxide (GO) annealed in argon and ammonia environment performed better because of the increase in conductivity and catalytic defect density of the material structure [14–16].

Graphene and graphene-based materials have been reported as efficient components for various parts of DSSCs. Mostly, graphene is used as a counterelectrode material in DSSC. Apart from that graphene-based materials that are used as a transparent conducting electrode, [13, 17] sensitizing materials along with dyes [18–20] are also used as an additive to the photo anode material to enhance the photo-electron transfer process. In this chapter, we are mainly presenting an overview of the recent advances in graphene-based nanocomposite materials for DSSC photoanode applications.

11.2 Graphene–TiO$_2$ Nanocomposite for Photoanode

As a photoanode, TiO$_2$ regulates the light-to-electricity conversion processes such as the electron transfer from the dye molecules to the TiO$_2$ conduction band (CB) and also the electron transfer from photoanode to the external circuit in the DSSCs. However, charge recombination is one of the main issues for the limitation of the performance of DSSC. In other words, back-electron transfer in the TiO$_2$ photoanode–electrolyte interface is supposed to be the major recombination pathway that limits the photogenerated electrons to reach the transparent conducting oxide (TCO) electrode resulting in reduced DSSC efficiency. There are several approaches used to prevent charge carrier recombination and improve electron transport such as composite semiconductor with different bandgaps, insertion of some doping elements in TiO$_2$, and incorporating charge carriers to direct the photogenerated electron toward external circuit. The following elements such as zirconium, aluminum, ruthenium, and niobium are mostly used as dopants for promoting charge transfer to enhance the DSSC efficiency. Further, there are several reports on incorporation of one-dimensional (1D) carbon nanotubes (CNTs) to metal oxide, but the efficiency of DSSCs is limited. Interestingly, graphene-based TiO$_2$ nanocomposites showed higher efficiency due to effective charge separation, excellent conductivity, and good contact with TiO$_2$. In addition, TiO$_2$ can attach to the graphene to form graphene bridges that reduce the TiO$_2$–TiO$_2$ connections and act as an electron transporter from the TiO$_2$ CB to the underlying collecting electrode [21].

Considerable efforts have been made to incorporate graphene into TiO$_2$ via various synthesis routes that showed morphology dependent properties. Graphene in different forms like ultra-thin layer, nanofibers as well as quantum dots were used. Apart from being used as the conductive substrate, graphene can also be mixed with TiO$_2$ to play as the photoanode in DSSCs. Graphene films have been explored as both the transparent electrode and the counterelectrode for DSSCs due to their high transparency and electron transfer mobility as well as their electrochemical activities. Furthermore, graphene layer can also be mixed with TiO$_2$ semiconductor to form a composite film and to enhance the electron transfer from TiO$_2$ to photoelectrode.

There are several factors such as the ratio of the graphene and the host semiconductor, the particle size TiO$_2$, size and thickness of the graphene film that determines the efficiency of the solar cell [22]. The effect of graphene loading in the nanocomposite was studied and proposed that ~0.7 wt% graphene addition led to the highest efficiency. The work clearly showed that the electron lifetime decreased from 17.6 to 6.4 ms by the addition of graphene and above 0.7 wt%, the excessive graphene involved in recombination processes along with TiO$_2$. Madhavan et al. [23] reported the electrospinning of TiO$_2$–graphene composite nanofibers to develop conductive nanofiber mats to increase the surface area. They observed a short circuit current of 16.2 mA/cm^2, an open circuit voltage of 0.71 V, a fill factor (FF) of 0.66, and an efficiency of 7.6%, compared with pure TiO$_2$ efficiency of 6.3%. Sun et al. found that by addition of graphene to the photoanode the PCE is 4.28%, which is an enhancement of 59% compare to the photoanode without graphene [24]. A systematic investigation of the incorporation of chemical exfoliation graphene sheets (GS) in TiO$_2$ nanoparticle films via molecular grafting method where the PCE for DSSC based on GS/TiO$_2$ composite films is more than five times higher than that of pure TiO$_2$. This enhancement is attributed to the improved conductivity of TiO$_2$ nanoparticle film and better dye loading of GS/TiO$_2$ film than that of TiO$_2$ film [25]. Incorporation of reduced graphene oxide (rGO) with metal oxides leads to fast charge recombination and transport rate between the conducting electrode and the semiconductor nanostructures. Graphene that possesses large surface area (~2600 m^2/g) and superior mobility of charge carriers (200 000 cm^2/V s) with metal oxides could promote the advantages in rGO-metal oxide composites where it can be suitably modified by chemical composition of metal oxide and rGO [26]. The main role of rGO in DSSC is to easily capture the photo-induced charge carriers from the dye molecules and to increase the electron transfer rate at the metal oxide semiconductor electrode and the TCO interface thus enhancing the solar cell efficiency. So far rGO and graphene have been successfully incorporated with various metal oxides such as TiO$_2$, ZnO, SnO$_2$, and PbO to form unique hybrid materials for photovoltaic and photoelectric devices.

Typically, graphene is dispersed in an aqueous medium, then precursor of the metal oxide or metal oxide nanoparticles directly mixed with graphene solution. After few minutes of stirring the solution, the colloidal solution was treated under high pressure through hydrothermal method for several hours. Finally, the product was filtered and dried in an oven. Then, the G-metal oxide composite prepared in the paste form for depositing on the conduction glass substrate [27]. A schematic illustration of the synthesis and fabrication process is shown in Figure 11.1.

To improve electron transport path toward photoanode using with and without graphene was represented as shown in Figure 11.2. Deposition of graphene layer between fluorinated tin oxide (FTO) and TiO$_2$ in the DSSC may hinder the direct contact between electrolyte solution and FTO electrode thereby effectively preventing the back-transport of electrons from the FTO electrode to the electrolyte and thus increasing open circuit voltage. The interfacial layer of graphene–TiO$_2$ with low roughness provided better adhesion between the FTO substrate and TiO$_2$ than without any interfacial layer resulting the DSSC efficiency increased from 4.9% to 5.2%.

As discussed above, the increasing DSSCs performance of rGO modified TiO$_2$ photoanode is mainly due to fast electron transport and less charge recombination. Figure 11.3 shows energy level diagram of the DSSCs with graphene-based photo electrode, the generated electrons from dye molecules by incident photons in the CB of TiO$_2$ (~4.2 eV) are accelerated into the external circuit by rGO, since the energy level of rGO (~4.4 eV) is lying between the TiO$_2$ and FTO

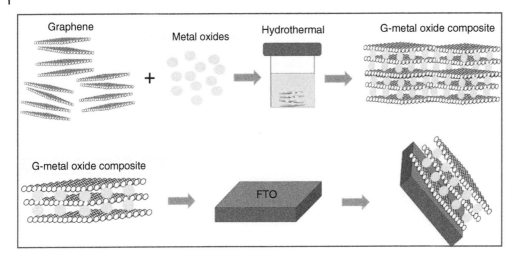

Figure 11.1 Illustrate the process for preparation of graphene or rGO-metal oxide nanocomposites.

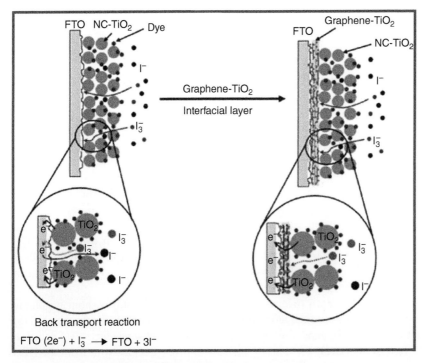

Figure 11.2 Schematic illustration of graphene–TiO$_2$ interfacial layer to prevent back-transport reaction of electrons. *Source:* Kim et al. 2009 [28]. Reproduced with permission of Elsevier.

($\sim$4.7 eV). Due to the suitable energy level of rGO, the captured electrons from TiO$_2$ are merely transferred to FTO without resistance.

There are several methods used for the preparation of the Graphene-TiO$_2$ composite photoanode for improving DSSCs performance. Wei et al. [30] reported improved DSSCs performance using a rGO/TiO$_2$ as photoanode. The rGO/TiO$_2$ thin film is deposited on FTO substrate through the screen-printing method. To improve electron transfer from TiO$_2$

Figure 11.3 Schematic illustration of electron transport in the interface of rGO/TiO₂ photoanodes in DSSC. *Source:* Wei et al. 2016 [29]. Reproduced with permission of The Royal Society of Chemistry.

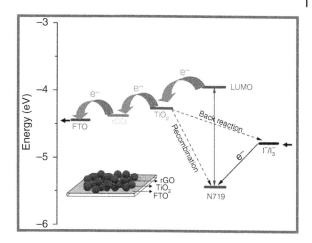

film to FTO substrate, graphene was used as a blocking layer, which led to the suppression of charge recombination rate, the enhancement of electron transport rate, and the electron collection efficiency was increased resulting in higher current density. The graphene was used with different concentrations from 0% to 0.2% with pristine TiO₂ nanoparticles. Moreover, optimum concentration of graphene was selected about 0.1% based on the device performance. The enhancement of device performance (~7.5%) using the graphene with TiO₂ was attributed to suppressed charge recombination rate, the enhanced electron transport rate, and increased the electron collection efficiency. Mehmood [31] have fabricated DSSCs and studied their performance using graphene/TiO₂ nanocomposite as a photoanode and pristine graphene as counter electrode. The graphene-based composite prepared by simply mixing graphene and TiO₂ past. They also prepared different weight percentage of the graphene from 0.04, 0.08, 0.12, and 0.16 (wt%). High efficiency was achieved at about 7.7% and 7.28% for TiO₂-G composite as photoanode with pt as counterelectrode and pristine graphene as counterelectrode, respectively. Figure 11.4a–c shows transmission electron microscopic (TEM) images of TiO₂–graphene composite and energy-dispersive X-ray spectroscopy (EDS) result of the corresponding sample. As can be seen in the figures, the TiO₂ particles were well dispersed on the graphene surface. The enhanced PCE is attributed to an improvement in charge collection efficiency of injected photoelectrons at composite photoanode.

Recently, the impact of preparation methods on TiO₂–rGO nanocomposite for photoanode application of DSSC was investigated by Nouri et al. [32]. The TiO₂–rGO composite was prepared by three different methods: (i) direct mixing of rGO and TiO₂, (ii) rGO mixed with intermediate stage of nanoparticulate titania formation and (iii) the rGO added into the TiO₂ precursor solution at the beginning before TiO₂ formation. For all the above methods, the concentration of rGO used was about 0.4%. In-situ doping of rGO with TiO₂ by addition of rGO at the beginning before titania formation had the most marked beneficial effects on the performance of DSSC devices resulting in the highest internal photocurrent efficiency (IPCE), current density, open circuit voltage, and efficiency of 8.62%. This was mainly attributed to the role of rGO in the TiO₂ films to suppress electron–hole recombination by facilitating electron transport. More recently, graphene/TiO₂ nanocomposite photoanode was used for DSSC with cosensitizers [33]. The dyes concentration was optimized about 0.2 mM N719/0.3 mM RK-1. Moreover, the concentration of the graphene was optimized at 0.1%. The graphene–TiO₂ composite prepared by mixing desired amount of the graphene solution (dispersed in ethanol) with TiO₂ past. The possible formation of the pristine TiO₂, graphene and graphene–TiO₂ structure investigated by simulation method is as shown in Figure 11.5. The improved DSSC performance

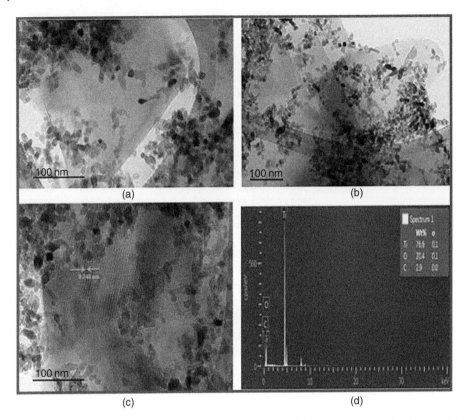

Figure 11.4 (a–c) TEM images with different magnifications of TiO_2–graphene composite and (d) EDS result of TiO_2–graphene nanocomposite. *Source:* Mehmood 2017 [31]. Reproduced with permission of Elsevier.

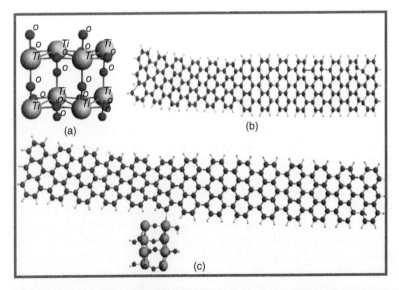

Figure 11.5 Simulated structures of (a) TiO_2, (b) graphene and (c) graphene doped TiO_2. *Source:* Mehmood et al. 2017 [33]. Reproduced with permission of Elsevier.

believed to occur due to increased light absorption, improved light harvesting efficiency and fast electron transport, which ultimately enhance the current density of solar cells. The PCE of 9.45% was achieved using the graphene–TiO$_2$ photoanode.

The nanocomposite of graphene with N-doped TiO$_2$ was also used as a photoanode where different concentration of graphene was mixed with N–TiO$_2$ paste directly and ultrasonicated until well dispersion. A maximum PCE of 9.32% was achieved for the 0.01 ml (graphene prepared 4.8 mg/ml concentration in ethanol) graphene solution dispersed in 0.6 g of the N-TiO$_2$ paste used photoanode DSSC, which is approximately 22% over that of DSSCs with none doped TiO$_2$ photoanode. The impressive PCE improvement is mainly due to increased dye loading, improved electron transport, and reduced recombination processes [34].

Apart from the TiO$_2$–graphene thin sheet, graphene quantum dots (GQDs) decorated on TiO$_2$ is one of the novel methods to improve DSSCs efficiency. Salam et al. [35] fabricated DSSCs using GQD decorated TiO$_2$ as photoanode. The GQD was prepared by top down approach, and the prepared GQD was decorated by electrophoretic filling method. The DSSC with a photoanode based on GQDs–TiO$_2$ exhibited high energy conversion efficiency of 6.22%, than the pure TiO$_2$ photoelectrode (4.81%), accompanied by an increment in both short-circuit photocurrent density and open circuit voltage. The enhancement of the DSSC performance was attributed to the better photogenerated electron transfer ability, reduced charge recombination, increased dye adsorption, and an effective harvesting of visible light by the addition of GQD with TiO$_2$. Table 11.1 shows a summary of the photovoltaic performance with respect to various methods applied for preparation of graphene–TiO$_2$ photoanode.

Mostly, TiO$_2$ and ZnO metal oxide nanostructures and their nanocomposites are being synthesised and investigated for the application as photoanode in DSSCs [56–58]. Hao et al. [59] prepared ZnO/TiO$_2$ nanocomposite via sol–gel method and reported the enhanced optical properties upon proper confinements between TiO$_2$ and ZnO. Chen et al. [60] demonstrated that the generated photocurrent of TiO$_2$/ZnO nanocomposite was largely enhanced with several orders of magnitude higher intensities than that of pristine nano-TiO$_2$ or ZnO. Some other reports also suggested about the enhancement of photocatalytic and optical properties of TiO$_2$/ZnO nanocomposites [61–63].

Except TiO$_2$ nanostructures, other n-type semiconductor metal oxides such as ZnO and SnO$_2$ also possess high electron mobility (115–155 and 240 cm^2/V s), suitable band gap (~3.4 and 3.6 eV) and are earth abundant materials. Because of the suitable band gap and superior electronic properties such as large excitation binding energy (60 meV), higher electron mobility, ZnO is found to be a good candidate for photoanode. Jayabal et al [64] fabricated DSSCs using ZnO photoanode and studied device performance for both with and without presence of graphene. Figure 11.6 shows schematic illustration of the electron transport on both photoanodes under illumination. The highoptical transparency of graphene allows more sunlight to the dye molecules and enhances the light harvesting in the DSSCs, results the DSSC efficiency was increased about two times than pure ZnO photoanode. Moreover, the presence of graphene with ZnO is suppressed by the electron–hole recombination, enhanced charge transport, and increased overall device performance.

To enhance photo-generated electron transport in DSSC, Xu et al. [65] used graphene as a scaffolds layer with ZnO thin film. The authors believed that the excellent electronically conductive graphene scaffolds incorporated into ZnO hierarchically structured nanoparticle (HSN) thin film layer could simultaneously capture and transport the photogenerated electrons injected into ZnO by excited dyes. By this new approach, the electron transport was improved by decreasing internal resistance and electron recombination loss. Ultimately, an impressively high PCE of 5.86% was achieved. The operation principle of the DSSC using graphene scaffold layer with ZnO phoanode thin film is shown in Figure 11.7. The reason

Table 11.1 Summary on the graphene–TiO$_2$ nanocomposite photoanodes.

Material	Synthesis technique	J_{SC} (mA/cm²)	V_{OC} (V)	Fill factor	η (%)	Reference cell (η% without graphene)	Reference
Reduced graphene oxide/TiO$_2$ blocking layer	Screen printing method	15.29	0.74	0.66	7.48	5.76	[30]
Graphene/TiO$_2$ nanocomposite	Dispersion of graphene in TiO$_2$	21.4	0.749	0.49	7.70	6.07	[31]
TiO$_2$–RGO nanocomposite		18.46	726	69.4	8.62	5.82	[32]
Graphene/TiO$_2$ nanocomposite	Ruthenizer and metal free organic photosensitizers	25.123	0.733	0.510	9.45	6.54	[33]
Graphene quantum dots decorated TiO$_2$	Electrospinning and electrophoretic deposition	11.72	0.68	78	6.22	4.81	[35]
TiO$_2$-reduced graphene oxide (TiO$_2$/rGO)	One-step hydrothermal method	16.75	0.74	0.65	7.57	5.52	[36]
Stacked rGO–TiO$_2$	Electrophoretic deposition	15.84	0.63	55.3	6.2	4.4	[37]
1D TiO$_2$-reduced graphene oxide	Hydrothermal method RGO with TiO$_2$ nanotubes and nanowires	10.7	0.78	0.639	5.33	4.0	[38]
TiO$_2$-rGO	Graphitic carbon nitride and reduced graphene oxide modified TiO$_2$	15.28	0.691	55.17	5.83	3.87	[39]
TiO$_2$ at reduced graphene oxide	Hydrothermal method	18.39	0.682	61.2	7.68	4.78	[40]
Nitrogen-reduced graphene oxide (N-rGO)–TiO$_2$	Solvothermal reduction	18.74	0.722	53.08	7.19	6.35	[41]
graphene–TiO$_2$ (from rGO) composites	One-pot hydrothermal synthesis	6.13	0.46	60.27	4.26	3.22	[42]
Graphene (from GO) with TiO$_2$ nanosheets	Solution based dispersion	16.8	0.606	0.567	5.77	4.61	[43]
Graphene-titania (TrGO)	Sol precipitation peptization	9.15	0.644	0.67	3.95	2.89	[44]

Material	Method						Ref.
Graphene (from GO) with TiO_2	Molecular grafting	6.67	0.56	0.45	1.68	0.32	[25]
Graphene–TiO_2	Solvothermal reaction	13.1	0.771	0.72	7.26	4.76	[45]
3D graphene network with TiO_2	CVD then solution-based dispersion	15.4			6.58	4.96	[46]
Graphene-embedded 3D TiO_2	Template infiltration	17.10			7.52	4.86	[47]
Graphene–TiO_2 composites	Simultaneous reduction–hydrolysis technique	13.93	0.70	73.4	7.1	5.3	[48]
Graphene–TiO_2	Solution dispersion	16.69	0.66	0.74	8.15	7.35	[49]
Graphene-P25 (TiO_2)	Ball milling method	10.284	0.616	0.637	5.09	4.43	[50]
TiO_2–graphene composite nanofibers	Electrospinning	16.2	0.71	0.66	7.6	6.3	[23]
Graphene and multiwall carbon nanotubes MWCNTs hybrid with TiO_2	Solution based dispersion	11.27	0.78	0.70	6.11	4.54	[51]
TiO_2/graphene	Single-layered graphene oxide sheets	14.64	0.66	73.02	7.06	5.89	[52]
Graphene integrated TiO_2	Hydrothermal synthesis	16.5	0.703	65.2	7.56	5.21	[53]
TiO_2 – N, F and S, co-doped graphene QDs	Microwave treatment in an ionic liquid	22.6	0.79	0.7	11.7	7.48	[54]
TiO_2-N-doped graphene (NDG)	Atmospheric pressure chemical vapor deposition (APCVD)	5.767	0.655	0.443	1.673	1.013	[55]

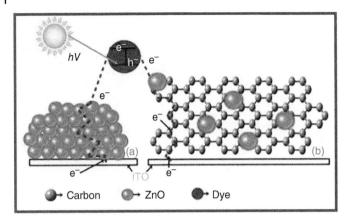

Figure 11.6 Schematic representation of electron transport mechanism in DSSCs for (a) ZnO and (b) ZnO–graphene photoanode. *Source:* Jayabal et al. 2014 [64]. Reproduced with permission of Springer Nature.

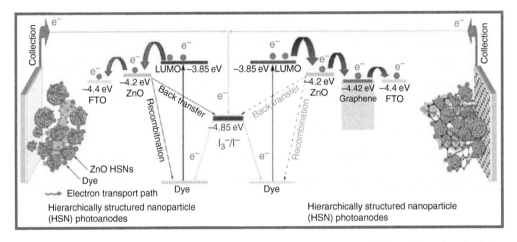

Figure 11.7 Schematic illustration of energy diagram and operation principle of the DSSCs with ZnO HSN photoanode and graphene/ZnOHSN photoanode, respectively. The diagram shows electron injection from excited dye into ZnO nanoparticles and then the electron transport of to the collection electrode. *Source:* Xu et al. 2013 [65]. Reproduced with permission of The American Chemical Society.

behind the improvement of performance shows that favorable band gap matching, graphene has a work function (−4.42 eV versus vacuum) similar to that of FTO (−4.4 eV versus vacuum).

More recently, GO–SnO$_2$ hybrid nanocomposite as photoanode was used in DSSC and the PCE investigated [66]. Remarkably, DSSC based on GO-SnO$_2$ hybrid nanocomposite photoanode showed PCE of 8.3% and it is about 12% higher than that of pure TiO$_2$ photoanode. The GO-SnO$_2$ nanocomposite prepared by directly mixing of GO (2 mg) with SnO$_2$ powder (2 mg) in 2 ml H$_2$O and the mixture was sonicated for one hour. A current density (J_{sc}) of 16.67 mA/cm^2, open circuit voltage (V_{oc}) of 0.77 V, and FF of 0.65 were achieved. Table 11.2 shows a summary of the photovoltaic performance with respect to various methods applied for the preparation of the graphene- and other metal oxides as photoanode. The obtained results clearly indicate that the hybrid nanocomposite is an appropriate and alternative photoanode material for DSSCs.

Table 11.2 Summary of the graphene-metal oxide (except TiO_2) nanocomposite photoanodes.

Material	Synthesis technique	J_{SC} (mA/cm²)	V_{OC} (V)	Fill factor	η (%)	Reference cell (η% without graphene)	References
Graphene quantum dots (GQD) functionalized ZnO photoanodes	In-situ solution	10.1	0.48	0.507	2.45	0.817	[67]
SnO₂-doped graphene (GO/SnO₂/TiO₂)	Ultra-sonication mixing	16.67	0.77	0.65	8.39	6.27	[66]
Poly[3-(2-hydroxyethyl)-2,5-thienylene] grafted reduced graphene oxide	In-situ polymerization	7.5	0.61	0.668	3.06	2.66	[68]
Graphene scaffolds incorporated into ZnO hierarchically structured nanoparticles	In-situ reduction of GO incorporated into ZnO HSN films	10.89	0.66	47.60	3.19	2.31	[65]
GO-ZnO	Aligned ZnO nanorods on hot filament chemical vapor deposition (HFCVD) grown graphene oxide	6.56	0.704	0.54	2.5	1.18	[17]
ZnO/graphene nanocomposite		0.00160	0.70	0.53	1.50	1.04	[64]

11.3 Conclusion and Remarks

This chapter highlighted the research on the use of graphene and graphene-based nanocomposite materials in the photoanode of DSSCs. Graphene-based materials, such as pristine graphene, graphene oxide, and reduced graphene oxide and graphene quantum dots possess attractive properties for various components of DSSC photoanode. The graphene-based nanocomposite materials showed different functionalities such as electron conducting layer, transparent conducting electrode, and sensitizer in the DSSC photoanode. When combined with other nanomaterials of metal, metal oxides, metal sulfides, etc., to form a nanocomposite due to synergetic effects few interesting properties are also emerged which enhanced the photovoltaic performances of DSSCs. From the above discussion, it is clear that graphene and its nanocomposites have the properties that are well suited for the purpose of making high-performance photoanode for DSSC. But, there is still lake of extensive studies on the graphene-based nanocomposite materials in the DSSC photoanode as compared to the applications toward counter electrode. Hence, an in-depth research on modification of DSSC photoanode employing graphenebased materials needs to be carried out. Particularly, the synthesis protocols for graphene-based materials with tunable morphology and adjustable properties and their better incorporation into other components to enhance the photoanode performance. On the other hand, loading of graphene-based material on to the host also greatly influence the performance of the cells, hence meticulous calculation on the amount of materials and characterization of their physical and chemical properties can play important roles in solving many issues limiting the performance of DSSC photoanodes. When graphene is used as a transparent conducting electrode, the major issue is to maintain the transparency. The unsatisfactory transparency is caused by multiple-layer graphene stacking and the high sheet resistance due to surface defects and oxidization. Better processing procedures are necessary to overcome these problems and enhance the chance for graphene to be used as feasible alternatives to TCOs in the DSSC photoanode. The strength and flexibility of graphene outperforms other flexible candidates. Further modification of graphene nanosheets may create a new generation of flexible electrodes. Considerable progress has been made on the preparation of graphene transparent conducting electrodes at the laboratory level; however, it remains a challenge to cost-effectively produce high-quality graphene on an industrial scale for the practical use of graphene in transparent conducting electrodes.

References

1 Mathew, S., Yella, A., Gao, P. et al. (2014). Dye-sensitized solar cells with 13% efficiency achieved through the molecular engineering of porphyrin sensitizers. *Nat. Chem.* 6: 242–247.

2 Kakiage, K., Aoyama, Y., Yano, T. et al. (2015). Highly-efficient dye-sensitized solar cells with collaborative sensitization by silyl-anchor and carboxy-anchor dyes. *Chem. Commun.* 51: 15894–15897.

3 Redmond, G., Fitzmaurice, D., and Graetzel, M. (1994). Visible light sensitization by cis-bis(thiocyanato) bis (2, 2′-bipyridyl-4, 4′-dicarboxylato) ruthenium(II) of a transparent nanocrystalline ZnO film prepared by sol-gel techniques. *Chem. Mater.* 6: 686–691.

4 Dinh, N.N., Bernard, M.-C., Hugot-Le Goff, A. et al. (2006). Photoelectrochemical solar cells based on SnO_2 nanocrystalline films. *C.R. Chim.* 9: 676–683.

5 Ou, J.Z., Rani, R.A., Ham, M.-H. et al. (2012). Elevated temperature anodized Nb_2O_5: a photoanode material with exceptionally large photoconversion efficiencies. *ACS Nano* 6: 4045–4053.

6 Gholamrezaei, S., Niasari, M.S., Dadkhah, M., and Sarkhosh, B. (2016). New modified sol–gel method for preparation $SrTiO_3$ nanostructures and their application in dye sensitized solar cells. *J. Mater. Sci. Mater. Electron.* 27: 118–125.

7 Tan, B., Toman, E., Li, Y., and Wu, Y. (2007). Zinc stannate (Zn_2SnO_4) dye-sensitized solar cells. *J. Am. Chem. Soc.* 129: 4162–4163.

8 Hara, K., Zhao, Z.-G., Cui, Y. et al. (2011). Nanocrystalline electrodes based on nanoporous-walled WO_3 nanotubes for organic-dye-sensitized solar cells. *Langmuir* 27: 12730–12736.

9 Ahmad, M.S., Pandey, A.K., and Rahim, N. (2017). Advancements in the development of TiO$_2$ photoanodes and its fabrication methods for dye sensitized solar cell (DSSC) applications. A review. *Renewable Sustainable Energy Rev.* 77: 89–108.

10 Novoselov, K.S., Geim, A.K., Morozov, S.V. et al. (2004). Electric field effect in atomically thin carbon films. *Science* 306: 666–669.

11 Zhou, K., Zhu, Y., Yang, X., and Li, C. (2010). One-pot preparation of graphene/Fe$_3$O$_4$ composites by a solvothermal reaction. *New J. Chem.* 34: 2950–2955.

12 Stankovich, S., Dikin, D.A., Piner, R.D. et al. (2007). Synthesis of graphene-based nanosheets via chemical reduction of exfoliated graphite oxide. *Carbon* 45: 1558–1565.

13 Wang, X., Zhi, L., and Mullen, K. (2008). Transparent, conductive graphene electrodes for dye-sensitized solar cells. *Nano Lett.* 8: 323–327.

14 Lee, J.S., Ahn, H.J., Yoon, J.C., and Jang, J.H. (2012). Three-dimensional nano-foam of few layer graphene grown by CVD for DSSC. *Phys. Chem. Chem. Phys.* 14: 7938–7943.

15 Yen, M.Y., Hsieh, C.K., Teng, C.C. et al. (2012). Metal-free, nitrogen-doped graphene used as a novel catalyst for dye-sensitized solar cell counter electrodes. *RSC Adv.* 2: 2725–2728.

16 Xue, Y., Liu, J., Chen, H. et al. (2012). Nitrogen-doped graphene foams as metal-free counter electrodes in high-performance DSSCs. *Angew. Chem. Int. Ed.* 51: 12124–12127.

17 Ameen, S., Akhtar, M.S., Song, M., and Shin, H.S. (2012). Vertically aligned ZnO nanorods on hot filament chemical vapor deposition grown graphene oxide thin film substrate: solar energy conversion. *ACS Appl. Mater. Interfaces* 4: 4405–4412.

18 Fang, X.L., Li, M.Y., Guo, K.M. et al. (2014). Graphene quantum dots optimization of dye-sensitized solar cells. *Electrochim. Acta* 137: 634–638.

19 Lee, E., Ryu, J., and Jang, J. (2013). Fabrication of graphene quantum dots via size-selective precipitation and their application in up-conversion-based DSSCs. *Chem. Commun.* 49: 9995–9997.

20 Zhu, S.J., Song, Y.B., Zhao, X.H. et al. (2015). The photoluminescence mechanism in carbon dots (graphene quantum dots, carbon nanodots, and polymer dots): current state and future perspective. *Nano Res.* 8: 355–381.

21 Eshaghi, A. and Aghaei, A.A. (2015). Effect of TiO$_2$–graphene nanocomposite photoanode on dye-sensitized solar cell performance. *Bull. Mater. Sci.* 38: 1177–1182.

22 Wang, H., Leonard, S.L., and Hu, Y.H. (2012). Promoting effect of graphene on dye-sensitized solar cells. *Ind. Eng. Chem. Res.* 51: 10613–10620.

23 Madhavan, A.A., Kalluri, S., and Chacko, D.K.e.a. (2012). Electrical and optical properties of electrospun TiO$_2$-graphene composite nanofibers and its application as DSSC photo-anodes. *RSC Adv.* 2: 13032–13037.

24 Sun, S.R., Gao, L., and Liu, Y.Q. (2010). Enhanced dye-sensitized solar cell using graphene-TiO$_2$ photoanode prepared by heterogeneous coagulation. *Appl. Phys. Lett.* 96: 083113.

25 Tang, Y.B., Lee, C.S., Xu, J. et al. (2010). Incorporation of graphenes in nanostructured TiO$_2$ films via molecular grafting for dye-sensitized solar cell application. *ACS Nano* 4: 3482–3488.

26 Castro Neto, A.H., Peres, N.M.R., Novoselov, K.S., and Geim, A.K. (2009). The electronic properties of graphene. *Rev. Mod. Phys.* 81 (1): 109–162.

27 Huang, X., Yin, Z., Wu, S. et al. (2011). Graphene-based materials: synthesis, characterization, properties, and applications. *Small* 7: 1876–1902.

28 Kim, S.R., Parvez, M.K., and Chhowalla, M. (2009). UV-reduction of graphene oxide and its application as an interfacial layer to reduce the back-transport reactions in dye-sensitized solar cells. *Chem. Phys. Lett.* 483: 124–127.

29 Wei, L., Chen, S., Yang, Y. et al. (2016). Reduced graphene oxide modified TiO_2 semiconductor materials for dye-sensitized solar cells. *RSC Adv.* 6: 100866–100875.

30 Wei, L., Wang, P., Yang, Y. et al. (2017). Enhanced performance of dye sensitized solar cells by using a reduced graphene oxide/TiO_2 blocking layer in the photoanode. *Thin Solid Films* 639: 12–21.

31 Mehmood, U. (2017). Efficient and economical dye-sensitized solar cells based on graphene/TiO2 nanocomposite as a photoanode and graphene as a Pt-free catalyst for counter electrode. *Org. Electron.* 42: 187–193.

32 Nouri, E., Mohammadi, M.R., and Lianos, P. (2016). Impact of preparation method of TiO_2-rGO nanocomposite photoanodes on the performance of dye-sensitized solar cells. *Electrochim. Acta* 219: 38–48.

33 Mehmood, U., Ahmad, S.H.A., Khan, A.H., and Qaiser, A.A. (2018). Co-sensitization of graphene/TiO_2 nanocomposite thin films with ruthenizer and metal free organic photosensitizers for improving the power conversion efficiency of dye-sensitized solar cells (DSSCs). *Sol. Energy* 170: 47–55.

34 Kim, S.B., Park, J.Y., Kim, C.S. et al. (2015). Effects of graphene in dye-sensitized solar cells based on nitrogen-doped TiO_2 composite. *J. Phys. Chem. C* 119 (29): 16552–16559.

35 Salam, Z., Vijayakumar, E., Subramania, A. et al. (2015). Graphene quantum dots decorated electrospun TiO_2 nanofibers as an effective photoanode for dye sensitized solar cells. *Sol. Energy Mater. Sol. Cells* 143: 250–259.

36 Zhang, H., Lv, Y., Yang, C. et al. (2018). One-step hydrothermal fabrication of TiO_2/reduced graphene oxide for high-efficiency dye-sensitized solar cells. *J. Electron. Mater.* 47: 1630–1637.

37 Peiris, D.S.U., Ekanayake, P., and Petra, M.I. (2018). Stacked rGO–TiO_2 photoanode via electrophoretic deposition for highly efficient dyesensitized solar cells. *Org. Electron.* 59: 399–405.

38 Cai, H., Li, J., Xu, X. et al. (2017). Nanostructured composites of one-dimensional TiO_2 and reduced graphene oxide for efficient dye-sensitized solar cells. *J. Alloys Compd.* 697: 132–137.

39 Lv, H., Hu, H., Cui, C. et al. (2017). Enhanced performance of dye-sensitized solar cells with layered structure graphitic carbon nitride and reduced graphene oxide modified TiO_2 photoanodes. *Appl. Surf. Sci.* 422 (2017): 1015–1021.

40 Cheng, G., Akhtar, M.S., Yang, O., and Stadler, F.J. (2013). Novel Preparation of anatase TiO_2@reduced graphene oxide hybrids for high-performance dye-sensitized solar cells. *ACS Appl. Mater. Interfaces* 5: 6635–6642.

41 Xiang, Z., Zhou, X., Wan, G. et al. (2014). Improving energy conversion efficiency of dye-sensitized solar cells by modifying TiO_2 photoanodes with nitrogen-reduced graphene oxide. *ACS Sustainable Chem. Eng.* 2: 1234–1240.

42 Anjusree, G.S., Nair, A.S., Nair, S.V., and Vadukumpully, S. (2013). One-pot hydrothermal synthesis of TiO_2/graphene nanocomposites for enhanced visible light photocatalysis and photovoltaics. *RSC Adv.* 3: 12933.

43 Fan, J., Liu, S., and Yu, J. (2012). Enhanced photovoltaic performance of dye-sensitized solar cells based on TiO_2 nanosheets/graphene composite films. *J. Mater. Chem.* 22: 17027–17036.

44 Siddick, S.Z., Lai, C.W., and Juan, J.C. (2018). An investigation of the dye-sensitized solar cell performance using graphene-titania (TrGO) photoanode with conventional dye and natural green chlorophyll dye. *Mater. Sci. Semicond. Process.* 74: 267–276.

45 He, Z., Phan, H., Liu, J. et al. (2013). Understanding TiO_2 size-dependent electron transport properties of a graphene-TiO_2 photoanode in dye-sensitized solar cells using conducting atomic force microscopy. *Adv. Mater.* 25: 6900–6904.

46 Tang, B., Hu, G., Gao, H., and Shi, Z. (2013). Three-dimensional graphene network assisted high performance dye sensitized solar cells. *J. Power Sources* 234: 60–68.

47 Kim, H., Yoo, H., and Moon, J.H. (2013). Graphene-embedded 3D TiO2 inverse opal electrodes for highly efficient dye-sensitized solar cells: morphological characteristics and photocurrent enhancement. *Nanoscale* 5: 4200–4204.

48 Chen, L., Zhou, Y., Tu, W. et al. (2013). Enhanced photovoltaic performance of a dye-sensitized solar cell using graphene–TiO_2 photoanode prepared by a novel in situ simultaneous reduction-hydrolysis technique. *Nanoscale* 5: 3481.

49 Sharma, G.D., Daphnomili, D., Gupta, K.S.V. et al. (2013). Enhancement of power conversion efficiency of dye-sensitized solar cells by co-sensitization of zinc-porphyrin and thiocyanate-free ruthenium(II)–terpyridine dyes and graphene modified TiO_2 photoanode. *RSC Adv.* 3: 22412.

50 Fang, X., Li, M., Guo, K. et al. (2012). Improved properties of dye-sensitized solar cells by incorporation of graphene into the photoelectrodes. *Electrochimica Acta* 65: 174–178.

51 Yen, M., Hsiao, M., Liao, S. et al. (2011). Preparation of graphene/multi-walled carbon nanotube hybrid and its use as photoanodes of dye-sensitized solar cells. *Carbon* 49: 3597–3606.

52 Kusumawati, Y., Koussi-Daoud, S., and Pauporte, T. (2016). TiO_2/graphene nanocomposite layers for improving the performances of dye-sensitized solar cells using a cobalt redox shuttle. *J. Photochem. Photobiol., A Chem.* 329: 54–60.

53 Liu, L., Zeng, B., Meng, Q. et al. (2016). Titanium dioxide/graphene anode for enhanced charge-transfer in dye-sensitized solar cell. *Synth. Met.* 222: 219–223.

54 Kundu, S., Sarojinijeeva, P., Karthick, R. et al. (2017). Enhancing the efficiency of DSSCs by the modification of TiO_2 photoanodes using N, F and S, co-doped graphene quantum dots. *Electrochim. Acta* 242: 337–343.

55 Joseph, E., Singh, B.S.M., Mohamed, N.M. et al. (2016). Investigating the performance of nitrogen-doped graphene photoanode in dyesensitized solar cells. *AIP Conf. Proc.* 1787: 040002-1–040002-6.

56 Samsuri, S.A.M., Rahman, M.Y.A., and Umar, A.A. (2017). Comparative study of the properties of TiO_2 nanoflower and TiO_2-ZnO composite nanoflower and their application in dye-sensitized solar cells. *Ionics* 23 (7): 1897–1902.

57 Liu, R., Yang, W.D., Qiang, L.S., and Liu, H.Y. (2012). Conveniently fabricated heterojunction ZnO/TiO_2 electrodes using TiO_2 nanotube arrays for dye-sensitized solar cells. *J. Power Sources* 220: 153–159.

58 Rajkumar, N., Kanmani, S.S., and Ramachandran, K. (2011). Performance of dye-sensitized solar cell based on TiO_2: ZnO nanocomposites. *Adv. Sci. Lett.* 4 (2): 627–633.

59 Hao, Y., Cheng-Chun, T., and Shou-Shan, F. (2001). Optical absorption of sol-gel derived ZnO/TiO_2 nanocomposite films. *Chin. Phys. Lett.* 18 (11): 1520.

60 Chen, D., Zhang, H., Hu, S., and Li, J. (2008). Preparation and enhanced photoelectrochemical performance of coupled bicomponent ZnO-TiO_2 nanocomposites. *J. Phys. Chem. C* 112 (1): 117–122.

61 Xiao, F.X. (2012). Construction of highly ordered ZnO-TiO_2 nanotube arrays (ZnO/TNTs) heterostructure for photocatalytic application. *ACS Appl. Mater. Interfaces* 4 (12): 7055–7063.

62 Mohamed, S.H., El-Hagary, M., and Althoyaib, S. (2012). Photocatalytic and optical properties of nanocomposite TiO_2-ZnO thin films. *Eur. Phys. J. Appl. Phys.* 57 (2): 20301.

63 Cheng, C., Amini, A., Zhu, C. et al. (2014). Enhanced photocatalytic performance of TiO_2-ZnO hybrid nanostructures. *Sci. Rep.* 4: 4181.

64 Jayabal, P., Gayathri, S., Sasirekha, V. et al. (2014). Preparation and characterization of ZnO/graphene nanocomposite for improved photovoltaic performance. *J. Nanopart. Res.* 16 (11): 2640–2647.

65 Xu, F., Chen, J., Wu, X. et al. (2013). Graphene scaffolds enhanced photogenerated electron transport in ZnO photoanodes for high-efficiency dye-sensitized solar cells. *J. Phys. Chem. C* 117: 8619–8627.

66 Sasikumar, R., Chen, T., Chen, S. et al. (2018). Developing the photovoltaic performance of dye-sensitized solar cells (DSSCs) using a SnO_2-doped graphene oxide hybrid nanocomposite as a photo-anode. *Opt. Mater.* 79: 345–352.

67 Sehgal, P. and Narula, A.K. (2018). Enhanced performance of porphyrin sensitized solar cell based on graphene quantum dots decorated photoanodes. *Opt. Mater.* 79: 435–445.

68 Chatterjee, S., Patra, A.K., Bhaumik, A., and Nandi, A.K. (2013). Poly[3-(2-hydroxyethyl)-2,5-thienylene] grafted reduced graphene oxide: an efficient alternate material of TiO_2 in dye sensitized solar cells. *Chem. Commun.* 49: 4646–4648.

12

Graphitic Carbon Nitride Based Nanocomposites as Photoanodes

T.S. Shyju[1,2,3], S. Anandhi[4], P. Vengatesh[3], C. Karthik Kumar[3], and M. Paulraj[2]

[1] Sathyabama Institute of Science and Technology, Centre for Nanoscience and Nanotechnology, Chennai, India
[2] University of Concepcion, Department of Physics, Faculty of Physical and Mathematical Sciences, Concepcion, Chile
[3] Sathyabama Institute of Science and Technology, Centre of Excellence for Energy Research, India
[4] Jeppiaar Maamallan Engineering College, Department of Physics, Sriperumbudur, Chennai, India

12.1 Introduction

Owing to their unique properties, nanocomposites, nanostructures, and nanomaterials are playing a key role in energy conversion and in energy storage applications. In recent years, carbon-based nanostructures are used to meet the future energy demands. Nanostructured carbon nitrides (C_3N_4) are very attractive candidates for energy-based devices due to its high hardness, low friction coefficient, and steadfast chemical inertness. It has a great potential in solving the issues related to energy and environmental applications. Graphitic carbon nitride (g-C_3N_4) is one among the carbon-based nanostructures which has attracted enormous attention in green technologies for arresting solar energy, energy storage, supercapacitor, fuel cells, electrocatalysis, and environmental remediation as well as for electronic and composite industry. g-C_3N_4 is a well-known polymeric materials mainly consisting of carbon, nitrogen, and is also one of the oldest material discovered in 1843. It is considered as an artificial polymer in the scientific literature [1]. Its structure is shown in Figure 12.1.

The history of C_3N_4 polymers has the embryonic form of melon, which is a linear polymer chain interconnected with tri-*s*-triazine through secondary nitrogen. It was first developed by Berzelius and named by Liebig [2]. Generally, it has seven phases and among these phases; graphitic carbon nitride is found to be the most stable phase in ambient conditions. g-C_3N_4 architecture shows an aromatic plane constructed from triazine and has a metal-free conjugated semiconductor photocatalyst for H_2 evolution and was first reported by Wang et al. [3] in 2009. Over all, polymeric materials have a general feature of low thermal stability and readily undergoes oxidization in oxidizing atmosphere, and this makes them unsuitable for practical applications. But g-C_3N_4 is found to have good thermal stability, is also environment friendly, and does not undergo degradation. In 2010, Junjiang Zhu et al. [4] reported the thermal stability of g-C_3N_4 at different conditions with the help of TGA curves, where it was found to be stable up to 600 °C in air, N_2 or O_2 atmosphere. Several researchers suggested that g-C_3N_4 crystals withstands high temperature (up to 600 °C), even in oxidizing atmosphere that leads to its potential application [5]. Moreover, g-C_3N_4 possess excellent chemical stability in different solvents such as alcohol, water, dimethylformamide, and toluene. It is reported that no changes were observed in the infrared (IR) spectra of g-C_3N_4 after being dipped into the abovementioned solvents even for over a period of month, this confirms its excellent durability and its stability.

Interfacial Engineering in Functional Materials for Dye-Sensitized Solar Cells, First Edition.
Edited by Alagarsamy Pandikumar, Kandasamy Jothivenkatachalam and Karuppanapillai B. Bhojanaa.

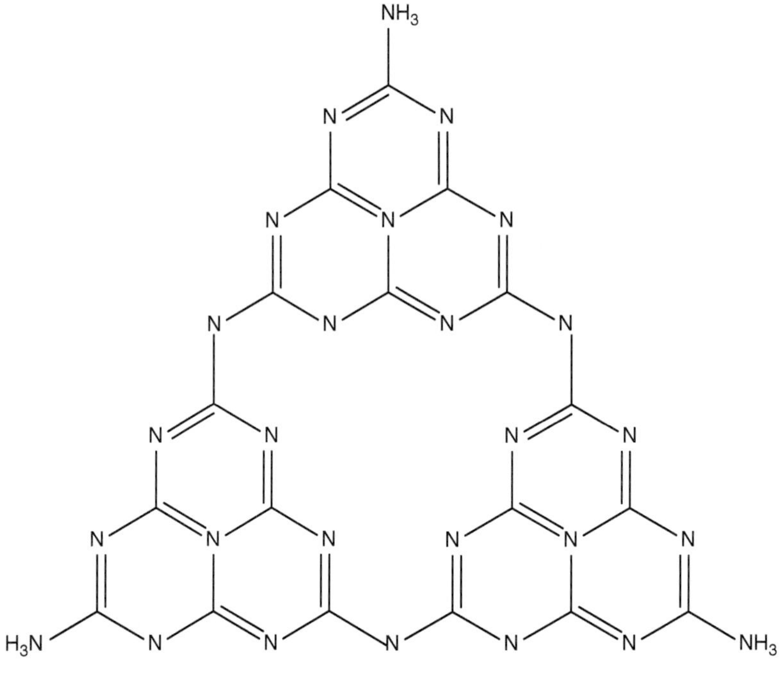

Figure 12.1 Overview of g-C$_3$N$_4$ structure.

12.2 Importance of Graphitic Carbon Nitride

In recent years, two-dimensional (thin film) g-C$_3$N$_4$ has provoked to be an interdisciplinary research fascination among the scientific community for its gorgeous properties such as electronic band structures, electrical properties and visible-light absorption, and high chemical and thermal stability. Various reports have been made so far on the discussion about the interfacial charge transfer of g-C$_3$N$_4$, physicochemical stability, and heterostructure properties, nano-composites employed as a photoanode in solar cell. Hence, this chapter is discussed in detail about the current progress in various dimensions of g-C$_3$N$_4$-based nanocomposites such as g-C$_3$N$_4$ nanotube, g-C$_3$N$_4$ nanofiber, g-C$_3$N$_4$ nanosheet, g-C$_3$N$_4$ nanosphere, owing to the enormous interest and excellent progress on the unitilization of g-C$_3$N$_4$-based nanocomposites. Most of the authors have discussed about the application of g-C$_3$N$_4$ in environmental applications such as pollutant degradation, heavy metal ions reduction, bacterial disinfection, and CO$_2$ reduction. However, this chapter discusses especially about the application of g-C$_3$N$_4$ in solar cells for energy conversion.

It is well-known that g-C$_3$N$_4$ is an analog of graphite which possess tectonic structure with repeating units of tri-*s*-triazine (shown in Figure 12.2.) The formation of g-C$_3$N$_4$ structure from various reactants with its processing temperature is also described in Figure 12.3. It is a yellow-colored material having good optical absorption observed using UV–Vis and photoluminescence spectroscopic studies, with a band gap of ~2.77 eV which is a unique optoelectronic properties. Moreover, the conduction band minimum (−1.12 eV versus normal hydrogen electrode (NHE)) of g-C$_3$N$_4$ is extremely negative, so the photogenerated electrons should have high reduction capability. Theoretical calculations showed that the absorption edge depends on structure, packing, and ad-atoms. Literature data confirms an optical absorption edge at 430 nm. In general, most common metal oxides such as ZnO, TiO$_2$, and Fe$_2$O$_3$

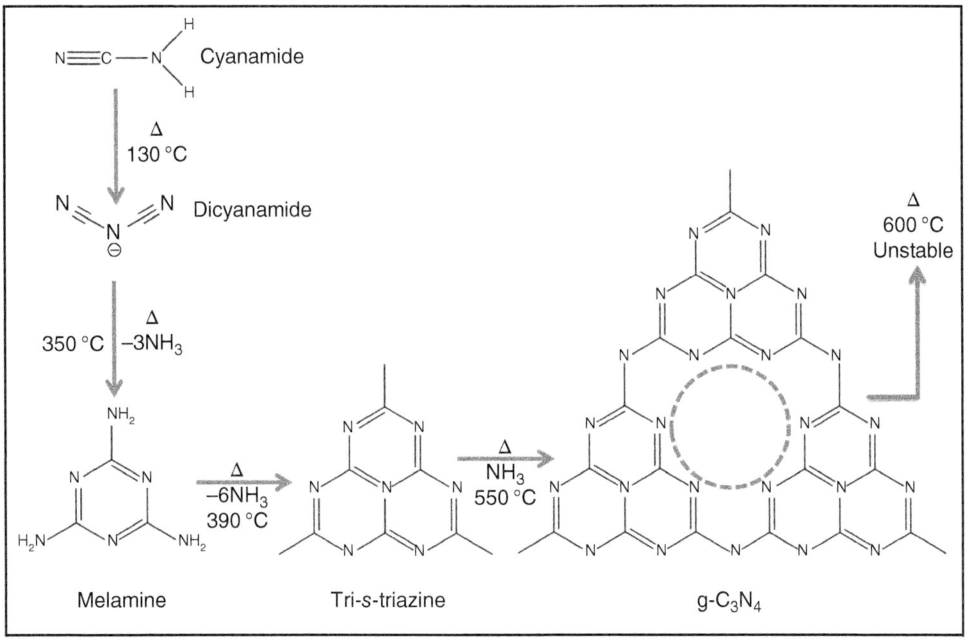

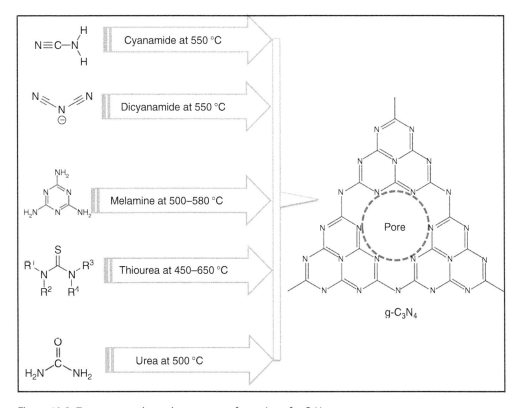

Figure 12.2 Phase transition of g-C$_3$N$_4$.

Figure 12.3 Temperature-dependent structure formation of g-C$_3$N$_4$.

are very promising aspirants, because they have a suitable band gap for different application (solar cell, water splitting, etc.). Grätzel et al. studied about the sensitizing semiconductors via immobilization of dyes to enhance the quantum yield of photogenerated electrons.

Carbon-based materials generally have a high recombination rate of photo-generated electrons (e^-) and holes (h^+) or exciton (e^-–h^+) pairs and small specific surface area. However, in pure g-C_3N_4, the radiative recombination rate of photogenerated exciton pairs is far above the ground state resulting in high fluorescence yield and thereby limiting its beneficial effects in photovoltaic (PV) device performance. One of the practical approaches to reduce the recombination rate of photogenerated excitons by coupling g-C_3N_4 with semiconductor materials such as: ZnO, CdS, ZnS, TiO_2, WO_3, $SrTiO_3$, with well-aligned band edge positions. The movement of excited state electrons from g-C_3N_4 to these semiconductors is more favored, since the conduction band edge of these semiconductors lies low as compared to the conduction band of g-C_3N_4, which in turn leads to the reduction in radiative annihilation of photogenerated excitons. Especially, tunable optical band gap and charge transport properties of ZnO and TiO_2 is well-known to design nanocomposite with g-C_3N_4 for solar energy harvesting. Optical and charge transport properties of oxides can be tuned by tailoring into various architectures such as nanotubes, nanorods, nanoparticles, nanosheets, nanodisks, nanospheres, etc. Jiao Men et al. [6] reported that g-C_3N_4 modified TiO_2 mesoporous photoelectrodes were used as photoanodes for quantum dot-sensitized solar cells (QDSSCs) and photodegradation of organic pollutant rhodamine B.

Surface to volume ratio of nanomaterials and their surface modifications using various chemical and bioconjugate reactions have enthused the interface among chemistry, materials science, and biology toward biomedical science. As discussed earlier, functionalizations or postfunctionalizations of g-C_3N_4 are often carried out to extend its applications to the broader research areas. For example, g-C_3N_4 can be doped in situ with boron [7], fluorine [8], sulfur [9], phosphor [10], etc. Various research articles deals with chemical dopant and enhancement of semiconductor electronic structures as well as the surface properties of the material by tuning the electronic structure, its ionic conductivity leading to improvement in performance. For example, Yuanjian Zhang et al. reported, doping with phosphor can improve the electronic conductivity by 4 orders of magnitude and photocurrent generation by 5 orders of magnitude, compared to the pure g-C_3N_4 [10]. This is an important step toward the photovoltaic applications of g-C_3N_4 thin layers. Many interesting physiochemical properties of g-C_3N_4 have also been documented [11]. It should be noted that [12, 13] surface modifications or functionalization are very useful for modifying the surface properties of g-C_3N_4, and these are important in concerning the optimization of its catalytic performance. Doping g-C_3N_4 retains most of the structural features, yet electronic features can be changed. To date, the combination of g-C_3N_4/TiO_2 nanocomposites presents a feasible and inspiring route toward improved charge separation in the electron transfer process.

12.3 Photoanodes for DSSC

Dye sensitized solar cell (DSSC) is a third-generation mesoporous photovoltaic device that has gained a great attention over the past two decades as an inexpensive alternate to silicon photovoltaics. The third-generation photovoltaic based on nanostructured materials are made of purely organic and a mixture of organic–inorganic components. These cells utilize some of the technologically important nano-compounds and are classified as nanocrystal, polymer, dye sensitized, perovskite, and concentrated solar cells. Figure 12.4 shows the overview of third-generation solar cells. Among them, DSSCs have been intensively investigated because

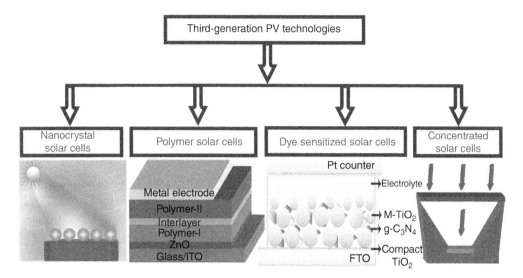

Figure 12.4 Overall view of third-generation photovoltaic devices.

they have shown excellent optical and electrical properties of semiconductor when quantum dots (QDs) were used. However, the power conversion efficiency (PCE) of DSSC is still far behind its theoretical value mainly due to the loss and recombination of photogenerated electrons during transport in circuit. A typical structure of DSSC includes a photoanode, a counter electrode, and an electrolyte solution. The photoanode as the crucial part of cell affords electron transport between sensitizer and external circuit. For enhancing PCE, numerous reports have been dedicated to the morphological control of TiO_2 (nanowires, nanotubes, nanorods, and nanoparticles), the most important semiconductor which is being utilized as photoanode in the configuration of DSSCs [6]. Among them, single crystalline TiO_2 nanorod array is one of the most desirable nanostructures for preparing photoanode in DSSC, due to its effective charge transfer property as well as excellent light harvesting ability. Also doping on TiO_2 lattice is an effective approach to improve the PCE of DSSC. In addition, the preparation of photoanodes based on nanocomposites plays a crucial role in DSSC in order to improve its PCE.

Semiconductor quantum dots or dye molecules harvest photon to produce electron–hole pairs, collects the electron to use in external circuit. Similarly, the photoanodes plays a critical role in collecting the electrons and transporting it to the transparent conducting oxide (TCO) substrates while blocking the holes and thereby reducing the recombination of photogenerated carriers. However, the control measures have been taken in order to avoid recombination and in turn to improve PCE. Introducing a suitable block layer outside of the photoanode material is one such method employed to reduce recombination and to promote the separation of exciton pairs which also forms type II band alignment between TiO_2 and such block layer material. $g\text{-}C_3N_4$ would be one such material as block layer due to its suitable behavior and also easy bonding with different class of materials as described in Figure 12.5.

Qiqian Gao et al. [14] prepared {001}-faceted TiO_2 photoanode via a one-step hydrothermal method and also prepared $g\text{-}C_3N_4$ by simple solid-state heat treatment. They modified the {001}-faceted TiO_2 photoanodes which is used as backbone scaffold of photoanodes and compared the PCE of both cells. The device with $g\text{-}C_3N_4/TiO_2$ nanocomposites as photoanode exhibited a maximum PCE of 4.23% with an open circuit voltage of 0.58 V, a short circuit current density of 15.5 mA/cm^2, and a fill factor (FF) of 0.48, giving 24.7% enhancement in the

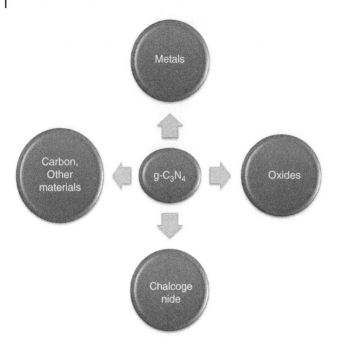

Figure 12.5 Formation of g-C$_3$N$_4$ nanocomposites using various materials.

cell efficiency. Holly F. Zarick et al. [15] demonstrated that the light trapping is enhanced in DSSCs with hybrid bimetallic gold core/silver shell nanostructures, enhancing the e$^-$/h$^+$ pairs generated in N719. Therefore, it is inferred that the electron injection into TiO$_2$ conduction band is accelerated by the metal nanostructures resulting in the minimal recombination. However, if the bimetallic nanostructures are mixed homogeneously throughout the active layer, the interparticle coupling between adjacent nanostructures is expected to happen that may promote long-range surface plasmon polariton modes, which would also contribute to the overall enhancements.

Owing to surface plasmon resonance (SPR) of Au/g-C$_3$N$_4$ nanocomposites, it exhibit high response to visible-light. On the other hand, the SPR effect of Au nanoparticle causes the intense local electromagnetic fields, which can speed the formation rate of holes and electrons within g-C$_3$N$_4$. Additionally, Fermi level of noble metal facilitates the separation of electrons and holes, which in turn enhances the quantum efficiency of g-C$_3$N$_4$, due to the intimate combination of the noble metal/g-C$_3$N$_4$ nanohybrids. Moreover, the transfer of electrons shifts the Fermi level to more negative potential, thereby improves the reducibility of electrons in the Fermi level close to the conduction band of g-C$_3$N$_4$. Moreover, the efficient utilization of sunlight can be realized due to SPR absorption in the visible, UV light response of the inter band transition of noble metal nanoparticles. In the Sections 12.2 and 12.3, we have briefly summarized the principles, mechanisms, various applications of g-C$_3$N$_4$, and its nanocomposites.

12.4 Preparation of Graphitic Carbon Nitride

The various researchers were discussed about the synthesis of g-C$_3$N$_4$ using various methods; herein we emphasis only few literatures and its features. Jinshui Zhang et al. [16] reported the synthesis of g-C$_3$N$_4$, briefly, 3.0 g melamine was heated up to 550 °C in air atmosphere and kept for four hours to produce the coarse g-C$_3$N$_4$. Two gram of final coarse C$_3$N$_4$ was transferred in a 50 ml flask containing 20 ml HNO$_3$, stirred at 80 °C for three hours. The final product was washed, dried, and grinded for further utilization. Jiao Men et al. [6] used the g-C$_3$N$_4$ powder to fabricate the films on fluorine-doped tin oxide (FTO) substrate for DSSC as a g-C$_3$N$_4$/TiO$_2$ photoanodes and sensitized with CdS QDs for fabricating DSSC.

Qiqian Gao et al. [14] reported the preparation of g-C$_3$N$_4$ based on previous works, briefly, 3 g melamine and 4 g urea were mixed in a crucible and heated at 550 °C for two hours using muffle furnace. The yellow color crystalline bulk g-C$_3$N$_4$ was obtained and grinded. The g-C$_3$N$_4$ paste was prepared by mixing 0.8 g of g-C$_3$N$_4$, 0.4 g of ethyl cellulose and 3.245 g of α-terpinol in 8.5 ml anhydrous ethanol and stirred for 24 hours. The g-C$_3$N$_4$ paste was spin-coated on the as-prepared TiO$_2$ nanorod. The as-received g-C$_3$N$_4$/TiO$_2$ nanorod photoanodes were subjected to a sintering process in air at 450 °C for 30 minutes.

The monolayer g-C$_3$N$_4$ was prepared using 2 g melamine placed in a crucible and calcinated at 550 °C (2 °C/min) for four hours in a muffle furnace. The obtained g-C$_3$N$_4$ was dissolved in (0.1 g) 10 ml of water under sonication. Simultaneously, 2 ml of NH$_3$·H$_2$O, 0.01 g of Na$_2$SO$_3$ and 10 ml of isooctane were added into the mixture. The mixture was heated in an oil bath at 95 °C for 10 minutes, and HAuCl$_4$·4H$_2$O (0.0011, 0.0021, or 0.0042 g) was added in drops. The mixture was stirred at 95 °C for one hour, followed by centrifuged and washed in water several times, and dried.

Yongmei Wu et al. [17] synthesized g-C$_3$N$_4$, briefly 6 g of urea were placed in an alumina crucible heated to 520 °C for four hours under N$_2$ environment to obtain g-C$_3$N$_4$ powder. The yellow-color product was washed in 0.1 M of nitric acid and distilled water to remove residual alkaline species adsorbed on the sample surface and dried at 80 °C overnight.

12.4.1 Bulk Graphitic Carbon Nitride

g-C$_3$N$_4$ was produced by the standard pyrolysis method using urea as the precursor, which was heated at 550 °C for four hours at 5 °C/min. About 120 mg of as-prepared g-C$_3$N$_4$ was acidified by 20 ml H$_2$SO$_4$ blended with 20 ml HNO$_3$ for 20 hours under sonication. The clear solution was diluted with 500 ml deionized water and a colloidal suspension was obtained, which was filtered with a 0.22 μm microporous membrane to remove the excess H$_2$SO$_4$ and HNO$_3$. The as-prepared product was then dried and dispersed in 20 ml of dichlorobenzene under sonication for two hours. Finally, the suspension was heated at 200 °C for 10 hours in a teflon-lined autoclave. Final solution was filtered with a 0.22 μm membrane, a clear and light yellow C$_3$N$_4$ was obtained. Although bulk g-C$_3$N$_4$ could be synthesized easily at large scale and at low cost, it is not an ideal candidate for sensing applications because of its low water dispersity and surface area.

12.4.2 Mesoporous Graphitic Carbon Nitrides

The meso-C$_3$N$_4$ [18] was prepared by directly dispersing 1.25 g of cyanamide in water, followed by the addition of 12 nm colloidal silica stirring at 100 °C overnight to remove water. The weight ratio between precursors and silica was always kept as 1 : 1. The resultant white powder was calcinationed at 550 °C and the obtained product were washed repeatedly with 4 M NH$_4$HF$_2$ solution to remove the silica template. Afterward, the product was sieved and washed repeatedly with water and absolute ethanol until neutral. Finally, the products were dried at 60 °C under vacuum overnight.

12.4.3 Doping in Graphitic Carbon Nitride

Homogeneous functionalization of g-C_3N_4 can be achieved at an atomic level by elemental doping and at molecular level by copolymerization of precursors and molecules with functional moieties, since doping play a vital role in tuning its properties. For sensitive sensing with a fast kinetic response, g-C_3N_4 with a large surface area, an abundance of active sites, and high water dispersity are desirable for the immobilization of recognition elements and the efficient mass transport. Generally, chemical doping is a current approach to modify the surface and electronic properties of semiconductors, for improving their device output. Therefore, various researches put their efforts on doping this material with boron and phosphorus using ionic liquids as doping agent.

12.4.4 Ag Deposited g-C_3N_4

A 50 g of urea was put into an alumina crucible under ambient pressure and then the crucible was heated using muffle furnace to 250 °C for one hour, 350 °C for two hours, and a final temperature 550 °C for two hours at a heating rate of 2 °C/min. The yellow color final product was washed using nitric acid (0.1 mol/l) and distilled water to remove any residual alkaline species adsorbed on the surface of the product and then the product was dried at 80 °C in a hot air oven for 12 hours. A total of 0.7 g of as-prepared g-C_3N_4 powder was dispersed in 90 ml of aqueous $AgNO_3$ solution with the concentration of 0.08, 0.4, 0.8, 1.6, and 4 mmol/l, respectively, to obtain Ag-deposited g-C_3N_4 materials with various Ag loadings (0.1, 0.5, 1, 2, and 5 wt%). The resulting solution was irradiated under a 300 W Xe lamp for two hours. The separated powder was washed with distilled water to remove the adsorbed Ag^+ ions and then the product was dried at 80 °C for 12 hours.

12.4.5 Chemical Doping

In general, chemical doping is an effective approach to transform the electronic structures of semiconductors as well as their surface properties. Yuanjian Zhang et al. [10] reported that the phosphorus doping of g-C_3N_4 not only significantly changed its morphology and surface property but also tuned its electronic structure and enhanced its ionic conductivity. They employed co-condensation strategy between dicyandiamide and phosphorus containing ionic liquid to dope phosphorus into polymeric g-C_3N_4. Such doping not only improved conductivity but also the photocurrent generation, which implies the significance of polymeric g-C_3N_4 in photovoltaic applications.

Doping of fluorine was also employed to modify the properties of graphite, activated carbons, and carbon nanotubes, Yong Wang et al. [8] synthesized directly by incorporating the different amount of NH_4F ($x = 0.05, 0.1, 0.5, 1.0, 2.0$ g) into polymeric carbon nitride solids by thermally induced condensation. X-ray diffraction is well-known for g-C_3N_4 where the typical diffraction shows at 27.4° corresponding to (002), $d = 0.33$ nm. They observed for as prepared fluorinated g-C_3N_4, that the (002) diffraction becomes broader and gradually reduce its intensity with increasing amount of NH_4F, which indicates incorporation of fluorine into the lattice. They also observed shifts in both the lowest unoccupied molecular orbitals (LUMOs) and highest occupied molecular orbitals (HOMOs) to higher energy values side. However, incorporation of F at the corner carbon shifts the LUMO to higher energy and the HOMO to lower energy values. This remarkable LUMO and HOMO energy changes are brought by fluorine doping with an extension of the spectrum toward absorption in the visible range. The optical band gap also decreased from 2.69 eV of g-C_3N_4 to 2.63 eV of carbon nano fibres (CNF) on fluorine doping.

12.5   Operation Principles of DSSC

This section particularly deals with the brief discussion about the operating principles of a DSSC based on a ruthenium complexes dye, TiO_2 photoanode or TiO_2 nanocomposite photoanode material and I^-/I^{3-} as redox couple and a counter electrode that are as follows and shown in Figure 12.6. The basic electron transfer processes occurring in such a DSSC, as well as the related potentials were presented in Figure 12.7. TiO_2 is a very good material that transmits the entire visible light into the device. In the initial process light strikes the dye molecules or sensitized semiconductor quantum dots present in the TiO_2 composites (g-C_3N_4 with CdS). In process 1 (Eq. (12.1)), photoexcitation of the dye molecules or g-C_3N_4 takes place resulting in the excitation of an electron into the conduction band.

$$S + hv \rightarrow S^* \tag{12.1}$$

$$S^* \rightarrow S^+ e^-_{(TiO_2)} \tag{12.2}$$

In process 2 (Eq. (12.2)), electron is injected into the conduction band of metal oxide (TiO_2). However, if the g-C_3N_4/TiO_2 nanocomposites are used as photoanodes, then the excited electron will be transferred to the conduction band of g-C_3N_4 and then it will be injected to the conduction band of TiO_2 which is lying just below the conduction band of g-C_3N_4. Thus injected electron travels through the semiconductor and finally reaches the transparent conductive substrate (process 3). In process 4, the dye is restored to its ground state by electron transfer from the electrolyte. In process 5, the I^{3-} ions formed by oxidation of I^- diffuse through the electrolyte to the cathode where the regenerative cycle is completed by electron transfer to reduce I^{3-} to I^-. Instead of dye molecules, we can prepare TiO_2 nanocomposites sensitized with semiconductor quantum dots where it will have higher conduction band minimum (g-C_3N_4) than that of TiO_2 and also (g-C_3N_4) effectively act as a blocking layer inhibiting the reverse transmission of electrons from TiO_2 to the electrolyte.

In general, the injection process occurs is accepted in the femtosecond time scale for a Ru-complex sensitizer attached to an oxide surface. For good device performance, the duration of the injection process should be compared with the decay of the excited state of the dye to the ground state (process 6). This is given by the excited state lifetime of the dye, which for typical Ru-complexes used in DSSCs are 20–60 ns. In process 7, the kinetics of the back-electron transfer reaction were discussed from the semiconductor conduction band to the oxidized sensitizer in Eq. (12.3).

$$S^+ + e^-_{(TiO_2)} \rightarrow S \tag{12.3}$$

$$I_3^- + 2e^-_{(TiO_2)} \rightarrow 3I^-_{(anode)} \tag{12.4}$$

In process 8, the recombination of electrons in TiO_2 with acceptors in the electrolyte, Eq. (12.4) is usually referred to as the electron lifetime (τ_n). Lifetime was observed with the I^-/I^{3-} are very long (1–20 ms under one sunlight).

By looking at the energetic levels of dye, semiconductor conduction band and electrolyte redox potential, it is clear that for an efficient operation of this kind of device, the LUMO of dye molecules must have an energy higher than oxide conduction band and the HOMO should be lower with respect to redox potential: for the above reasons. Sensitizers have to be designed and synthesized in the appropriate way in order to match the energy requirements of semiconductor and redox couple. The voltage generated under illumination corresponds to the difference between the electrochemical potentials of the electrons at the two contacts, which is generally for DSSCs, the difference between the Fermi level of the mesoporous TiO_2 layer and the

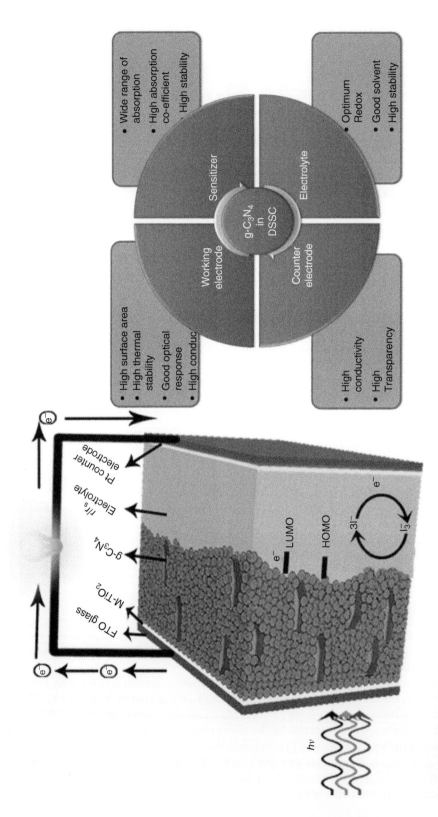

Figure 12.6 Typical DSSC structure with g-C₃N₄/TiO₂ nanocomposites as photoanode and its different properties and applications.

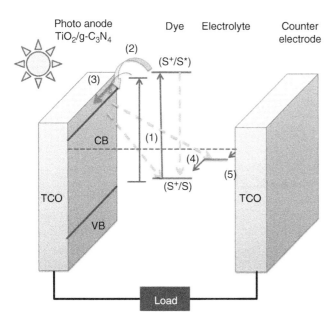

Figure 12.7 Energy level diagram for a DSSC based on Ru-complexes dye, TiO_2 photoanode, I^-/I^{3-} redox couple and Pt counter electrode.

redox potential of the electrolyte is taken into consideration. Electrical properties of the PV cell is a built-in electric field which provides the needed voltage to drive the current through an external load.

12.5.1 Nanostructured Graphitic Carbon Nitride in DSSC

Todate, $g-C_3N_4$ nanostructures have been combined with various materials such as inorganic semiconductors [14], metals [19], metal organic frame works [10, 20], conducting polymers [11, 21], sensors [22], and photocatalysis [5, 8]. These nanocomposites were synthesized by bottom-up and posttreatment methods.

The photocurrent generation in these nanocomposites involves three steps: first, being the creation of exciton pairs in $g-C_3N_4$; second, the passage of these excitons to the material surface, and third, the reaction of these excitons with solution-solubilized electron acceptors or donors. Nanostructured $g-C_3N_4$ commonly exhibits a rapid recombination rate of its photogenerated electron–hole pairs, leading to the low photocurrent output. In general, the nanocomposites are employed to improve carrier separation that facilitate charge rectification and promote charge migration. When high-conductivity materials such as noble metal nanoparticles, conducting polymers and carbonaceous nanomaterials were deposited on $g-C_3N_4$ nanostructures, they can act as a reservoir for electrons from the conduction band of $g-C_3N_4$, promotes interfacial electron transfer processes, and improves photocurrent signal. When semiconductors and dyes were deposited on $g-C_3N_4$ nanostructures, g C_3N_4 nanostructures act as either electron donors to transfer photoinduced electrons to a semiconductor with a large band gap such as TiO_2 and ZnO or as an electron acceptor of photoinduced electrons from a dye/semiconductor with a narrow band gap. It is worth mentioning that intimate contact between a uniform $g-C_3N_4$ film and conductive substrate is critical for the generation of the sensor with high photocurrent density. Currently, physical and chemical are the main methods used for the preparation of $g-C_3N_4$ thin film. The poor adhesion of the $g-C_3N_4$ nanoarchitectures to the surface leads to poor stability and performance. So researchers perform extensive efforts to fabricate the dense and uniform g-C$_3$N$_4$ film.

The conduction band minimum of g-C$_3$N$_4$ lying higher than that of TiO$_2$ allows the g-C$_3$N$_4$ layer to effectively act as a blocking layer inhibiting the reverse transmission of electrons from TiO$_2$ to the electrolyte. Besides, the lower band gap of g-C$_3$N$_4$ (2.7 eV), compared to TiO$_2$ (3.2 eV), widens the spectra absorption range, thus improving the cell performance. Recently, Wang et al. [23] fabricated the composites of g-C$_3$N$_4$ and conductive carbon black by a simple ball-milling process and used as the counter electrode of DSSCs. Yang and coworkers [21] applied g-C$_3$N$_4$ in polymer solar cells (PSCs) for the first time by doping solution-processable g-C$_3$N$_4$ quantum dots in the active layer, leading to a dramatic efficiency enhancement.

Xiang Chen et al. [21] says, g-C$_3$N$_4$ can be used in PSCs, which represent a promising renewable energy source featuring because of low manufacturing cost, flexibility, and easy fabrication. g-C$_3$N$_4$ graphene has been extensively applied in bulk-heterojunction PSCs due to its extraordinary electronic and optical properties such as high carrier mobility at room temperature and good optical transparency. Qiqian Gao et al. [14] introduce two-dimensional g-C$_3$N$_4$ layer in the single crystal TiO$_2$ nanorod compared with pure TiO$_2$ photoanodes. The g-C$_3$N$_4$ modified photoanodes show a noticeable enhancement in cell performances. This is due to the equivalent conduction and valence bands of g-C$_3$N$_4$ and TiO$_2$, which greatly enhanced the separation and transfer of the photogenerated electrons and holes, effectively suppressed interfacial recombination (see Figure 12.8).

Tridip Ranjan Chetia et al. [24] performed systematic investigations on g-C$_3$N$_4$ that reveals light harvesting ability of the solar cell by impeding photo-induced electron interception to the redox couple and injecting population to the conduction band of semiconductor. Also, g-C$_3$N$_4$ acts as a barrier for photoinduced electron interception at the working electrode/electrolyte interface and boosts the solar cell efficiency.

Jian Xu et al. [25] fabricated DSSCs by employing g-C$_3$N$_4$/TiO$_2$ nanosheets as photoanodes and found the significant enhancement in PCE of DSSCs (nearly 28%). At first, TiO$_2$ nanosheets were prepared and mixed with urea to obtain a thin layer of g-C$_3$N$_4$ on the surface of TiO$_2$ nanosheets. This photoanode (g-C$_3$N$_4$/TiO$_2$) in DSSC functioned effectively as the blocking layer to suppress the recombination.

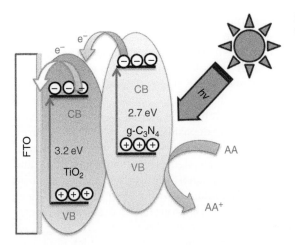

Figure 12.8 Mechanism of electron injection from g-C$_3$N$_4$/TiO$_2$ nanocomposites to FTO.

12.6 Graphitic Carbon Nitride in Polymer Films Solar Cell

Two-dimensional nanomaterials have attracted extensive attention due to their unique mechanical, thermal, optical, and electronic properties and its prospective application in electronics and optoelectronics, such as organic field-effect transistors, polymer light-emitting diodes and PSCs. As compared to bulk C_3N_4, the highly anisotropic 2D-nanosheets possessed a high specific surface area, a larger band gap, improved electron transport ability along the in-plane direction, and increased lifetime of photoexcited charge carriers owing to the quantum confinement effect. The utilization of two-dimensional g-C_3N_4 nanosheets as a cathode interfacial layer in inverted PSCs is also demonstrated. The device with C_3N_4 as cathode interfacial layer exhibited a remarkable improvement in PCE, indicating that C_3N_4 can effectively modify the bulk-heterojunction/indium-doped tin oxide (ITO) interface to facilitate efficient electron collection. The results indicate that the solution exfoliated g-C_3N_4 is a new and promising electron transport material for solution-processed organic optoelectronic devices [26]. A thin film of C_3N_4 on ITO substrates can effectively lower the work function of ITO, which is favorable for electron transport and can prohibit charge recombination near the cathode.

As compared to bulk g-C_3N_4, the highly anisotropic 2D-nanosheets possessed a high specific surface area, a larger band gap, improved electron transport ability along the in-plane direction, and increased lifetime of photoexcited charge carriers owing to the quantum confinement effect. Theoretical calculations based on density functional theory have revealed that g-C_3N_4 nanosheets exhibited unique electronic and optical properties. It reveals that the sheets of g-C_3N_4 have the semiconductor properties with a greater band gap than graphene and the wave functions of the valence band and the conduction band derives from nitrogen pz orbitals and carbon pz orbitals, respectively.

12.7 Preparation of Carbon Nitride Counter Electrode

Carbon nitride thin films were developed on a commercial available conductive FTO glass substrate by radio frequency magnetron sputtering process. Herein, high-purity (99.99%) carbon target was used as a source material. During sputtering process nitrogen gas (N_2) were fed into the chamber as reactive gas to form the corresponding carbon nitride coating. N_2 and Ar flow rates were accurately controlled by mass flow controllers at 2 and 20 sccm, respectively. Also sputtering power, deposition time, base pressure, and working pressure were optimized in order to get good quality, uniform carbon nitride thin films [27]. Wentao Xu et al. [28] developed carbon nitride films onto pure Si(111)substrates by reactive radio frequency magnetron sputtering combined with nitrogen gas at 500 °C. Optimized the film deposition, parameters for normal reactive sputtering using various gas flow and powers also concluded that the nitrogen radical beam produced an obvious increase in the nitrogen concentration through a slight increase in the hardness of the carbon nitride thin film. Logothetidis S. et al. [29] deposited nitrogenated α-C films by reactive RF magnetron sputtering using carbon target in plasma of Ar/N (4.4 at.% N) mixtures on Si substrates. The films are composed of two phases, sp^3 bonded C_3N_4 and sp^2 bonded amorphous carbon nitride (CN). Nitrogen incorporation into α-C films impart the sp^3 bonding essential in the entire films, but it is largely dependent on deposition conditions.

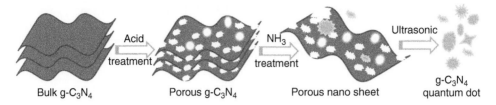

Figure 12.9 Synthesis of single layered g-C$_3$N$_4$ quantum dots from bulk g-C$_3$N$_4$.

12.8 Quantum Dot Graphitic Carbon Nitride

Recently, Zhang et al. [30] found that single-layered g-C$_3$N$_4$ quantum dots exhibit typical two-photon absorption characteristics and emit bright green emission after simultaneously absorbing two near-infrared photons. Furthermore, the photoluminescence intensity of g-C$_3$N$_4$ in nanocomposites can be modulated by photo-induced electron transfer between g-C$_3$N$_4$ nanostructure and the other substances in the nanocomposite. g-CN nanostructures can act as electron donors in the photo-induced electron transfer process. Compared to bulk g-C$_3$N$_4$, nanoscaled g-C$_3$N$_4$ usually exhibits a blue-shifted fluorescence (FL) peak with a higher intensity. A blue-shift indicates a wide band gap originating from quantum size effect. The higher FL intensity is caused by defects and termination sites in the nanoscaled g-CN, which promotes electron delocalization and thereby increases the recombination probability of photogenerated electron–hole pairs (see Figure 12.9).

12.9 Porous Graphitic Carbon Nitride

The preparation of mesoporous graphitic carbon nitride is discussed in detail in the above Section 12.4.2, Xiaofei Yang [18] performed photocatalysis activity for bulk and mesoporous C$_3$N$_4$, and they concluded that, the final oxygen amount was similar, but the reaction rate was faster for the M-C$_3$N$_4$ system. Yongmei Wu et al. [17] studied the novel ternary composite of heterostructured mesoporous g-C$_3$N$_4$/TiO$_2$ material prepared via a facile calcination approach with commercially available urea and synthesized amphourous TiO$_2$ precursors, and its composition, morphology and optical properties that were well characterized. This novel mesoporous TiO$_2$ covered with g-C$_3$N$_4$ exhibits best cell efficiency in near future.

12.10 Summary

In this chapter, we highlighted the importance of g-C$_3$N$_4$, preparation methods for g-C$_3$N$_4$ nanocomposites system and incorporation of g-C$_3$N$_4$ nanocomposites for the various applications. We also sketched out the performance, such as high adsorption, light absorption, charge separation, transportation, and charge carriers lifetime. The present chapter mainly represents the fabrication, structure, and photovoltaic performance, as well as mechanisms of g-C$_3$N$_4$ coupled with metal oxide, sulfides, composite oxide, etc. Interesting properties may be explored by combining novel metal oxides with g-C$_3$N$_4$ composites for better understanding of the fundamental properties and rapid development of advanced new nanomaterial. Also present investigations are mainly focused on an overall performance of the device efficiency. In addition to that, we need to understand charge generation, separation, and transport properties at nanoscale interfaces. The detailed mechanism studies of the charge transfer process will be necessary to further advance the field. Therefore, more studies are required to promote the general understanding of the enhancement mechanism of g-C$_3$N$_4$-based nanocomposites, especially employing advanced in situ techniques. Also, it is necessary to develop a uniform method to assess the photovoltaic performance as the current evaluation methods are diverse and energy-related global issues could be overcome in the near future.

Acknowledgment

One of the authors T.S. Shyju would like to thank CONICYT, FONDECYT (No. 3160445) Government of Chile. Authors thank Mr. Suresh Udaiyappan, CNSNT, SIST for the scientific contribution.

References

1 Zhao, Z., Sun, Y., and Dong, F. (2015). Graphitic carbon nitride based nanocomposites: a review. *Nanoscale* 7: 15–37. https://doi.org/10.1039/C4NR03008G.

2 Liebig, J. (1834). *Ann. Pharm.* 10: 1–47. https://doi.org/10.1002/jlac.18340100102.

3 Wang, X., Maeda, K., Thomas, A. et al. (2009). A metal-free polymeric photocatalyst for hydrogen production from water under visible light. *Nat. Mater.* 8: 76–80. https://doi.org/10.1038/nmat2317.

4 Zhu, J., Wei, Y., Chen, W. et al. (2010). Graphitic carbon nitride as a metal-free catalyst for NO decomposition. *Chem. Commun.* 46: 6965–6967. https://doi.org/10.1159/000101629.

5 Zhu, J., Xiao, P., Li, H., and Carabineiro, S.A.C. (2014). Graphitic carbon nitride: synthesis, properties, and applications in catalysis. *ACS Appl. Mater. Interfaces* 6: 16449–16465. https://doi.org/10.1021/am502925j.

6 Men, J., Gao, Q., Sun, S. et al. (2017). *Carbon Nitride Doped TiO$_2$ Photoelectrodes for Photocatalysts and Quantum Dot Sensitized Solar Cells. Mater. Res. Bull.*, 85: 209–215. Elsevier Ltd. https://doi.org/10.1016/j.materresbull.2016.09.023.

7 Wang, Y., Li, H., Yao, J. et al. (2011). Synthesis of boron doped polymeric carbon nitride solids and their use as metal-free catalysts for aliphatic C–H bond oxidation. *Chem. Sci.* 2: 446–450. https://doi.org/10.1039/c0sc00475h.

8 Wang, Y., Di, Y., Antonietti, M. et al. (2010). Excellent visible-light photocatalysis of fluorinated polymeric carbon nitride solids. *Chem. Mater.* 22: 5119–5121. https://doi.org/10.1021/cm1019102.

9 Liu, G., Niu, P., Sun, C. et al. (2010). Unique electronic structure induced high photoreactivity of sulfur-doped graphitic C$_3$N$_4$. *J. Am. Chem. Soc.* 132: 11642–11648. https://doi.org/10.1021/ja103798k.

10 Zhang, Y., Mori, T., Ye, J., and Antonietti, M. (2010). Phosphorus-doped carbon nitride solid enhanced electrical conductivity. *J. Am. Chem. Soc.* 132 (5): 6294–6295. https://doi.org/10.1002/asia.200900685.

11 Wang, X., Blechert, S., and Antonietti, M. (2012). Polymeric graphitic carbon nitride for heterogeneous photocatalysis. *ACS Catal.* 2: 1596–1606. https://doi.org/10.1021/cs300240x.

12 Zheng, Y., Liu, J., Liang, J. et al. (2012). Graphitic carbon nitride materials: controllable synthesis and applications in fuel cells and photocatalysis. *Energy Environ. Sci.* 5: 6717–6731. https://doi.org/10.1039/c2ee03479d.

13 Li, X.H. and Antonietti, M. (2013). Metal nanoparticles at mesoporous N-doped carbons and carbon nitrides: functional Mott–Schottky heterojunctions for catalysis. *Chem. Soc. Rev.* 42: 6593–6604. https://doi.org/10.1039/c3cs60067j.

14 Gao, Q., Sun, S., Li, X. et al. (2016). Enhancing performance of CdS quantum dot-sensitized solar cells by nanorods. *Nanoscale Res. Lett.*: 1–9. https://doi.org/10.1186/s11671-016-1677-1.

15 Zarick, H.F., Erwin, W.R., Boulesbaa, A. et al. (2016). Improving light harvesting in dye-sensitized solar cells using hybrid bimetallic nanostructures. *ACS Photonics* 3: 385–394. https://doi.org/10.1021/acsphotonics.5b00552.

16 Zhang, J., Zhang, M., Lin, L., and Wang, X. (2015). Sol processing of conjugated carbon nitride powders for thin-film fabrication. *Angew. Chem. Int. Ed.* 54: 6297–6301. https://doi.org/10.1002/anie.201501001.

17 Wu, Y., Chen, S., Zhao, J. et al. (2016). Mesoporous graphitic carbon nitride and carbon-TiO$_2$ hybrid composite photocatalysts with enhanced photocatalytic activity under visible light irradiation. *J. Environ. Chem. Eng.* 4: 797–807. https://doi.org/10.1016/j.jece.2015.10.023.

18 Yang, X., Chen, Z., Xu, J. et al. (2015). Tuning the morphology of g-C$_3$N$_4$ for improvement of Z-scheme photocatalytic water oxidation. *ACS Appl. Mater. Interfaces* https://doi.org/10.1021/acsami.5b02649.

19 Jiang, L., Yuan, X., Pan, Y. et al. (2017). Applied catalysis B: environmental doping of graphitic carbon nitride for photocatalysis: a review. *Appl. Catal., B* 217: 388–406. https://doi.org/10.1016/j.apcatb.2017.06.003.

20 Yang, Y., Guo, Y., Liu, F. et al. (2013). Preparation and enhanced visible-light photocatalytic activity of silver deposited graphitic carbon nitride plasmonic photocatalyst. *Appl. Catal., B* 142–143: 828–837. https://doi.org/10.1016/j.apcatb.2013.06.026.

21 Chen, X., Liu, Q., Wu, Q. et al. (2016). Incorporating graphitic carbon nitride (g-C$_3$N$_4$) quantum dots into bulk-heterojunction polymer solar cells leads to efficiency enhancement. *Adv. Funct. Mater.*: 1719–1728. https://doi.org/10.1002/adfm.201505321.

22 Chen, L. and Song, J. (2017). Tailored graphitic carbon nitride nanostructures: synthesis, modification, and sensing applications. *Adv. Funct. Mater.* 27: 1–15. https://doi.org/10.1002/adfm.201702695.

23 Wang, G., Zhang, J., and Hou, S. (2016). g-C$_3$N$_4$/conductive carbon black composite as Pt-free counter electrode in dye-sensitized solar cells. *Mater. Res. Bull.* 76: 454–458. https://doi.org/10.1016/j.materresbull.2016.01.001.

24 Chetia, T.R., Ansari, M.S., and Qureshi, M. (2016). Graphitic carbon nitride as a photovoltaic booster in quantum dot sensitized solar cells: a synergistic approach for enhanced charge separation and injection. *J. Mater. Chem. A* 4: 5528–5541. https://doi.org/10.1039/c6ta00761a.

25 Xu, J., Wang, G., Fan, J. et al. (2015). g-C$_3$N$_4$ modified TiO$_2$ nanosheets with enhanced photoelectric conversion efficiency in dye-sensitized solar cells. *J. Power Sources* 274: 77–84. https://doi.org/10.1016/j.jpowsour.2014.10.033.

26 Zhou, L., Xu, Y., Yu, W. et al. (2016). Ultrathin two-dimensional graphitic carbon nitride as a solution-processed cathode interfacial layer for inverted polymer solar cells. *J. Mater. Chem. A* 4: 8000–8004. https://doi.org/10.1039/c6ta01894g.

27 Wu, C., Li, G., Cao, X. et al. (2017). Carbon nitride transparent counter electrode prepared by magnetron sputtering for a dye-sensitized solar cell. *Green Energy Environ.* 2: 302–309. https://doi.org/10.1016/j.gee.2017.06.002.

28 Xu, W., Wang, L., Kojima, I. et al. (2006). Carbon nitride thin films prepared by radio-frequency magnetron sputtering combined with a nitrogen radical beam source carbon nitride thin films prepared by radio-frequency magnetron sputtering combined with a nitrogen radical beam source. *J. Appl. Phys.* 94, 7345 (2003) https://doi.org/10.1063/1.1625412.

29 Logothetidis, S., Lefakis, H., and Gioti, M. (1998). Carbon nitride thin films prepared by reactive r.f. magnetron sputtering. *Carbon* 36: 757–760. https://doi.org/10.1016/S0008-6223(98)00073-6.

30 Zhang, X., Wang, H., Wang, H. et al. (2014). Single-layered graphitic-C_3N_4 quantum dots for two-photon fluorescence imaging of cellular nucleus. *Adv. Mater.* 26: 4438–4443. https://doi.org/10.1002/adma.201400111.

Index

Interfacial Engineering in Functional Materials for Dye-Sensitized Solar Cells, First Edition.
Edited by Alagarsamy Pandikumar, Kandasamy Jothivenkatachalam and Karuppanapillai B. Bhojanaa.
© 2020 John Wiley & Sons, Inc. Published 2020 by John Wiley & Sons, Inc.